Edward Y. Chang

Foundations of Large-Scale Multimedia Information Management and Retrieval

Mathematics of Perception

Edward Y. Chang

Foundations of Large-Scale Multimedia Information Management and Retrieval

Mathematics of Perception

With 108 figures, 21 of them in color

Author
Edward Y. Chang
Ph.D., Director of Research
Google Inc. Mountain View
CA 94306, USA
Email: edchang@google.com
http://infolab.stanford.edu/~echang/

ISBN 978-7-302-24976-4
Tsinghua University Press, Beijing

ISBN 978-3-642-20428-9   e-ISBN 978-3-642-20429-6
Springer Heidelberg Dordrecht London New York

Library of Congress Control Number: 2011924542

Printed on acid-free paper

Springer is part of Springer Science+Business Media (www.springer.com)

To my family
   Lihyuarn, Emily, Jocelyn, and Rosalind.

Foreword

The last few years have been transformative time in information and communication technology. Possibly this is one of the most exciting period after Gutenberg's moveable print revolutionized how people create, store, and share information. As is well known, Gutenberg's invention had tremendous impact on human societal development. We are again going through a similar transformation in how we create, store, and share information. I believe that we are witnessing a transformation that allows us to share our experiences in more natural and compelling form using audio-visual media rather than its subjective abstraction in the form of text. And this is huge.

It is nice to see a book on a very important aspect of organizing visual information by a researcher who has unique background in being a sound academic researcher as well as a contributor to the state of art practical systems being used by lots of people. Edward Chang has been a research leader while he was in academia, at University of California, Santa Barbara, and continues to apply his enormous energy and in depth knowledge now to practical problems in the largest information search company of our time. He is a person with a good perspective of the emerging field of multimedia information management and retrieval.

A good book describing current state of art and outlining important challenges has, enormous impact on the field. Particularly, in a field like multimedia information management the problems for researchers and practitioners are really complex due to their multidisciplinary nature. Researchers in computer vision and image processing, databases, information retrieval, and multimedia have approached this problem from their own disciplinary perspective. The perspective based on just one discipline results in approaches that are narrow and do not really solve the problem that requires true multidisciplinary perspective. Considering the explosion in the volume of visual data in the last two decades, it is now essential that we solve the urgent problem of managing this volume effectively for easy access and utilization. By looking at the problem in multimedia information as a problem of managing information about the real world that is captured using different correlated media, it is possible to make significant progress. Unfortunately, most researchers do not have time and interest to look beyond their disciplinary boundaries to understand the real problem and address it. This has been a serious hurdle in the progress in multimedia information management.

I am delighted to see and present this book on a very important and timely topic by an eminent researcher who has not only expertise and experience, but also energy and interest to put together an in depth treatment of this interdisciplinary topic. I am not aware of any other book that brings together concepts and techniques in this emerging field in a concise book. Moreover, Prof. Chang has shown his talent in pedagogy by organizing the book to consider needs of undergraduate students as well as graduate students and researchers. This is a book that will be equally useful for people interested in learning about the state of the art in multimedia information management and for people who want to address challenges in this transformative field.

Ramesh Jain
Irvine, February 2011

Preface

The volume and accessibility of images and videos is increasing exponentially, thanks to the sea-change of imagery captured from film to digital form, to the availability of electronic networking, and to the ubiquity of high-speed network access. The tools for organizing and retrieving these multimedia data, however, are still quite primitive. One such evidence is the lack of effective tools to-date for organizing personal images or videos. Another clue is that all Internet search engines today still rely on the keyword search paradigm, which knowingly suffers from the semantic aliasing problem. Existing organization and retrieval tools are ineffective partly because they fail to properly model and combine "content" and "context" of multimedia data, and partly because they fail to effectively address the scalability issues. For instance, today, a typical content-based retrieval prototype extracts some signals from multimedia data instances to represent them, employs a poorly justified distance function to measure similarity between data instances, and relies on a costly sequential scan to find data instances similar to a query instance. From feature extraction, data representation, multimodal fusion, similarity measurement, feature-to-semantic mapping, to indexing, the design of each component has mostly not been built on solid scientific foundations. Furthermore, most prior art focuses on improving one single component, and demonstrates its effectiveness on small datasets. However, the problem of multimedia information management and retrieval is inherently an interdisciplinary one, and tackling the problem must involve collaboration between fields of machine learning, multimedia computing, cognitive science, and large-scale computing, in addition to signal processing, computer vision, and databases. This book presents an interdisciplinary approach to first establish scientific foundations for each component, and then address interactions between components in a scalable manner in terms of both data dimensionality and volume.

This book is organized into twelve chapters of two parts. The first part of the book depicts a multimedia system's key components, which together aims to comprehend semantics of multimedia data instances. The second part presents methods for scaling up these components for high-dimensional data and very large datasets. In part one we start with providing an overview of the research and engineering challenges in Chapter 1. Chapter 2 presents feature extraction, which obtains useful signals from multimedia data instances. We discuss both model-based and data-driven, and then a hybrid approach. In Chapter 3, we deal with the problem of formulating users' query concepts, which can be complex and subjective. We show how active learning and kernel methods can be used to work effectively with both keywords and perceptual features to understand a user's query concept with minimal user feedback. We argue that only after a user's query concept can be thoroughly comprehended, it is then possible to retrieve matching objects. Chapters 4 and 5 address the problem of distance-function formulation, a core subroutine of information retrieval for measuring similarity between data instances. Chapter 4 presents Dynamic Partial function and its foundation in cognitive psychology. Chapter 5 shows how an effective function can also be learned from examples in a data-driven way. Chapters 6, 7 and 8 describe methods that fuse metadata of multiple modalities. Multimodal fusion is important to properly integrate perceptual features of various kinds (e.g., color, texture, shape; global, local; time-invariant, time-variant), and to properly combine metadata from multiple sources (e.g., from both content and context). We present three techniques: super-kernel fusion in Chapter 6, fusion with causal strengths in Chapter 7, and combinational collaborative filtering in Chapter 8.

Part two of the book tackles various scalability issues. Chapter 9 presents the problem of imbalanced data learning where the number of data instances in the target class is significantly out-numbered by the other classes. This challenge is typical in information retrieval, since the information relevant to our queries is always the minority in the dataset. The chapter describes algorithms to deal with the problem in vector and non-vector spaces, respectively. Chapters 10 and 11 address the scalability issues of kernel methods. Kernel methods are a core machine learning technique with strong theoretical foundations and excellent empirical successes. One major shortcoming of kernel methods is its cubic computation time required for training and linear for classification. We present parallel algorithms to speed up the training time, and fast indexing structures to speed up the classification time. Finally, in Chapter 12, we present our effort in speeding up Latent Dirichlet Allocation (LDA), a robust method for modeling texts and images. Using distributed computing primitives, together with data placement and pipeline techniques, we were able to speed up LDA $1,500$ times when using $2,000$ machines.

Although the target application of this book is multimedia information retrieval, the developed theories and algorithms are applicable to analyze data of other domains, such as text documents, biological data and motion patterns.

This book is designed for researchers and practitioners in the fields of multimedia, computer vision, machine learning, and large-scale data mining. We expect the reader to have some basic knowledge in Statistics and Algorithms. We recommend that the first part (Chapters 1 to 8) to be used in an upper-division undergraduate course, and the second part (Chapters 9 to 12) in a graduate-level course. Chapters 1 to 6 should be read sequentially. The reader can read Chapters 7 to 12 in selected order. Appendix lists our open source sites.

Edward Y. Chang
Palo Alto, February 2011

Acknowledgements

I would like to thank contributions of my Ph.D students and research colleagues (in roughly chronological order): Beitao Li, Simon Tong, Kingshy Goh, Yi Wu, Navneet Panda, Gang Wu, John R. Smith, Bell Tseng, Kevin Chang, Arun Qamra, Wei-Cheng Lai, Kaihua Zhu, Hongjie Bai, Hao Wang, Jian Li, Zhihuan Qiu, Wen-Yen Chen, Dong Zhang, Xiance Si, Hongji Bao, Zhiyuan Liu, Maosong Sun, Dingyin Xia, Zhiyu Wang, and Shiqiang Yang. I would also like to thank the funding supported by three NSF grants: NSF Career IIS-0133802, NSF ITR IIS-0219885, and NSF IIS-0535085.

Contents

Chapter 1
Introduction — Key Subroutines of Multimedia Data Management

Abstract This chapter presents technical challenges that multimedia information management faces. We enumerate five key subroutines required to work together effectively so as to enable robust and scalable solutions. We provide pointers to the rest of the book, where in-depth treatments are presented.
Keywords: Mathematics of perception, multimedia data management, multimedia information retrieval.

1.1 Overview

The tasks of multimedia information management such as clustering, indexing, and retrieval, come up against technical challenges in at least three areas: data representation, similarity measurement, and scalability. First, data representation builds layers of abstraction upon raw multimedia data. Next, a distance function must be chosen to properly account for similarity between any pair of multimedia instances. Finally, from extracting features, measuring similarity, to organizing and retrieving data, all computation tasks must be performed in a scalable fashion with respect to both data dimensionality and data volume. This chapter outlines design issues of five essential subroutines, and they are:

1. Feature extraction,
2. Similarity (distance function formulation),
3. Learning (supervised and unsupervised),
4. Multimodal fusion, and
5. Indexing.

1.2 Feature Extraction

Feature extraction is fundamental to all multimedia computing tasks. Features can be classified into two categories, *content* and *context*. Content refers directly to raw imagery, video, and mucic data such as pixels, motions, and tones, respectively, and their representations. Context refers to metadata collected or associated with content when a piece of data is acquired or published. For instance, EXIF camera parameters and GPS location are contextual information that some digital cameras can collect. Other widely used contextual information includes surrounding texts of an image/photo on a Web page, and social interactions on a piece of multimedia data instance. Context and content ought to be fused synergistically when analyzing multimedia data [1].

Content analysis is a subject studied for more than a couple of decades by researchers in disciplines of computer vision, signal processing, machine learning, databases, psychology, cognitive science, and neural science. Limited progress has been made in each of these disciplines. Many researchers now are convinced that interdisciplinary research is essential to make ground breaking advancements. In Chapter 2 of this book, we introduce a model-based and data-driven hybrid approach for extracting features. A promising model-based approach was pioneered by neural scientist Hubel [2], who proposed a feature learning pipeline based on human visual system. The principal reason behind this approach is that human visual system can function so well in some challenging conditions where computer vision solutions fail miserably. Recent neural-based models proposed by Lee [3] and Serre [4] show that such model can effectively deal with viewing of different positions, scales, and resolutions. Our empirical study confirmed that such model-based approach can recognize objects of rigid shapes, such as watches and cars. However, for objects that do not have invariant features such as pizzas of different toppings, and cups of different colors and shapes, the model-based approach loses its advantages. For recognizing these objects, the data-driven approach can depict an object by collecting a representative pool of training instances. When combining model-based and data-driven, the hybrid approach enjoys at least three advantages:

1. *Balancing feature invariance and selectivity*. To achieve feature selectivity, the hybrid approach conducts multi-band, multi-scale, and multi-orientation convolutions. To achieve invariance, it keeps signals of sufficient strengths via pooling operations.
2. *Properly using unsupervised learning to regularize supervised learning*. The hybrid approach introduces unsupervised learning to reduce features so as to prevent the subsequent supervised layer from learning trivial solutions.
3. *Augmenting feature specificity with diversity*. A model-based only approach cannot effectively recognize irregular objects or objects with diversified patterns; and therefore, we must combine such with a data-driven pipeline.

Chapter 2 presents the detailed design of such a hybrid model involving disciplines of neural science, machine learning, and computer vision.

1.3 Similarity

At the heart of data management tasks is a distance function that measures *similarity* between data instances. To date, most applications employ a variant of the *Euclidean distance* for measuring similarity. However, to measure similarity meaningfully, an effective distance function ought to consider the idiosyncrasies of the application, data, and user (hereafter we refer to these factors as contextual information). The quality of the distance function significantly affects the success in organizing data or finding relevant results.

In Chapters 4 and 5, we present two methods, first an unsupervised in Chapter 4 and then a supervised in Chapter 5, to quantify similarity. Chapter 4 presents Dynamic Partial Function (DPF), which we formulated based on what we learned from some intensive data mining on large image datasets. Traditionally, similarity is a measure of all respects. For instance, the Euclidean function considers all features in equal importance. One step forward was to give different features different weights. The most influential work is perhaps that of Tversky [5], who suggests that similarity is determined by matching features of compared objects. The weighted Minkowski function and the quadratic-form distances are the two representative distance functions that match the spirit. The weights of the distance functions can be learned via techniques such as relevance feedback, principal component analysis, and discriminative analysis. Given some similar and some dissimilar objects, the weights can be adjusted so that similar objects can be better distinguished from the other objects.

However, the assumption made by these distance functions, that all similar objects are similar in the same respects [6], is questionable. We propose that *similarity is a process that provides respects for measuring similarity*. Suppose we are asked to name two places that are similar to England. Among several possibilities, Scotland and New England could be two reasonable answers. However, the respects England is similar to Scotland differ from those in which England is similar to New England. If we use the shared attributes of England and Scotland to compare England and New England, the latter pair might not be similar, and vice versa. This example depicts that objects can be similar to the query object in different respects. A distance function using a fixed set of respects cannot capture objects that are similar in different sets of respects. Murphy and Medin [7] provide early insights into how similarity works in human perception: "The explanatory work is on the level of determining which attributes will be selected, with similarity being at least as much a consequence as a cause of a concept coherence." Goldstone [8] explains that similarity is the process that determines the respects for measuring similarity. In other words, a distance function for measuring a pair of objects is formulated only after the objects are compared, not before the comparison is made. The respects for the comparison are activated in this formulation process. The activated respects are more likely to be those that can support coherence between the compared objects. DPF activates different features for different object pairs. The activated features are those with minimum differences — those which provide coherence between the objects. If coherence can be maintained (because sufficient a number of features are similar), then the objects paired are perceived as similar. Cognitive psychology seems able to explain much of the effectiveness of DPF.

Whereas DPF learns similar features in an unsupervised way, Chapter 5 presents a supervised method to learn a distance function from contextual information or user feedback. One popular method is to weight the features of the Euclidean distance (or more generally, the L_p-norm) based on their importance for a target task [9, 10, 11]. For example, for answering a *sunset* image-query, color features should be weighted higher. For answering an *architecture* image-query, shape and texture features may be more important. Weighting these features is equivalent to performing a *linear* transformation in the space formed by the features. Although linear models enjoy the twin advantages of simplicity of description and efficiency of computation, this same simplicity is insufficient to model similarity for many real-world data instances. For example, it has been widely acknowledged in the image/video retrieval domain that a query concept is typically a nonlinear combination of perceptual features (color, texture, and shape) [12, 13]. Chapter 5 presents a *nonlinear* transformation on the feature space to gain greater flexibility for mapping features to semantics.

At first it might seem that capturing nonlinear relationships among contextual information can suffer from high computational complexity. We avoid this concern by employing the *kernel trick*, which has been applied to several algorithms in statistics, including Support Vector Machines and kernel PCA. The kernel trick lets us generalize distance-based algorithms to operate in the *projected space*, usually nonlinearly related to the *input space*. The *input space* (denoted as $\mathscr{I}$) is the original space in which data vectors are located, and the *projected space* (denoted as $\mathscr{P}$) is that space to which the data vectors are projected, linearly or nonlinearly. The advantage of using the *kernel trick* is that, instead of explicitly determining the coordinates of the data vectors in the projected space, the distance computation in $\mathscr{P}$ can be efficiently performed in $\mathscr{I}$ through a kernel function.

Through theoretical discussion and empirical studies, Chapters 4 and 5 show that when similarity measures have been improved, data management tasks such as clustering, learning, and indexing can perform with marked improvements.

1.4 Learning

The principal design goal of a multimedia information retrieval system is to return data (images or video clips) that accurately match users' queries (for example, a search for pictures of a deer). To achieve this design goal, the system must first comprehend a user's query concept (i.e., a user's perception) thoroughly, and then find data in the low-level input space (formed by a set of perceptual features) that match the concept accurately. Statistical learning techniques can assist achieving the design goal via two complementary avenues: semantic annotation and query-concept learning.

Both semantic annotation and query-concept learning can be cast into the form of a supervised learning problem, which consists of three steps. First, a representative set of perceptual features is extracted from each training instance. Second, each training feature-vector (other representations are possible) is assigned semantic labels. Third, a classifier is trained by a supervised learning algorithm, based on the labeled instances, to predict the class labels of a query instance. Given a query instance represented by its features, the semantic labels can be predicted. In essence, these steps learn a mapping between the perceptual features and a human perceived concept or concepts.

Chapter 3 presents the challenges of semantic annotation and query-concept learning. To illustrate, let D denote the number of low-level features (extracted by methods presented in Chapter 2), N the number of training instances, N^+ the number of positive training instances, and N^- the number of negative training instances ($N = N^+ + N^-$). Two major technical challenges arise:

1. *Scarcity of training data.* The features-to-semantics mapping problem often comes up against the $D > N$ challenge. For instance, in the query-concept learning scenario, the number of low-level features that characterize an image (D) is greater than the number of images a user would be willing to label (N) during a relevance feedback session. As pointed out by David Donoho, the theories underlying "classical" data analysis are based on the assumptions that $D < N$, and N approaches infinity. But when $D > N$, the basic methodology which was used in the classical situation is not similarly applicable.
2. *Imbalance of training classes.* The target class in the training pool is typically outnumbered by the non-target classes ($N^- \gg N^+$). For instance, in a k-class classification problem where each class has about the same number of training instances, the target class is outnumbered by the non-target classes by a ratio of k:1. The class boundary of imbalanced training classes tends to skew toward the target class when k is large. This skew makes class prediction less reliable.

To address these challenges, Chapter 3 presents a small sample, active learning algorithm, which also adjusts its sampling strategy in a concept-dependent way. Chapter 9 presents a couple of approaches to deal with imbalanced training classes. When conducting annotation, the computation task faces the challenge of dealing with a substantially large N. From Chapter 10 to Chapter 12 , we discuss parallel algorithms, which can employ thousands of CPUs to achieve near-linear speedup, and indexing methods, which can substantially reduce retrieval time.

1.5 Multimodal Fusion

Multimedia metadata can be collected from multiple channels or sources. For instance, a video clip consists of visual, audio, and caption signals. Besides, a Web page where the video clip is embedded, and the users who have viewed the video can provide contextual signals for analyzing that clip. When mapping features extracted from multiple sources to semantics, a fusion algorithm must incorporate useful information while removing noise. Chapters 6, 7, and 8 are devoted to address multimodal fusion.

Chapter 6 focuses on addressing two questions: (1) what are the *best* modalities? and (2) how can we optimally fuse information from multiple modalities? Suppose we extract l, m, n features from the visual, audio, and caption tracks of videos. At one extreme, we could treat all these features as one modality and form a feature vector of $l + m + n$ dimensions. At the other extreme, we could treat each of the $l + m + n$ features as one modality. We could also regard the extracted features from each media-source as one modality, formulating a visual, audio, and caption modality with l, m, and n features, respectively. Almost all prior multimodal-fusion work in the multimedia community employs one of these three approaches. But, can any of these feature compositions yield the optimal result?

Statistical methods such as principle component analysis (PCA) and independent component analysis (ICA) have been shown to be useful for feature transformation and selection. PCA is useful for denoising data, and ICA aims to transform data to a space of independent axes (components). Despite their best attempt under some error-minimization criteria, PCA and ICA do not guarantee to produce independent components. In addition, the created feature space may be of very high dimensions and thus be susceptible to the *curse of dimensionality*. Chapter 6 first presents an *independent modality analysis* scheme, which identifies independent modalities, and at the same time, avoids the curse-of-dimensionality challenge. Once a good set of modalities has been identified, the second research challenge is to fuse these modalities in an optimal way to perform data analysis (e.g., classification). Chapter 6 presents the *super-kernel fusion* scheme to fuse individual modalities in a non-linear way. The *super-kernel fusion* scheme finds the best combination of modalities through supervised training.

Chapter 6 addresses the problem of fusing multiple modality of multimedia data *content*. Chapter 7 addresses the problem of fusing *context* with *content*. Semantic labels can be roughly divided into two categories: wh labels and non-wh labels. Wh-semantics include time (when), people (who), location (where), landmarks (what), and event (inferred from when, who, where, and what). Providing the when and where information is trivial. Already cameras can provide time, and we can easily infer an approximate location from GPS or CellID. However, determining the what and who requires contextual information in addition to time, location, and photo content. More precisely, contextual information can include time, location, camera parameters, user profile, and even social graphs. Content of images consists of perceptual features, which can be divided into holistic features (e.g., color, shape and texture characteristics of an image), and local features (edges and salient points of regions or objects in an image). Besides context and content, another important source of information (which has been largely ignored) is the relationships between semantic labels (which we refer to as semantic ontology). To explain the importance of having a semantic ontology, let us consider an example with two semantic labels: outdoor and sunset. When considering contextual information alone, we may be able to infer the outdoor label from camera parameters: focal length and lighting condition. We can infer sunset from time and location. Notice that inferring outdoor and sunset do not rely on any common contextual modality. However, we can say that a sunset photo is outdoor with certainty (but not the other way). By considering semantic relationships between labels, photo annotation can take advantage of contextual information in a "transitive" way.

To fuse context, content, and semantic ontology in a synergistic way, Chapter 7 presents EXTENT, an inferencing framework to generate semantic labels for photos. EXTENT uses an influence diagram to conduct semantic inferencing. The variables on the diagram can either be decision variables (i.e., causes) or chance variables (i.e., effects). For image annotation, decision variables include time, location, user profile, and camera parameters. Chance variables are semantic labels. However, some variables may play both roles. For instance, time can affect some camera parameters (such as exposure time and flash on/off), and hence these camera parameters are both decision and chance variables. Finally, the influence diagram connects decision variables to chance variables with arcs weighted by causal strength.

To construct an influence diagram, we rely on both domain knowledge and data. In general, learning such a probabilistic graphical model from data is an NP hard problem. Fortunately, for image annotation, we have abundant prior knowledge about the relationships between context, content, and semantic labels, and we can use them to substantially reduce the hypothesis space to search for the right model. For instance, time, location, and user profile, are independent of each other. Camera parameters such as exposure time and flash on/off depend on time, but are independent of other modalities. The semantic ontology provides us the relationships between words. The only causal relationships that we must learn from data are those between context/content and semantic labels (and their causal strengths).

Once causal relationships have been learned, causal strengths must be accurately accounted for. Traditional probabilistic graphical models such as Bayesian networks use conditional probability to quantify the correlation between two variables. Unfortunately, conditional probability characterizes *covariation*, not *causation* [14, 15, 16]. A basic tenet of classical statistics is that correlation does not imply causation. Instead, we use recently developed *causal-power* theory [17] to account for causation. We show that fusing context and content using causation achieves superior results over using correlation.

Finally, Chapters 8 presents a fusion model called Combinational Collaborative Filtering (CCF) using a latent layer. CCF views a community of common interests from two simultaneous perspectives: *a bag of users* and *a bag of multimodal features*. A community is viewed as a bag of participating users; and at the same time, it is viewed as a bag of multimodal features describing that community. Traditionally, these two views are independently processed. Fusing these two views provides two benefits. First, by combining *bags of features* with *bags of users*, CCF can perform *personalized* community recommendations, which the *bags of features* alone model cannot. Second, augmenting *bags of users* with *bags of features*, CCF improves information density to perform more effective recommendations. Though the chapter uses community recommendation as an application, one can use the CCF scheme for recommending any objects, e.g., images, videos, and songs.

1.6 Indexing

With the vast volume of data available for search, indexing is essential to provide scalable search performance. However, when data dimension is high (higher than 20 or so), no nearest-neighbor algorithm can be significantly faster than a linear scan of the entire dataset. Let n denote the size of a dataset and d the dimension of data, the theoretical studies of [18, 19, 20, 21] show that when $d \gg \log n$, a linear search will outperform classic search structures such as k-d-trees [22], SR-trees [23], and SS-trees [24]. Several recent studies (e.g., [19, 20, 25]) provide empirical evidence, all confirming this phenomenon of *dimensionality curse*.

Nearest neighbor search is inherently expensive, especially when there are a large number of dimensions. First, the search space can grow exponentially with the number of dimensions. Second, there is simply no way to build an index on disk such that all nearest neighbors to any query point are physically adjacent on disk. The prohibitive nature of exact nearest-neighbor search has led to the development of *approximate nearest-neighbor search* that returns instances approximately similar to the query instance [18, 26]. The first justification behind approximate search is that a feature vector is often an approximate characterization of an object, so we are already dealing with approximations [27]. Second, an approximate set of answers suffices if the answers are relatively close to the query concept. Of late, three approximate indexing schemes, *locality sensitive hashing* (LSH) [28], M-trees [29], and clustering [27] have been employed in applications such as image-copy detection [30] and bio-sequence-data matching [31]. These approximate indexing schemes speed up similarity search significantly (over a sequential scan) by slightly lowering the bar for accuracy.

In Chapter 11, we present our *hypersphere indexer*, named SphereDex, to perform approximate nearest-neighbor searches. First, the indexer finds a roughly central instance among a given set of instances. Next, the instances are partitioned based on their distances from the central instance. SphereDex builds an *intra-partition* (or local) index within each partition to efficiently prune out irrelevant instances. It also builds an *inter-partition* index to help a query to identify a good starting location in a neighboring partition to search for nearest neighbors. A search is conducted by first finding the partition to which the query instance belongs. (The query instance does not need to be an existing instance in the database.) SphereDex then searches in this and the neighboring partitions to locate nearest neighbors of the query. Notice that since each partition has just two neighboring partitions, and neighboring partitions can largely be sequentially laid out on disks, SphereDex can enjoy sequential IO performance (with a tradeoff of transferring more data) to retrieve candidate partitions into memory. Even in situations (e.g., after a large batch of insertions) when one sequential access might not be feasible for retrieving all candidate partitions, SphereDex can keep the number of non-sequential disk accesses low. Once a partition has been retrieved from the disk, SphereDex exploits geometric properties to perform intelligent intra-partition pruning so as to minimize the computational cost for finding the top-k approximate nearest neighbors. Through empirical studies on two very large, high-dimensional datasets, we show that SphereDex significantly outperforms both LSH and M-trees in both IO and CPU time. Though we mostly present our techniques for approximate nearest-neighbor queries, Chapter 11 also briefly describes the extensibility of SphereDex to support farthest-instance queries, especially hyperplane queries to support key data-mining algorithms like Support Vector Machines (SVMs).

1.7 Scalability

Indexing deals with retrieval scalability. We must also address scalability of learning, both supervised and unsupervised. Since 2007, we have parallelized five mission-critical algorithms including SVMs [32], Frequent Itemset Mining [33], Spectral Clustering [34], Probabilistic Latent Semantic Analysis (PLSA) [35], and Latent Dirichlet Allocation (LDA) [36]. In this book, we present Parallel Support Vector Machines (PSVM) in Chapter 10 and an enhanced PLDA+ in Chapter 12.

Parallel computing has been an active subject in the distributed computing community over several decades. In PSVM, we use Incomplete Cholesky Factorization to approximate a large matrix so as to reducing the memory use substantially. For speeding up LDA, we employ data placement and pipeline processing techniques to substantially reduce the communication bottleneck. We are able to achieve $1,500$ speedup when $2,000$ machines are simultaneously used: i.e., a two-month computation task on a single machine can now be completed in an hour. These parallel algorithms have been released to the public via Apache open source (please check out the Appendix).

1.8 Concluding Remarks

As we stated in the beginning of this chapter, multimedia information management research is multidisciplinary. In feature extraction and distance function formulation, the disciplines of computer vision, psychology, cognitive science, neural science, and database have been involved. In indexing and scalability, distributed computing and database communities have contributed a great deal. In devising learning algorithms to bridge the semantic gap, machine learning and neural science are the primary forces behind recent advancements. Together, all these communities are increasingly working together to develop robust and scalable algorithms. In the remainder of this book, we detail the design and implementation of these key subroutines of multimedia data management.

References

1. Chang, E.Y. Extent: Fusing context, content, and semantic ontology for photo annotation. In *Proceedings of ACM Workshop on Computer Vision Meets Databases (CVDB) in conjunction with ACM SIGMOD*, pages 5–11, 2005.
2. Hubel, D.H., Wiesel, T.N. Receptive fields and functional architecture of monkey striate cortex. *Journal of Physiology*, 195(1):215–243, 1968.
3. Lee, H., Grosse, R., Ranganath, R., Ng, A. Convolutional deep belief networks for scalable unsupervised learning of hierarchical representations. In *Proceedings of International Conference on Machine Learning (ICML)*, 2009.
4. Serre, T. *Learning a dictionary of shape-components in visual cortex: comparison with neurons, humans and machines*. PhD Thesis, Massachusetts Institute of Technology, 2006.
5. Tversky, A. Feature of similarity. *Psychological Review*, 84:327–352, 1977.
6. Zhou, X.S., Huang, T.S. Comparing discriminating transformations and svm for learning during multimedia retrieval. In *Proc. of ACM Conf. on Multimedia*, pages 137–146, 2001.
7. Murphy, G., Medin, D. The role of theories in conceptual coherence. *Psychological Review*, 92:289–316, 1985.
8. Goldstone, R.L. Similarity, interactive activation, and mapping. *Journal of Experimental Psychology: Learning, Memory, and Cognition*, 20:3–28, 1994.
9. Aggarwal, C.C. Towards systematic design of distance functions for data mining applications. In *Proceedings of ACM SIGKDD*, pages 9–18, 2003.
10. Fagin, R., Kumar, R., Sivakumar, D. Efficient similarity search and classification via rank aggregation. In *Proceedings of ACM SIGMOD Conference on Management of Data*, pages 301–312, June 2003.
11. Wang, T., Rui, Y., Hu, S.M., Sun, J.Q. Adaptive tree similarity learning for image retrieval. *Multimedia Systems*, 9(2):131–143, 2003.
12. Rui, Y., Huang, T. Optimizing learning in image retrieval. In *Proceedings of IEEE CVPR*, pages 236–245, June 2000.
13. Tong, S., Chang, E. Support vector machine active learning for image retrieval. In *Proceedings of ACM International Conference on Multimedia*, pages 107–118, October 2001.
14. Heckerman, D. A bayesian approach to learning causal networks. In *Proceedings of the Conference on Uncertainty in Artificial Intelligence*, pages 107–118, 1995.
15. Pearl, J. *Causality: Models, Reasoning and Inference*. Cambridge University Press, 2000.
16. Pearl, J. Causal inference in the health sciences: A conceptual introduction. *Special issue on causal inference, Kluwer Academic Publishers, Health Services and Outcomes Research Methodology*, 2:189–220, 2001.
17. Novick, L.R., Cheng, P.W. Assessing interactive causal influence. *Psychological Review*, 111(2):455–485, 2004.
18. Arya, S., Mount, D., Netanyahu, N., Silverman, R., Wu, A. An optimal algorithm for approximate nearest neighbor searching in fixed dimensions. In *Proceedings of the 5th SODA*, pages 573–82, 1994.
19. Indyk, P., Motwani, R. Approximate nearest neighbors: towards removing the curse of dimensionality. In *Proceedings of VLDB*, pages 604–613, 1998.
20. Kleinberg, J.M. Two algorithms for nearest-neighbor search in high dimensions. In *Proceedings of the 29th STOC*, 1997.
21. Weber, R., Schek, H.J., Blott, S. A quantitative analysis and performance study for similarity-search methods in high-dimensional spaces. In *Proc. 24th Int. Conf. Very Large Data Bases VLDB*, pages 194–205, 1998.
22. Bentley, J. Multidimensional binary search trees used for associative binary searching. *Communications of ACM*, 18(9):509–517, 1975.
23. Katayama, N., Satoh, S. The SR-tree: an index structure for high-dimensional nearest neighbor queries. In *Proceedings of ACM SIGMOD Int. Conf. on Management of Data*, pages 369–380, 1997.

24. White, D.A., Jain, R. Similarity indexing with the SS-Tree. In *Proceedings of IEEE ICDE*, pages 516–523, 1996.
25. Kushilevitz, E., Ostrovsky, R., Rabani, Y. Efficient search for approximate nearest neighbor in high dimensional spaces. In *Proceedings of the 30th STOC*, pages 614–23, 1998.
26. Clarkson, K. An algorithm for approximate closest-point queries. In *Proceedings of the 10th SCG*, pages 160–64, 1994.
27. Li, C., Chang, E., Garcia-Molina, H., Wilderhold, G. Clindex: Approximate similarity queries in high-dimensional spaces. *IEEE Transactions on Knowledge and Data Engineering (TKDE)*, 14(4):792–808, July 2002.
28. Gionis, A., Indyk, P., Motwani, R. Similarity search in high dimensions via hashing. *VLDB Journal*, pages 518–529, 1999.
29. Ciaccia, P., Patella, M. Pac nearest neighbor queries: Approximate and controlled search in high-dimensional and metric spaces. In *Proceedings of IEEE ICDE*, pages 244–255, 2000.
30. Qamra, A., Meng, Y., Chang, E.Y. Enhanced perceptual distance functions and indexing for image replica recognition. *IEEE Transactions on Pattern Analysis and Machine Intelligence (PAMI)*, 27(3), 2005.
31. Buhler, J. Efficient large-scale sequence comparison by locality-sensitive hashing. *Bioinformatics*, 17:419–428, 2001.
32. Chang, E.Y., Zhu, K., Wang, H., Bai, H., Li, J., Qiu, Z., Cui, H. Parallelizing support vector machines on distributed computers. In *Proceedings of NIPS*, 2007.
33. Li, H., Wang, Y., Zhang, D., Zhang, M., Chang, E.Y. PFP: Parallel fp-growth for query recommendation. In *Proceedings of ACM RecSys*, pages 107–114, 2008.
34. Song, Y., Chen, W., Bai, H., Lin, C.J., Chang, E.Y. Parallel spectral clustering. In *Proceedings of ECML/PKDD*, pages 374–389, 2008.
35. Chen, W., Zhang, D., Chang, E.Y. Combinational collaborative filtering for personalized community recommendation. In *Proceedings of ACM KDD*, pages 115–123, 2008.
36. Wang, Z., Zhang, Y., Chang, E.Y., Sun, M. PLDA+ parallel latent dirichlet allocation with data placement and pipeline processing. *ACM Transactions on Intelligent System and Technology*, 2(3), 2011.

Chapter 2
Perceptual Feature Extraction

Abstract In this chapter[†], we present a deep model-based and data-driven hybrid architecture (DMD) for feature extraction. First, we construct a deep learning pipeline for progressively learning image features from simple to complex. We mix this deep model-based pipeline with a data-driven pipeline, which extracts features from a large collection of unlabeled images. Sparse regularization is then performed on features extracted from both pipelines in an unsupervised way to obtain representative patches. Upon obtaining these patches, a supervised learning algorithm is employed to conduct object prediction. We present how DMD works and explain why it works more effectively than traditional models from both aspects of neuroscience and computational learning theory.

Keywords: Data driven, deep learning, DMD , feature extraction, model-based.

2.1 Introduction

Extracting useful features from a scene is an essential step of any computer vision and multimedia analysis tasks. Though progress has been made in past decades, it is still quite difficult for computers to accurately recognize an object or analyze the semantics of an image. In this chapter, we study two extreme approaches of feature extraction, *model-based* and *data-driven*, and then evaluate a hybrid scheme.

One may consider model-based and data-driven to be two mutually exclusive approaches. In practice; however, they are not. Virtually all *model* construction relies on some information from data; all data-driven schemes are built upon some *models*, simple or complex. The key research questions for the feature-extraction task of an object recognition or image annotation application are:

[†] ©ACM, 2010. This chapter is a minor revision of the author's work with Zhiyu Wang and Dingyin Xia [1] published in VLS-MCMR'10. Permission to publish this chapter is granted under copyright license #2587600190581.

1. *Can more data help a model*? Given a *model*, can the availability of more training
   data improve feature quality, and hence improve annotation accuracy?
2. *Can an improve model help data-driven*? Give some fixed amount of data, can
   a model be enhanced to improve feature quality, and hence improve annotation
   accuracy?

We first closely examine a model-based deep-learning scheme, which is neuro-science-motivated. Strongly motivated by the fact that the human visual system can effortlessly conduct these tasks, neuroscientists have been developing vision models based on physiological evidences. Though such research may still be in its infancy and several hypotheses remain to be validated, some widely accepted theories have been established. This chapter first presents such a model-based approach. Built upon the pioneer neuroscience work of Hubel [2], all recent models are founded on the theory that visual information is transmitted from the primary visual cortex (V1) over extrastriate visual areas (V2 and V4) to the inferotemporal cortex (IT). IT in turn is a major source of input to the prefrontal cortex (PFC), which involves in linking perception to memory and action [3]. The pathway from V1 to IT, which is called the *visual frontend* [4], consists of a number of simple and complex layers. The lower layers attain simple features that are invariant to scale, position and orientation at the pixel level. Higher layers detect complex features at the object-part level. Pattern reading at the lower layers are unsupervised; whereas recognition at the higher layers involves supervised learning. Computational models proposed by Lee [5] and Serre [6] show such a multi-layer generative approach to be effective in object recognition.

Our empirical study compared features extracted by a neuroscience-motivated deep-learning model and those extracted by a data-driven scheme through an application of image annotation. For the data-driven scheme, we employed features of some widely used pixel aggregates such as shapes and color/texture patches. These features construct a feature space. Given a previously unseen data instance, its annotation is determined through some nearest-neighbor scheme such as k-NN or kernel methods. The assumption of the data-driven approach is that if the features of two data instances are close ("similar") in the feature space, then their target semantics would be same. For a data-driven scheme to work well, its feature space must be densely populated with training instances so that unseen instances can find sufficient number of nearest neighbors as their references.

We made two observations from the results of our experimental study. First, when the number of training instances is small, the model-based deep-learning scheme outperforms the data-driven. Second, while both feature sets commit prediction errors, each does better on certain objects. Model-based tends to do well on objects of a regular, rigid shape with similar interior patterns; whereas the data-driven model performs better in recognizing objects of variant perceptual characteristics. These observations establish three guidelines for feature design.

1. *Recognizing objects with similar details*. For objects that have regular features,
   invariance should be the top priority of feature extraction. A feature-extraction
   pipeline that can be invariant to scale, position and orientation requires only a

handful of training instances to obtain a good set of features for recognizing this class of objects. For this class of objects, model-based works very well.

2. *Recognizing objects with different details.* Objects with variant features, such as strawberries in different orientations and environmental settings, or dalmatians with their varying patterns, do not have invariant features to extract. For recognizing such an object class, *diversity* is the design priority of feature extraction. To achieve diversity, a learning algorithm requires a large number of training instances to collect abundant samples. Therefore, data-driven works better for this class of objects.

3. *Recognizing objects within abstracts.* Classifying objects of different semantics such as whales and lions being mammals, or tea cups and beer mugs being cups, is remote to percepts. Abstract concept classification requires a WordNet-like semantic model.

The first two design principles confirm that feature extraction must consider both feature invariance and feature diversity; but how? A feedforward pathway model designed by Poggio's group [7] holds promises in obtaining invariant features. However, additional signals must be collected to enhance the diversity aspect. As Serre [6] indicates, feedback signals are transmitted back to V1 to pay attention to details. Biological evidences suggest that a feedback loop in visual system instructs cells to "see" local details such as color-based shapes and shape-based textures. These insights lead to the design of our hybrid model DMD , which combines a deep model-based pipeline with a data-driven pipeline to form a six-layer hierarchy. While the model-based pipeline faithfully models a deep learning architecture based on visual cortex's feedforward path [8], the data-driven pipeline extracts augmented features in a heuristic-based fashion. The two pipelines join at an unsupervised middle layer, which clusters low-level features into feature patches. This unsupervised layer is a critical step to effectively regularize the feature space [9, 10] for improving subsequent supervised learning, making object prediction both effective and scalable. Finally, at the supervised layer, DMD employs a traditional learning algorithm to map patches to semantics. Empirical studies show that DMD works markedly better than traditional models in image annotation. DMD 's success is due to (1) its simple to complex deep pipeline for balancing invariance and selectivity, and (2) its model-based and data-driven hybrid approach for fusing feature specificity and diversity.

In this chapter, we show that a model-based pipeline encounters limitations. As we have explained, a data-driven pipeline is necessary for recognizing objects of different shapes and details. DMD employs both approaches of *deep* and *hybrid* to achieve improved performance for the following reasons:

1. *Balancing feature invariance and selectivity.* DMD implements Serre's method [8] to achieve a good balance between feature invariance and selectivity. To achieve feature selectivity, DMD conducts multi-band, multi-scale, and multi-orientation convolutions. To achieve invariance, DMD only keeps signals of sufficient strengths via pooling operations.

2. *Properly using unsupervised learning to regularize supervised learning.* At the second, the third, and the fifth layers, DMD introduces unsupervised learning to reduce features so as to prevent the subsequent supervised layer from learning trivial solutions.

3. *Augmenting feature specificity with diversity.* Through empirical study, we identified that a model-based only approach cannot effectively recognize irregular objects or objects with diversified patterns; and therefore, fuse into DMD a data-driven pipeline. We point out subtle pitfalls in combining model-based and data-driven and propose a remedy for noise reduction.

2.2 DMD Algorithm

DMD consists two pipelines of six steps. Given a set of training images, the *model-based* pipeline feeds training images to the edge detection step. At the same time, the *data-driven* pipeline feeds training images directly to the step of sparsity regularization. We first discuss the *model-based pipeline* of DMD in Section 2.2.1, and then its *data-driven* pipeline in Section 2.2.2.

2.2.1 Model-Based Pipeline

Figure 2.1 depicts that visual information is transmitted from the primary visual cortex (V1) over extrastriate visual areas (V2 and V4) to the inferotemporal cortex (IT). Physiological evidences indicate that the cells in V1 largely conduct *selection* operations, and cells in V2 and V4 *pooling* operations. Based on such, M. Riesenhuber and T. Poggio's theory of feedforward path of object recognition in the cortex [7] establishes a qualitative way to model the ventral stream in the visual cortex. Their model suggests that the visual system consists of multiple layers of computational units where *simple S* units alternate with *complex C* units. The S units deal with signal selectivity, whereas the C units deal with invariance. Lower layers attain features that are invariant to scale, position, and orientation at the pixel level. Higher layers detect features at the object-part level. Pattern reading at the lower layers are largely unsupervised, whereas recognition at the higher layers involves supervised models. Recent advancements in *deep learning* [10] have led to mutual justifications that a model-based, hierarchical model enjoys these computational advantages:

- Deep architectures enjoy advantages over shallow architectures (please consult [11] for details), and
- Unsupervised initiated deep architectures can enjoy better generalization performance [12].

Motivated by both physiological evidences [8, 14] and computational learning theories [5], we designed DMD's model-based pipeline with six steps:

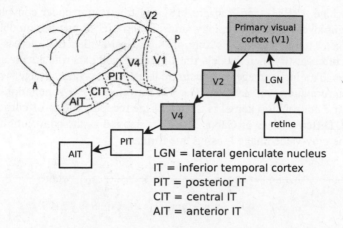

Fig. 2.1: Information flow in the visual cortex. (See the brain structure in [13].)

1. *Edge selection* (Section 2.2.1.1). This step corresponds to the operation conducted by cells in V1 and V2 [15], which detect edge signals at the pixel level.
2. *Edge pooling* (Section 2.2.1.2). This step also corresponds to cells in V1 and V2. The primary operation is to pool strong, representative edge signals.
3. *Sparsity regularization* (Section 2.2.1.3). To prevent too large a number of features, which can lead to dimensionality curse, or too low a level, which may lead to trivial solutions, DMD uses this unsupervised step to group edges into patches.
4. *Part selection* (Section 2.2.1.4). There is not yet strong physiological evidence, but it is widely believed that V2 performs part selection and then feeds signals directly to V4. DMD models this step to look for image patches matching those prototypes (patches) produced in the previous step.
5. *Part pooling* (Section 2.2.1.5). Cells in V4 [16], which have larger receptive fields than V1, deal with parts. Because of their larger receptive fields, V4's selectivity is preserved over translation.
6. *Supervised learning* (Section 2.2.1.6). Learning occurs at all steps and certainly at the level of inferotemporal (IT) cortex and prefrontal cortex (PFC). This topmost layer employs a supervised learning algorithm to map a patch-activation vector to some objects.

2.2.1.1 Edge Selection

In this step, computational units model the classical simple cells described by Hubel and Wiesel in the primary visual cortex (V1) [17]. A simple selective operation is performed by V1 cells. To model this operation, Serre [8] uses Gabor filters to perform a 2D convolution, Lee [5] suggests using a convolutional restricted Boltz-

mann machine (RBM), and Ranzato [18] constructs an encoder convolution. We
initially employed T. Serre's strategy since T. Serre [6, 8] justifies model selection
and parameter tuning based on strong physiological evidences, whereas computer
scientists often justify their models through contrived experiments on a small set
of samples. The input image is transmitted into a gray-value image, where only
the *edge* information is of interest. The 2D convolution is a summation-like operation, whose convolution kernel is to model the receptive fields of cortical simple
cells [19]. Different sizes of Gabor filters are applied as the convolution kernel to
process the gray-value image $\mathbf{I}$, using this format:

$$\mathbf{F}_s(x,y) = \exp(-\frac{x_0^2 + \gamma^2 y_0^2}{2\sigma^2}) \times \cos(\frac{2\pi}{\lambda}x_0), \text{ where} \qquad (2.1)$$

$$x_0 = x\cos\theta + y\sin\theta \text{ and } y_0 = -x\sin\theta + y\cos\theta.$$

In Eq. (2.1), γ is the aspect ratio and θ is the orientation, which takes values $0°$,
$45°$, $90°$, and $135°$. Parameters σ and λ are the effective width and wave length,
respectively. The Gabor filter forms a 2D matrix with the value at position (x,y) to
be $\mathbf{F}_s(x,y)$. The matrix size $(s \times s)$ or the Gabor filter size ranges from 7×7 to $37 \times$
37 pixels in intervals of two pixels. Thus there are 64 (16 scales $\times 4$ orientations)
different receptive field types in total. With different parameters, Gabor filters can
cover different orientations and scales and hence increase selectivity. The output of
the edge-selection step is produced by 2D convolutions (conv2) of the input image
and $n_b \times n_s \times n_f = 64$ Gabor filters of

$$\mathbf{I}_{S\_edge(i_b,i_s,i_f)} = conv2(\mathbf{I}, \mathbf{F}_{i_F}), \text{ where} \qquad (2.2)$$

$$i_F = (i_b \times n_s + i_s) \times n_f + i_f.$$

2.2.1.2 Edge Pooling

In the previous step, several edge-detection output matrices are produced, which
sufficiently support selectivity. At the same time, there is clearly some redundant
or noisy information produced from these matrices. Physiological evidences on cats
show that a MAX-like operation is taken in complex cells [15] to deal with redundancy and noise. To model this MAX-like operation, Serre's, Lee's, and Ranzato's
work all agree on applying a MAX operation on outputs from the simple cells. The
response $\mathbf{I}_{edge(i_b,i_f)}$ of a complex unit corresponds to the response of the strongest of
all the neighboring units from the previous edge-selection layer. The output of this
edge-pooling layer is as follows:

$$\mathbf{I}_{edge(i_b,i_f)}(x,y) = \max_{i_s \in \mathbf{v}_s, m \in \mathcal{N}(x,y)} \mathbf{I}_{S\_edge(i_b,i_s,i_f)}(x_m,y_m),$$

where (x_m, y_m) stands for edge-selection results at position (x, y). The max is taken over the two scales within the same spatial neighborhood of the same orientation, justified by the experiments conducted by the work of Serre [20].

Input: image data $\mathbf{I}$
Output: edge feature $\{\mathbf{I}_{edge(i_b, i_f)} | i_b, i_f\}$

Initialization:
1: Calculate normalize parameters $\mathbf{I}_{Norm}$
2: $n_b \leftarrow 8$ // total # of bands
3: $n_s \leftarrow 2$ // total # of scales
4: $n_f \leftarrow 4$ // total # of orientations

Edge Selection:
1: **for** $i_b = 1$ to n_b **do**
2:     **for** $i_s = 1$ to n_s **do**
3:         **for** $i_f = 1$ to n_f **do**
4:             $\mathbf{I}_{S\_edge(i_b, i_s, i_f)} \leftarrow \mathbf{conv2}(\mathbf{I}, \mathbf{Gabor}(i_b, i_s, i_f))$
5:         **end for**
6:     **end for**
7: **end for**
8: Normalize $\mathbf{I}_{S\_edge}$ by $\mathbf{I}_{Norm}$

Edge Pooling:
1: **for** $i_b = 1$ to n_b **do**
2:     **for** $i_f = 1$ to n_f **do**
3:         **for** each x, y in dimensions of $I_{edge(i_b, i_f)}$ **do**
4:             $\mathbf{I}_{edge(i_b, i_f)}(x, y) \leftarrow \max_{i_s} \mathbf{I}_{s\_edge(i_b, i_s, i_f)}(x, y)$
5:         **end for**
6:     **end for**
7: **end for**

Fig. 2.2: DMD Model-based pipeline, steps 1 and 2

2.2.1.3 Sparsity Regularization

A subtle and important step of a deep architecture is to perform proper initialization between layers. The edge-pooling step may produce a huge number of edges. With such a large-sized output, the next layer may risk learning trivial solutions at the pixel level. Both Serre [8] and Ekanadham [21] suggest to sparsify the output of V2 (or input to V4).

To perform the sparsification, we form pixel *patches* via sampling. In this way, not only the size of the input to the part-selection step is reduced, but patches larger than pixels can regularize the learning at the upper layers. The regularization effect is achieved by the fact that parts are formed by neighboring edges, not edges at random positions. Thus, there is no reason to conduct learning directly on the edges.

A patch is a region of pixels sampled at a random position of a training image at four orientations. An object can be fully expressed if enough representative patches have been sampled. It is important to note that this sampling step can be performed incrementally when new training images are available. The result of this unsupervised learning step is n_p prototype patches, where n_p can be set initially to be a large value, and then trimmed back by the part-selection step.

In Section 2.2.2 we show that the data-driven pipeline also produces patches by sampling a large number of training instances. Two pipelines join at this unsupervised regularization step.

2.2.1.4 Part Selection

So far, DMD has generated patches via clustering and sampling. This part-selection step finds out which patches may be useful and of what patches an object part is composed. Part selection units describe a larger region of objects than the edge detection, by focusing on parts of the objects. Similar to our approach, Serre's S_2 units behave as radial basis function (RBF) units, Lee uses a convolutional deep belief network (CDBN), and Ranzato's algorithm implements a convolutional operation for the decoder. All are consistent with well-known response properties of neurons in the primate inferotemporal cortex (IT).

Serre proposes using Gaussian-like Euclidean distance to measure similarity between an image and the pre-calculated prototypes (patches). Basically, we would like to find out what patches an object consists of. Analogically, we are constructing a map from object-parts to an object using the training images. Once the mapping has been learned, we can then classify an unseen image.

To perform part selection, we have to examine if patches obtained in the regularization step appear frequently enough in the training images. If a patch appears frequently, that patch can be selected as a part; otherwise, that patch is discarded for efficiency. For each training image, we match its edge patches with the n_p prototyped patches generated in the previous step. For the i_b^{th} band of an image's edge detection output, we obtain for the i_p^{th} patch a measure as follows:

$$\mathbf{I}_{S\_part(i_b,i_p)} = \exp(-\beta||\mathbf{X}_{i_b} - \mathbf{P}_{i_p}||^2), \qquad (2.3)$$

where β is the sharpness of the tuning and $\mathbf{P}_{i_p}$ is one of the n_p patches learned during sparsity regularization. $\mathbf{X}_{i_b}$ is a transformation of the $\mathbf{I}_{edge(i_b,i_f)}$ with all n_f orientations merged to fit the size of $\mathbf{P}_{i_p}$. We obtain n_b measurements of the image for each prototype patch. Hence the total number of measurements that this part-selection step makes is the number of patches times the number of bands, or $n_p \times n_b$.

2.2.1.5 Part Pooling

Each image is measured against n_p patches, and for each patch, n_b measurements are performed. To aggregate n_b measurements into one, we resort to the part-pooling units. The part-pooling units correspond to visual cortical V4 neurons. It has been discovered that a substantial fraction of the neurons takes the maximum input as output in visual cortical V4 neurons of rhesus monkeys (macaca mulatta) [16], or

$$\mathbf{v}_{part(i_P)} = \min_{i_b} \mathbf{I}_{S\_part(i_b,i_P)}. \tag{2.4}$$

The MAX operation (maximizing similarity is equivalent to minimizing distance) can not only maintain feature invariance, but also scale down feature-vector size. The output of this stage for each training image is a vector of n_p values as depicted by the pseudo code in Fig. 2.3.

Input: edge features $\{\mathbf{I}_{edge(i_b,i_f)} | i_b, i_f\}$, patches $\{\mathbf{P}_{i_P} | i_P\}$
Output: part features $\mathbf{v}_{part}$

Initialization:
1: $n_P \leftarrow size\ of\ \{\mathbf{P}_{i_P} | i_P\}$ // total # of patches
2: $n_b \leftarrow 8$ // total # of bands, the same as before
Part Selection:
1: **for** $i_P = 1$ to n_P **do**
2:    merge $\mathbf{I}_{edge(i_b,i_f)}$ for all i_f to get $\mathbf{I}_{edge(i_b)}$
3:    $\mathbf{I}_{S\_part(i_b,i_P)} \leftarrow distance(\mathbf{I}_{edge(i_b)}, \mathbf{P}_{i_P})$
4: **end for**
Part Pooling:
1: **for** $i_P = 1$ to n_P **do**
2:    $\mathbf{v}_{part(i_P)} \leftarrow \min_{i_b} \mathbf{I}_{S\_part(i_b,i_P)}$
3: **end for**

Fig. 2.3: DMD steps 4 and 5

2.2.1.6 Supervised Learning

At the top layer, DMD performs part-to-object mapping. At this layer, any traditional *shallow* learning algorithm can work reasonably well. We employ SVMs to perform the task. The input to SVMs is a set of vector representations of image patches produced by this model-based pipeline and by the data-driven pipeline, which we present next. Each image is represented by a vector of real values each depicting the image's perceptual strength matched by a prototype patch.

2.2.2 Data-Driven Pipeline

The key advantage of the model-based pipeline is feature invariance. For objects that have a rigid body of predictable patterns, such as a watch or a phone, the model-based pipeline can obtain invariant features from a small number of training instances. Indeed, our experimental results presented in Section 2.3 show that it takes just five training images to effectively learn the features of a watch and to recognize it. Unfortunately, for objects that can have various appearances such as pizzas with different toppings, the model-based pipeline runs into limitations. The features it learned from the toppings of one pizza cannot help recognize a pizza with different toppings. The key reason for this is that invariance may cause overfitting, and that hurts selectivity.

To remedy the problem, DMD adds a data-driven pipeline. The principal idea is to collect enough examples of an object so that feature selectivity can be improved. By collecting signals from a large number of training data, it is also likely to collect signals of different scales and orientations. In other words, instead of relying solely on a model-based pipeline to deal with invariance, we can collect enough examples to ensure with high probability that the collected examples can cover most transformations of features.

Another duty that the data-driven pipeline can fulfill is to augment a key shortcoming of the model-based pipeline, i.e., it considers only the feedforward pathway of the visual system. It is well understood that some complex recognition tasks may require recursive predictions and verifications. Backprojection models [22, 23] and attention models [24] are still in early stage of development, and hence there is no solid basis of incorporating feedback. DMD uses heuristic-based signal processing subroutines to extract patches for the data-driven pipeline. The extracted patches are merged with those learned in the sparse-regularization step of the model-based pipeline.

We extracted patches in multiple resolutions to improve invariance [25, 26]. We characterized images by two main features: color and texture. We consider shapes as attributes of these main features.[1]

2.2.2.1 Color Patches

Although the wavelength of visible light ranges from 400 nanometers to 700 nanometers, research effort [28] shows that the colors that can be named by all cultures are generally limited to eleven. In addition to *black* and *white*, the discernible colors are *red, yellow, green, blue, brown, purple, pink, orange* and *gray*.

We first divided color into 12 color bins including 11 bins for culture colors and one bin for outliers [26]. At the coarsest resolution, we characterized color using a

---

[1] *Disclaimer*: We do not claim these heuristic-based features to be novel. Other heuristic-based features [27] may also be useful. What we consider to be important is that these features can augment model-based features to improve diversity before a principled theory can be formulated by neuroscientists to model cortex feedback/feedforward recursive signals.

color mask of 12 bits. To recorded color information at finer resolutions, we record eight additional features for each color. These eight features are color histograms, color means (in H, S and V channels), color variances (in H, S and V channel), and two shape characteristics: elongation and spread. Color elongation characterizes the shape of a color and spreadness characterizes how that color scatters within the image [29]. We categorize color features by coarse, medium and fine resolutions.

2.2.2.2 Texture Patches

Texture is an important cue for image analysis. Studies [30, 31, 32, 33] have shown that characterizing texture features in terms of structuredness, orientation, and scale (coarseness) fits well with models of human perception. A wide variety of texture analysis methods have been proposed in the past. We chose a discrete wavelet transformation (DWT) using quadrature mirror filters [31] because of its computational efficiency.

Each wavelet decomposition on a 2D image yields four subimages: a $\frac{1}{2} \times \frac{1}{2}$ scaled-down image of the input image and its wavelets in three orientations: horizontal, vertical and diagonal. Decomposing the scaled-down image further, we obtain a tree-structured or wavelet packet decomposition. The wavelet image decomposition provides a representation that is easy to interpret. Every subimage contains information of a specific scale and orientation and also retains spatial information. We obtain nine texture combinations from subimages of three scales and three orientations. Since each subimage retains the spatial information of texture, we also compute elongation and spreadness for each texture channel.

2.2.2.3 Feature Fusion

Now, given an image, we can extract the above color and texture information to produce some clusters of features. These clusters are similar to those patches generated by the model-based pipeline. All features, generated by the model-based or data-driven pipeline are inputs to the sparsity regularization step, depicted in Section 2.2.1.3, to conduct subsequent processing. In Section 2.3.3 we discuss where fusion can be counter-productive and propose a remedy to reduce noise.

2.3 Experiments

Our experiments were designed to answer three questions:

1. How does a model-based approach compare to a data-driven approach? Which model performs better? Where and why?
2. How does DMD perform compared to an individual approach, model-based or data-driven?

3. How much does the unsupervised regularization step help?

To answer these questions we conducted three experiments:

1. Model-based versus data-driven model.
2. DMD versus individual approaches.
3. Parameter tuning at the regularization step.

2.3.1 Dataset and Setup

To ensure that our experiments can cover object appearances of different character-istics (objects of similar details, different details, and within abstracts, as depicted in Section 2.1), we collected training and testing data from ImageNet [34]. We se-lected 10,885 images of 100 categories to cover the above characteristics. We fol-lowed the two pipelines of DMD to extract model-based features and data-driven features. For each image category, we cross-validated by using 15 or 30 images for training, and the remainder for testing to compute annotation accuracy. Because of the small training-data size, using linear SVMs turned out to be competitive to using advanced kernels. We thus employed linear SVMs to conduct all experiments.

2.3.2 Model-Based vs. Data-Driven

This experiment was designed to evaluate individual models and explain where and why model-based or data-driven is more effective.

2.3.2.1 Overall Accuracy

Table 2.1 summarizes the average annotation accuracy of the model-based-only ver-sus data-driven-only method with 15 and 30 training images, respectively. The table shows that the model-based method to be more effective. (We will shortly explain this result to be not-so-meaningful.) We next looked into individual categories to examine the reasons why.

Table 2.1: Model-based vs. data-driven

Training Size	Model-based	Data-driven
15	22.53%	16.32%
30	28.06%	20.61%

2.3.2.2 Where Model-Based Works Better

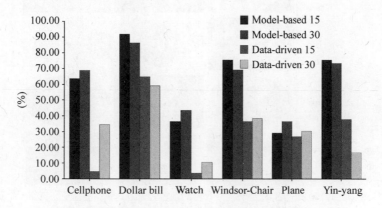

Fig. 2.4: Model-based outperforms data-driven (See color insert)

Figure 2.4 shows a set of six categories (example images shown in Table 2.2) where the model-based method performs much better than the data-driven in class-prediction accuracy (on the y-axis).

First, on some categories such as 'dollar bills', 'Windsor-chair' and 'yin-yang', increasing training data from 15 to 30 does not improve annotation accuracy. This is because these objects exhibit precise features (e.g., all dollar bills are the same), and as long as a model-based pipeline can deal with scale/position/orientation invariance, the feature-learning process requires only a small number of examples to capture their idiosyncrasies. The data-driven approach on these six categories performs quite poorly because its lacking the ability to deal with feature invariance. For instance, the data-driven pipeline cannot recognize watches of different sizes and colors, or Windsor-chairs of different orientations.

2.3.2.3 Where Data-Driven Works Better

Data-driven works better than model-based when objects exhibit patterns that are similar but not necessarily identical nor scale/position/orientation invariant. Data-driven can work effectively when an object exhibit some consistent characteristics such as apples are red or green, and dalmatians have black patches of irregular shapes.

Figure 2.5 shows a set of six categories where data-driven works substantially better than model-based in class-prediction accuracy.

Table 2.2: Images with a rigid body (See color insert)

Category	Images
Cellphone	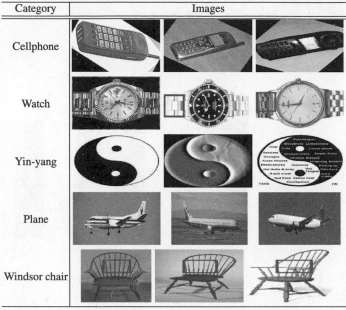
Watch	
Yin-yang	
Plane	
Windsor chair	

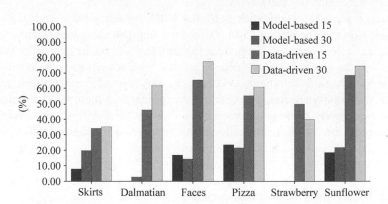

Fig. 2.5: Data-driven outperforms model-based (See color insert)

Table 2.3 display example images from four categories, 'strawberry', 'sun-flower', 'dalmatian', and 'pizza'. Both 'strawberry' and 'sunflower' exhibit strong color characteristics, whereas 'dalmatian' and 'pizza' exhibit predictable texture features. For these categories, once after sufficient amount of samples can be collected, semantic-prediction can be performed with good accuracy. The model-based

Table 2.3: Images of color or texture rich (See color insert)

Class	Category	Images
Visual words (color)	Strawberry	
	Sunflower	
Visual words (texture)	Dalmatian	
	Pizza	

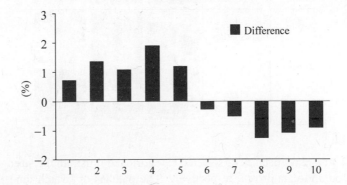

pipeline performs poorly in this case because it does not consider color nor can it find invariant patterns (e.g., pizzas have different shapes and toppings).

2.3.2.4 Accuracy vs. Training Size

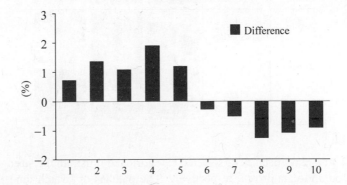

Fig. 2.6: Accuracy comparison (x-axis for the training numbers and y-axis for the accuracy model-based minus data-driven)

We varied the number of training instances for each category from one to ten to further investigate the strengths of model-based and data-driven. The x-axis of Figure 2.6 indicates the number of training instances, and the y-axis the accuracy of model-base subtracted by the accuracy of data-driven. A positive margin means that the model-based outperforms the data-driven in class prediction, whereas a negative means that the data-driven outperforms. When the size of training data is below five, the model-based outperforms data-driven. As we have observed from the previous results, model-based can do well with a small number of training instances on objects of invariant features. On the contrary, a data-driven approach cannot be productive when the number of training instances is scarce, as its name suggests. Also as we have explained, even the features of an object class can be quite predictable, varying camera and environmental parameters can produce images of different scales, orientations, and colors. A data-driven pipeline is not expected to do well unless it can get ample samples for capturing these variations.

2.3.2.5 Discussion

Table 2.4: Texture examples of some images (See color insert)

Class	Category	Images
Rigid body	Watch	
	Windsor chair	
Texture	Dalmatian	
	Pizza	

Table 2.4 displays patches of selected categories to illustrate the strengths of the model-based and data-driven, respectively. The patches of 'watch' and 'Windsor-chair' show patterns of almost identical edges. Model-based thus works well with these objects.

The 'dalmatian' and 'pizza' images tell a different story. The patches of these objects are similar but not identical. Besides, these patterns do not have strong edges. The color and texture patterns extracted by the data-driven pipeline can be more productive for recognizing these objects when sufficient samples are available.

Note: Let us revisit the result presented in Table 2.1. It is easy to make the data-driven look better by adding more image categories favoring data-driven to the testbed. Or the bias of a dataset can favor one approach over the other. Thus, evaluating which model works better cannot be done by looking only at the average accuracy. On a dataset of a few hundred or thousand categories, evaluation should be performed on individual categories.

2.3.3 DMD vs. Individual Models

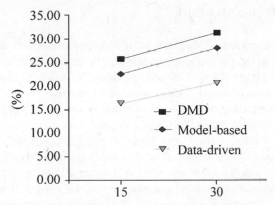

Fig. 2.7: DMD versus individual models

Fusing model-based and data-driven seems like a logical approach. Figure 2.7 shows that DMD outperforms individual models in classification accuracy on training sizes of both 15 and 30. Table 2.5 reports four selected categories on which we also performed training on up to 700 training instances.

Table 2.5: Fusion accuracy with large training pool

Category	Model-based		Data-driven		DMD	
	30	Large [2]	30	Large [2]	30	Large [2]
Plane	36.60%	**97.17%**	30.41%	90.57%	29.51%	96.23%
Watch	33.97%	**60.87%**	8.61%	43.48%	24.40%	55.07%
Skirts	19.82%	53.66%	35.14%	58.54%	29.73%	**65.85%**
Sunflower	18.18%	25.00%	60.00%	65.00%	63.64%	**70.00%**

2.3.3.1 Where Fusion Helps

For objects that the data-driven model works better, adding more features from the model-based pipeline can be helpful. The data-driven model needs a large number of examples to characterize an object (typically deformable as we have discussed), and those features produced by the model-based pipeline can help. On 'skirts' and 'sunflower', DMD outperforms data-driven when 100 images were used for training 'skirts', and 65 for 'sunflower'. (These are the maximum amount of training data that can be obtained from ImageNet for these categories.) On these deformable objects, we have explained why the data-driven approach is more productive. Table 2.5 shows that the model-based features employed by DMD can help further improve class-prediction accuracy.

2.3.3.2 Where Fusion May Hurt

For objects that the model-based can achieve better prediction accuracy, adding more features form the data-driven pipeline can be counter-productive. For all object categories in Figure 2.4 with 30 training instances, adding data-driven features reduces classification accuracy. This is because for those objects where the model-based performs well, additional features may contribute noise. For instance, the different colors of watches can make watch recognition harder when color is considered as a feature. Table 2.5 shows that when 30 training instances were used, DMD 's accuracy on 'watch' was degraded by 25 percentile. However, the good news is that when ample samples were provided to capture almost all possible colors of watches, the prediction accuracy improved. When we used 170 watch images to train the 'watch' predictor, the accuracy of DMD trails data-driven by just five percentile. On the 'plane' category where we were able to get a hold of 700 training instances, both DMD and model-base enjoy the same degree of accuracy.

2.3.3.3 Strength-Dominant Fusion

One key lesson learned from this experiment is that the MAX operator in the pooling steps of DMD is indeed effective in telling features from noise. Therefore, DMD can consider amplifying the stronger signals from either the model-based or the data-driven pipeline. When the signals extracted from an image well match some patches generated by the model-based pipeline, the model-based classifier should dominate the final class-prediction decision. This simple adjustment to DMD can improve its prediction accuracy on objects of rigid bodies, and hence the average prediction accuracy. Another key lesson learned (though obvious) is that the amount of training instances must be large to make the data-driven pipeline effective. When the training size is small and model-based can be effective, the data-driven pipeline should be turned off.

2.3.4 Regularization Tuning

This experiment examined the effect of the realization step. Recall that regularization employs unsupervised schemes to achieve both feature selection and feature reduction. Table 2.6 shows the effect of the number of prototype patches n_p on the final prediction accuracy of DMD . When n_p is set between $2,000 \times 4$ and $2,500 \times 4$, the prediction accuracy is the best. A small n_p may cause underfitting and too large an n_p may cause overfitting. Parameter n_p decides selectivity, and its best setting must be tuned through an empirical process like this.

Table 2.6: Accuracy of different number of patches

$n_p/4$	Training Size		$n_p/4$	Training Size	
	15	30		15	30
200	21.01%	26.09%	1500	23.62%	29.04%
300	22.08%	27.82%	2000	**23.75%**	29.21%
500	22.53%	28.54%	2500	23.30%	**29.61%**
1000	23.00%	28.82%	3000	23.27%	29.42%

2.3.5 Tough Categories

Table 2.7 presents six categories where DMD cannot perform effectively with a small amount of training data. For instance, the best prediction accuracy with 30 training images on 'barrel' and 'cup' is 12% and 17%, respectively. These objects exhibit such diversified perceptual features, and neither model-based nor data-driven with a small set of training data can capture their characteristics. Furthermore, the 'cup' category is more than percepts, as liquid containers of different shapes and colors can be classified as a cup. To improve class-prediction accuracy on these challenging categories, more training data ought to be collected to make DMD effective.

2.4 Related Reading

The CBIR community has been striving to bridge the *semantic gap* [35] between low-level features and high-level semantics for over a decade. (For a comprehensive survey, please consult [27].) Despite the progress has been made on both feature extraction and computational learning, these algorithms are far from being completely successful. On feature extraction, scale-invariant feature transform (SIFT) was considered a big step forward in the last decade for extracting scale-invariant features.

Table 2.7: Tough categories (See color insert)

Category	Images
Beaver	
Barrel	
Cup	
Mayfly	
Wild cat	
Gramophone	

SIFT may have improved feature invariance, but it does not effectively deal with feature selectivity. Indeed, SIFT has shown to be effective in detecting near-replicas of images [36], but it alone has not been widely successful in more general recognition tasks. As for computational learning, discriminative models such as linear SVMs [37] and generative models such as latent Dirichlet Allocation (LDA) [38] have been employed to map features to semantics. However, applying these and similar models directly to low-level features are considered to be *shallow* learning, which may be too limited for modeling complex vision problems. Yoshua Bengio [9] argues in theory that some functions simply cannot be represented by architectures that are too shallow.

If shallow learning suffers from limitations, then why have't deep learning been widely adopted? Indeed, neuroscientists have studied how the human vision system works for over 40 years [2]. Ample evidences [39, 40, 41] indicate that the human visual system is a pipeline of multiple layers: from sensing pixels and detecting edges to forming patches, recognizing parts, and then composing parts into objects. Physiological evidences strongly suggest that *deep* learning, rather than *shallow*, is appropriate. Unfortunately, before the work of Hinton in 2006 [10] deep models were not fully embraced partly because of their high intensity of computation and partly because of the well known problem of local optima. Recent advancements in neuroscience [3] motivated computer scientists to revisit deep learning in two respects:

1. *Unsupervised learning in lower layers.* The first couple of layers of the visual system are unsupervised, whereas supervised learning is conducted at the latter layers.
2. *Invariance and selectivity tradeoff.* Layers in the visual system deal with invariance and selectivity alternately.

These insights have led to recent progress in deep learning. First, Salakhutdinow et al. [42] show that using an unsupervised learning algorithm to conduct pretraining at each layer, a deep architecture can achieve much better results. Second, Serre [8] shows that by alternating pooling and sampling operations between layers, a balance between feature invariance and selectivity can be achieved. The subsequent work of Lee [5] confirms these insights.

2.5 Concluding Remarks

In this chapter, we first conducted empirical study to compare the model-based and the data-driven approach to image annotation. From our experimental results, we learned insights to design DMD , a hybrid architecture of deep model-based and data-drive learning. We showed the usefulness of unsupervised learning at three steps: edge-pooling, sparse regularization, and part-pooling. Unsupervised learning plays a pivotal role in making good tradeoffs between invariance and selectivity, and between specificity and diversity. We also showed that the data-driven pipeline can always be helped by the model-based. However, the other way may introduce noise when the amount of training data is scarce. DMD makes proper adjustments in making model-based and data-driven complement each other to achieve good performance.

Besides perceptual features that can be directly extracted from images, researchers have also considered camera parameters, textual information surrounding images, and social metadata, which can be characterized as contextual features. Nevertheless, perceptual feature extraction remains to be a core research topic of computer vision and image processing. Contextual features can complement image content but cannot substitute perceptual features. How to combine context and content belongs to the topic of multimodal fusion, which we address in Chapters 7, 8 and 9.

References

1. Wang, Z., Xia, D., Chang, E.Y. A deep model-based and data-driven hybrid architecture for image annotation. In *Proceedings of ACM International Workshop on Very-Large-Scale Multimedia Corpus, Mining and Retrieval*, pages 13–18, 2010.
2. Hubel, D.H., Wiesel, T.N. Receptive fields and functional architecture of monkey striate cortex. *Journal of Physiology*, 195(1):215–243, 1968.

3. Miller, E. The prefrontal cortex and cognitive control. *Nature Reviews Neuroscience*, 1(1):59–66, 2000.
4. Potamianos, G., Neti, C., Luettin, J., Matthews, I. Audio-visual automatic speech recognition: An overview. In *Issues in Visual and Audio-Visual Speech Processing*. MIT Press, 2004.
5. Lee, H., Grosse, R., Ranganath, R., Ng, A. Convolutional deep belief networks for scalable unsupervised learning of hierarchical representations. In *Proceedings of International Conference on Machine Learning (ICML)*, 2009.
6. Serre, T. *Learning a dictionary of shape-components in visual cortex: comparison with neurons, humans and machines*. PhD Thesis, Massachusetts Institute of Technology, 2006.
7. Riesenhuber, M., Poggio, T. Are Cortical Models Really Bound by the Binding Problem. *Neuron*, 24(1):87–93, 1999.
8. Serre, T., Wolf, L., Bileschi, S., Riesenhuber, M., Poggio, T. Robust object recognition with cortex-like mechanisms. *IEEE Trans Pattern Anal Mach Intell*, 29(3):411–426, 2007.
9. Bengio, Y. *Learning Deep Architectures for AI*. Now Publishers Inc, 2009.
10. Hinton, G., Osindero, S., Teh, Y. A fast learning algorithm for deep belief nets. *Neural Computation*, 18(7):1527–1554, 2006.
11. Bengio, Y., LeCun, Y. Scaling learning algorithms towards AI. In *Large-Scale Kernel Machines*, pages 321–360. The MIT Press, 2007.
12. Loosli, G., Canu, S., Bottou, L. Training invariant support vector machines using selective sampling. In *Large Scale Kernel Machines*, pages 301–320. The MIT Press, 2007.
13. Yasuda, M., Banno, T., Komatsu, H. Color selectivity of neurons in the posterior inferior temporal cortex of the macaque monkey. *Cerebral Cortex*, 20(7):1630–1646, 2009.
14. Tsunoda, K., Yamane, Y., Nishizaki, M., Tanifuji, M. Complex objects are represented in macaque inferotemporal cortex by the combination of feature columns. *Nature Neuroscience*, 4:832–838, 2001.
15. Lampl, I., Ferster, D., Poggio, T., Riesenhuber, M. Intracellular measurements of spatial integration and the MAX operation in complex cells of the cat primary visual cortex. *Journal of Neurophysiology*, 92(5):2704, 2004.
16. Gawne, T., Martin, J. Responses of primate visual cortical V4 neurons to simultaneously presented stimuli. *Journal of Neurophysiology*, 88(3):1128, 2002.
17. Hubel, D., Wiesel, T. Receptive fields, binocular interaction and functional architecture in the cat's visual cortex. *Journal of Physiology*, 160(1):106, 1962.
18. Ranzato, M., Huang, F., Boureau, Y., LeCun, Y. Unsupervised learning of invariant feature hierarchies with applications to object recognition. In *Proceedings of IEEE CVPR*, 2007.
19. Jones, J., Palmer, L. An evaluation of the two-dimensional Gabor filter model of simple receptive fields in cat striate cortex. *Journal of Neurophysiology*, 58(6):1233, 1987.
20. Serre, T., Riesenhuber, M. Realistic modeling of simple and complex cell tuning in the hmax model, and implications for invariant object recognition in cortex. *MIT Technical Report*, 2004.
21. Ekanadham, C., Reader, S., Lee, H. Sparse deep belief net models for visual area V2. *Proceedings of NIPS*, 2008.
22. Olshausen, B., Anderson, C., Van Essen, D. A neurobiological model of visual attention and invariant pattern recognition based on dynamic routing of information. *Journal of Neuroscience*, 13(11):4700, 1993.
23. Walther, D., Serre, T., Poggio, T., Koch, C. Modeling feature sharing between object detection and top-down attention. *Journal of Vision*, 5(8):1041, 2005.
24. Chikkerur, S., Serre, T., Poggio, T. A Bayesian inference theory of attention: neuroscience and algorithms. *MIT Technical Report MIT-CSAIL-TR-2009-047*, 2009.
25. Chang, E., Li, B., Li, C. Toward perception-based image retrieval. *IEEE Content-Based Access of Image and Video Libraries*, pages 101–105, 2000.
26. Tong, S., Chang, E.Y. Support vector machine active learning for image retrieval. In *Proceedings of ACM International Conference on Multimedia*, pages 107–118. ACM New York, NY, USA, 2001.
27. Datta, R., Joshi, D., Li, J., Wang, J. Image retrieval: Ideas, influences, and trends of the new age. *ACM Computing Surveys (CSUR)*, 40(2):1–60, 2008.

28. Coren, S., Ward, L.M., Enns, J.T. *Sensation and Perception* 6th *edition*. John Wiely & Sons, 2003.

29. Leu, J. Computing a shape's moments from its boundary. *Pattern Recognition*, 24(10):949–957, 1991.

30. Tamura, H., Mori, S., Yamawaki, T. Textural features corresponding to visual perception. *IEEE Transactions on Systems, Man and Cybernetics*, 8(6):460–473, 1978.

31. Smith, J., Chang, S. Automated image retrieval using color and texture. *IEEE Trans. Pattern Anal. Mach. Intell.*, 1996.

32. Wu, P., Manjunath, B., Newsam, S., Shin, H. A texture descriptor for browsing and similarity retrieval. *Signal Processing: Image Communication*, 16(1-2):33–43, 2000.

33. Ma, W., Zhang, H. Benchmarking of image features for content-based retrieval. In *Proceedings of Asilomar Conference on Signals Systems and Computers*, pages 253–260, 1998.

34. Deng, J., Dong, W., Socher, R., Li, L., Li, K., Li., F.F. Imagenet: A large-scale hierarchical image database. In *Proceedings of IEEE CVPR*, pages 156–161, 2009.

35. Smeulders, A., Worring, M., Santini, S., Gupta, A., Jain, R. Content-based image retrieval at the end of the early years. *IEEE Trans. Pattern Anal. Mach. Intell.*, 22(12):1349–1380, Dec 2000.

36. Ke, Y., Sukthankar, R. PCA-SIFT: A more distinctive representation for local image descriptors. In *Proceedings of IEEE CVPR*, pages 506–513, 2004.

37. Chapelle, O., Haffner, P., Vapnik, V. SVMs for histogram-based image classification. *IEEE Transactions on Neural Networks*, 10(5):1055, 1999.

38. Blei, D., Ng, A., Jordan, M. Latent dirichlet allocation. *Journal of Machine Learning Research*, 3:993–1022, 2003.

39. Serre, T., L., W., Poggio, T. Object recognition with features inspired by visual cortex. In *Proceedings of IEEE CVPR*, 2005.

40. Gross, C. Visual functions of inferotemporal cortex. *Handbook of Sensory Physiology*, 7(3), 1973.

41. Gross, C., Rocha-Miranda, C., Bender, D., . Visual properties of neurons in inferotemporal cortex of the macaque. *Journal of Neurophysiology*, 35(1):96–111, 1972.

42. Salakhutdinov, R., Mnih, A., Hinton, G. Restricted Boltzmann machines for collaborative filtering. In *Proceedings of International Conference on Machine Learning (ICML)*, pages 791–798, 2007.

Chapter 3
Query Concept Learning

Abstract With multimedia databases it is difficult to specify queries directly and explicitly. Relevance feedback interactively *learns* a user's desired output or *query concept* by asking the user whether certain proposed multimedia objects (e.g., images, videos, and songs) are relevant or not. For a learning algorithm to be effective, it must *learn* a user's query concept accurately and quickly, while also asking the user to label only a small number of data instances. In addition, the concept-learning algorithm should consider the *complexity* of a concept in determining its learning strategies. This chapter[†] presents the use of support vector machines active learning in a concept-dependent way (SVM^{CD}_{Active}) for conducting relevance feedback. A concept's complexity is characterized using three measures: *hit-rate*, *isolation* and *diversity*. To reduce concept complexity so as to improve concept learnability, a multimodal learning approach is designed to use the semantic labels of data instances to intelligently adjust the *sampling strategy* and the *sampling pool* of SVM^{CD}_{Active}. Empirical study on several datasets shows that active learning outperforms traditional passive learning, and concept-dependent learning is superior to concept-independent relevance-feedback schemes.
Keywords: Relevance feedback, active learning, kernel methods, query concept learning.

3.1 Introduction

One key design task, when constructing multimedia databases, is the creation of an effective relevance feedback component. While it is sometimes possible to arrange multimedia data instances (e.g., images) by creating a hierarchy, or by hand-labeling each data instance with descriptive words, these methods are time-consuming, costly, and subjective. Alternatively, requiring an end-user to specify a multime-

---

[†] ©ACM, 2004. This chapter is written based on the author's work with Simon Tong [1], Kingshy Goh, and Wei-Cheng Lai [2]. Permission to publish this chapter is granted under copyright licenses #2587600971756 and #2587601214143.

dia query in terms of low-level features (such as color and spatial relationships for images) is challenging to the end-user, because a query for images, videos, or music is hard to articulate, and articulation can vary from one user to another.

Thus, we need a way for a user to implicitly inform a database of his or her desired output or *query concept*. To address this requirement, relevance feedback can be used as a query refinement scheme to derive or learn a user's query concept. To solicit feedback, the refinement scheme displays a few media-data instances and the user labels each instance as "relevant" or "not relevant." Based on the responses, another set of data instances from the database is presented to the user for labeling. After a few such querying rounds, the refinement scheme returns a number of instances from the database that seem to fit the needs of the user.

The construction of such a query refinement scheme (hereafter called a *query-concept learner* or *learner*) can be regarded as a machine learning task. In particular, it can be seen as a case of *pool-based active learning* [3, 4]. In pool-based active learning the learner can access to a pool of unlabeled data and can request the user's label for a certain number of instances in the pool. In the image, video, or music retrieval domain, the unlabeled pool would be the entire database. An instance would be an image, a video clip, or a piece of music segment, and the two possible labelings for each media-data instance would be "relevant" or "not relevant." The goal for the learner is to learn the user's *query concept* — in other words, to label each data instance within the database in such a manner that the learner's labeling and the user's labeling will agree.

The main issue with active learning is finding a method for choosing informative data instances within the pool to ask the user to label. We call such a request for the label of a data instance a *pool-query*. Most machine learning algorithms are *passive* in the sense that they are generally applied using a randomly selected training set. The key idea with *active* learning is that it should choose its next pool-query based upon the past answers to previous pool-queries.

In general, such a learner must meet two critical design goals. First, the learner must learn target concepts accurately. Second, the learner must grasp a concept quickly, with only a small number of labeled instances, since most users do not wait around to provide a great deal of feedback. *Support vector machine active learner* (SVM_{Active}^{CD}) is effective to achieve these goals. SVM_{Active}^{CD} combines active learning with support vector machines (SVMs). SVMs [5, 6] have met with significant success in numerous real-world learning tasks. However, like most machine learning algorithms, SVMs use a randomly selected training set, which is not very useful in the relevance feedback setting. Recently, general purpose methods for active learning with SVMs have been independently developed by a number of researchers [7, 8, 9]. In the first part of this chapter, we use the theoretical motivation of [9] on active learning with SVMs to extend the use of support vector machines to the task of relevance feedback for image databases. (Readers can also consult [10] for the use of SVM active learning on music retrieval.)

When conducting a pool-query, SVM_{Active}^{CD} should consider the *complexity* of the target concept to adjust its *sampling strategies* for choosing informative data instances within the pool to ask user to label. In the second part of this chapter, we characterize *concept complexity* using three parameters: *scarcity*, *isolation* and *diversity*. Each of these parameters affects the learnability of a concept in a different way. The *hit-rate* (scarcity) is the percentage of instances matching the target concept in a database. When the *hit-rate* is low, it is difficult for active learning to find enough relevant instances to match the query. When *isolation* of a concept is poor, or several concepts are not well isolated from one another in the input space[1], the learner might confuse the target concept with its neighboring concepts. Finally, when a concept's *diversity* is high, the algorithm needs to be more explorative in the input space, rather than narrowly focused, so as to find the relevant instances that are scattered in the input space. We investigate how these complexity factors affect concept learnability, and then present a multimodal learning approach that uses the semantic labels (keywords) of media-data instances to guide our *concept-dependent active-learning* process. SVM_{Active}^{CD} adjusts the *sample pool* and *sampling strategy* according to concept complexity: it carefully selects the sample pool to improve *hit-rate* and *isolation* of the query concept, and it enhances the concept's learnability by adapting its sampling strategy to the concept's *diversity*.

Intuitively, SVM_{Active}^{CD} works by combining the following four ideas:

1. SVM_{Active}^{CD} regards the task of learning a target concept as one of learning an SVM binary classifier. An SVM captures the query concept by separating the relevant data instances from irrelevant ones with a hyperplane in a projected space, usually a very high-dimensional one. The projected points on one side of the hyperplane are considered relevant to the query concept and the rest irrelevant.
2. SVM_{Active}^{CD} learns the classifier quickly via active learning. The active part of SVM_{Active}^{CD} selects the most informative instances with which to train the SVM classifier. This step ensures fast convergence to the query concept in a small number of feedback rounds.
3. SVM_{Active}^{CD} learns the classifier in a concept-dependent way. With multimodal information from keywords and media-data features, SVM_{Active}^{CD} improves concept learnability for effectively capturing the target concept.
4. Once the classifier is trained, SVM_{Active}^{CD} returns the top-k most relevant data instances. These are the k data instances farthest from the hyperplane on the query concept side.

3.2 Support Vector Machines and Version Space

We shall consider SVMs in the binary classification setting. We are given training data $\{x_1 \ldots x_n\}$ that are vectors in some space $X \subseteq \mathbb{R}^d$. We are also given their labels $\{y_1 \ldots y_n\}$ where $y_i \in \{-1, 1\}$. In their simplest form, SVMs are hyperplanes that separate the training data by a maximal margin (see Figure 3.1). All vectors lying on

---

[1] The *input space* (denoted as X in machine learning and statistics literature) is defined as the original space in which the data vectors are located, and the *feature space* (denoted as F) is the space into which the data are projected, either linearly or non-linearly.

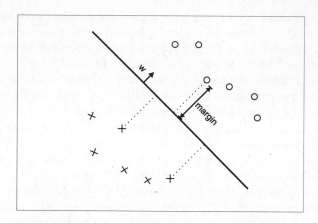

Fig. 3.1: A simple linear Support Vector Machine

one side of the hyperplane are labeled as -1, and all vectors lying on the other side are labeled as 1. The training instances that lie closest to the hyperplane are called *support vectors*. More generally, SVMs allow us to project the original training data in space X to a higher dimensional feature space F via a Mercer kernel operator K. In other words, we consider the set of classifiers of the form: $f(\mathbf{x}) = \sum_{i=1}^{n} \alpha_i K(\mathbf{x}_i, \mathbf{x})$. When $f(\mathbf{x}) \geq 0$ we classify $\mathbf{x}$ as $+1$, otherwise we classify $\mathbf{x}$ as -1.

When K satisfies Mercer's condition [5] we can write: $K(\mathbf{u}, \mathbf{v}) = \Phi(\mathbf{u}) \cdot \Phi(\mathbf{v})$ where $\Phi : \mathsf{X} \to \mathsf{F}$ and "$\cdot$" denotes an inner product. We can then rewrite f as:

$$f(\mathbf{x}) = \mathbf{w} \cdot \Phi(\mathbf{x}), \text{ where } \mathbf{w} = \sum_{i=1}^{n} \alpha_i \Phi(\mathbf{x}_i). \tag{3.1}$$

Thus, by using K we are implicitly projecting the training data into a different (often higher dimensional) feature space F. The SVM then computes the α_is that correspond to the maximal margin hyperplane in F. By choosing different kernel functions we can implicitly project the training data from X into space F. (Hyperplanes in F correspond to more complex decision boundaries in the original space X.)

Two commonly used kernels are the polynomial kernel $K(\mathbf{u}, \mathbf{v}) = (\mathbf{u} \cdot \mathbf{v} + 1)^p$, which induces polynomial boundaries of degree p in the original space X, and the radial basis function kernel $K(\mathbf{u}, \mathbf{v}) = (e^{-\gamma(\mathbf{u}-\mathbf{v}) \cdot (\mathbf{u}-\mathbf{v})})$, which induces boundaries by placing weighted Gaussians upon key training instances. In the remainder of this section we will assume that the modulus of the training data feature vectors are constant, i.e., for all training instances $\mathbf{x}_i$, $\|\Phi(\mathbf{x}_i)\| = \varphi$ for some fixed φ. The quantity $\|\Phi(\mathbf{x}_i)\|$ is always constant for radial basis function kernels, and so the assumption has no effect for this kernel. For $\|\Phi(\mathbf{x}_i)\|$ to be constant with the polynomial kernels we require that $\|\mathbf{x}_i\|$ be constant. It is possible to relax this constraint on $\Phi(\mathbf{x}_i)$. We shall discuss this option at the end of Section 3.3.

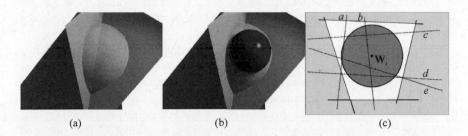

      (a)   (b)   (c)

Fig. 3.2: **(a)** Version space duality. The surface of the hypersphere represents unit weight vectors. Each of the two hyperplanes corresponds to a labeled training instance. Each hyperplane restricts the area on the hypersphere in which consistent hypotheses can lie. Here version space is the surface segment of the hypersphere closest to the camera. **(b)** An SVM classifier in version space. The dark embedded sphere is the largest radius sphere whose center lies in version space and whose surface does not intersect with the hyperplanes. The center of the embedded sphere corresponds to the SVM, its radius is the margin of the SVM in F and the training points corresponding to the hyperplanes that it touches are the support vectors. **(c)** Simple Margin Method [9].

Given a set of labeled training data and a Mercer kernel K, there is a set of hyperplanes that separate the data in the induced feature space F. We call this set of consistent hyperplanes or *hypotheses* the *version space* [11]. In other words, hypothesis f is in version space if for every training instance $\mathbf{x}_i$ with label y_i we have that $f(\mathbf{x}_i) > 0$ if $y_i = 1$ and $f(\mathbf{x}_i) < 0$ if $y_i = -1$. More formally:

Definition 3.1. Our set of possible hypotheses is given as:

$$\mathsf{H} = \left\{ f \mid f(\mathbf{x}) = \frac{\mathbf{w} \cdot \Phi(\mathbf{x})}{\|\mathbf{w}\|}, \text{ where } \mathbf{w} \in \mathsf{W} \right\},$$

where our *parameter space* W is simply equal to F. The *Version space*, V is then defined as:

$$\mathsf{V} = \{ f \in \mathsf{H} \mid \forall i \in \{1, 2 \dots n\}, \ y_i f(\mathbf{x}_i) > 0 \}.$$

Notice that since H is a set of hyperplanes, there is a bijection (an exact correspondence) between unit vectors $\mathbf{w}$ and hypotheses f in H. Thus we will redefine V as:

$$\mathsf{V} = \{ \mathbf{w} \in \mathsf{W} \mid \|\mathbf{w}\| = 1, \ y_i(\mathbf{w} \cdot \Phi(\mathbf{x}_i)) > 0, i = 1, 2 \dots n \}.$$

Note that a version space exists only if the *training* data are linearly separable in the feature space. Thus, we require linear separability of the training data in the feature space. This restriction is much less harsh than it might at first seem. First, the feature space often has a very high dimension and so in many cases it results in the data set being linearly separable. Second, as noted by [12], it is possible to modify any kernel so that the data in the newly induced feature space is linearly separable.

This is done by redefining all training instances $\mathbf{x}_i$: $K(\mathbf{x}_i, \mathbf{x}_i) \leftarrow K(\mathbf{x}_i, \mathbf{x}_i) + v$ where v is a positive regularization constant. The effect of this modification is to permit linear non-separability of the training data in the original feature space.

There exists a duality between the feature space F and the parameter space W [13, 14] which we shall take advantage of in the next section: points in F correspond to hyperplanes in W and *vice versa*.

Clearly, by definition, points in W correspond to hyperplanes in F. The intuition behind the converse is that observing a training instance $\mathbf{x}_i$ in feature space restricts the set of separating hyperplanes to ones that classify $\mathbf{x}_i$ correctly. In fact, we can show that the set of allowable points $\mathbf{w}$ in W is restricted to lie on one side of a hyperplane in W. More formally, to show that points in F correspond to hyperplanes in W, suppose we are given a new training instance $\mathbf{x}_i$ with label y_i. Then any separating hyperplane must satisfy $y_i(\mathbf{w} \cdot \Phi(\mathbf{x}_i)) > 0$. Now, instead of viewing $\mathbf{w}$ as the normal vector of a hyperplane in F, think of $y_i\Phi(\mathbf{x}_i)$ as being the normal vector of a hyperplane in W. Thus $y_i(\mathbf{w} \cdot \Phi(\mathbf{x}_i)) = \mathbf{w} \cdot y_i\Phi(\mathbf{x}_i) > 0$ defines a half-space in W. Furthermore $\mathbf{w} \cdot y_i\Phi(\mathbf{x}_i) = 0$ defines a hyperplane in W that acts as one of the boundaries to version space V. Notice that version space is a connected region on the surface of a hypersphere in parameter space. See Figure 3.2(a) for an example.

SVMs find the hyperplane that maximizes the margin in feature space F. One way to pose this is as follows:

$$\text{maximize}_{\mathbf{w} \in F} \quad \min_i\{y_i(\mathbf{w} \cdot \Phi(\mathbf{x}_i))\}$$
$$\text{subject to:} \quad \|\mathbf{w}\| = 1$$
$$y_i(\mathbf{w} \cdot \Phi(\mathbf{x}_i)) > 0 \quad i = 1 \ldots n.$$

By having the conditions $\|\mathbf{w}\| = 1$ and $y_i(\mathbf{w} \cdot \Phi(\mathbf{x}_i)) > 0$ we cause the solution to lie in version space. Now, we can view the above problem as finding the point $\mathbf{w}$ in version space that maximizes the distance $\min_i\{\mathbf{w} \cdot y_i\Phi(\mathbf{x}_i)\}$. From the duality between feature and parameter space, and since $\|\Phi(\mathbf{x}_i)\| = 1$, then each $y_i\Phi(\mathbf{x}_i)$ is a unit normal vector of a hyperplane in parameter space and each of these hyperplanes delimits the version space. Thus we want to find the point in version space that maximizes the minimum distance to any of the delineating hyperplanes. That is, SVMs find the center of the largest radius hypersphere whose center can be placed in version space and whose surface does not intersect with the hyperplanes corresponding to the labeled instances, as in Figure 3.2(b). It can be easily shown that the hyperplanes touched by the maximal radius hypersphere correspond to the support vectors and that the radius of the hypersphere is the margin of the SVM.

3.3 Active Learning and Batch Sampling Strategies

SVM_{Active}^{CD} performs the following two steps for each round of relevance feedback until the process is terminated by the user.

- *Sampling*. Select a batch of data instances and ask the user to label them as either "relevant" or "irrelevant" to the query concept, and
- *Learning*. Learn an SVM on data labeled thus far.

After the relevance feedback rounds have been performed, SVM_{Active}^{CD} retrieves the top-k most relevant data instances. The final SVM boundary separates "relevant" data instances from irrelevant ones. The top-k most "relevant" data instances are the k data instances farthest from the SVM boundary on the relevant side. The key step of SVM_{Active}^{CD} is its *sampling* step in which it selects most useful data instances to solicit user feedback. In the remainder of this section, we present the theoretical foundation of SVM_{Active}^{CD} (Section 3.3.1), and then four sampling strategies (Section 3.3.2).

3.3.1 Theoretical Foundation

In pool-based active learning we have a pool of unlabeled instances. It is assumed that the instances $\mathbf{x}$ are independently and identically distributed according to some underlying distribution $F(\mathbf{x})$, and the labels are distributed according to some conditional distribution $P(y \mid \mathbf{x})$.

Given an unlabeled pool U, an *active learner* ℓ has three components: (f, q, L). The first component is a classifier, $f : \text{L} \rightarrow \{-1, 1\}$, trained on the current set of labeled data L (and possibly unlabeled instances in U too). The second component $q(\text{L})$ is the querying function that, given a current labeled set L, decides which instance in U to query next. The active learner can return a classifier f after each pool-query (*online learning*) or after some fixed number of pool-queries.

The main difference between an active learner and a regular passive learner is the querying component q. This brings us to the issue of how to choose the next unlabeled instance in the pool to query. We use an approach that queries such instances in order to reduce the size of the version space as much as possible. We need one more definition before we can proceed:

Definition 3.2. *Area*(V) is the surface area that the version space V occupies on the hypersphere $\|\mathbf{w}\| = 1$.

We wish to reduce the version space as fast as possible. Intuitively, one good way of doing this is to choose a pool-query that halves the version space. More formally, we can use the following lemma to motivate which instances to use as our pool-query:

Lemma 3.1. *(Tong & Koller, 2000) Suppose we have an input space X, finite dimensional feature space F (induced via a kernel K), and parameter space W. Suppose active learner ℓ^* always queries instances whose corresponding hyperplanes in parameter space W halves the area of the current version space. Let ℓ be any other active learner. Denote the version spaces of ℓ^* and ℓ after i pool-queries as V_i^* and V_i respectively. Let P denote the set of all conditional distributions of y given $\mathbf{x}$. Then,*

$$\forall i \in \mathbb{N}^+ \sup_{P \in \mathsf{P}} E_P[Area(\mathsf{V}_i^*)] \le \sup_{P \in \mathsf{P}} E_P[Area(\mathsf{V}_i)],$$

with strict inequality whenever there exists a pool-query $j \in \{1 \ldots i\}$ by ℓ that does not halve version space V_{j-1}.

This lemma says that, for any given number of pool-queries, ℓ^* minimizes the maximum expected size of the version space, where the maximum is taken over all conditional distributions of y given $\mathbf{x}$.

Now, suppose $\mathbf{w}^* \in \mathsf{W}$ is the unit parameter vector corresponding to the SVM that we would have obtained had we known the actual labels of *all* of the data in the pool. We know that $\mathbf{w}^*$ must lie in each of the version spaces $\mathsf{V}_1 \supset \mathsf{V}_2 \supset \mathsf{V}_3 \ldots$, where V_i denotes the version space after i pool-queries. Thus, by shrinking the size of the version space as much as possible with each pool-query we are reducing as fast as possible the space in which $\mathbf{w}^*$ can lie. Hence, the SVM that we learn from our limited number of pool-queries will lie close to $\mathbf{w}^*$.

This discussion provides motivation for an approach in which we query instances that split the current version space into two equal parts insofar as possible. Given an unlabeled instance $\mathbf{x}$ from the pool, it is not practical to explicitly compute the sizes of the new version spaces V^- and V^+ (i.e., the version spaces obtained when $\mathbf{x}$ is labeled as -1 and $+1$ respectively). There is a way of approximating this procedure as noted by [9]:

Simple Method. Recall from Section 3.2 that, given data $\{\mathbf{x}_1, \mathbf{x}_2 \ldots \mathbf{x}_i\}$ and labels $\{y_1, y_2 \ldots y_i\}$, the SVM unit vector $\mathbf{w}_i$ obtained from this data is the center of the largest hypersphere that can fit inside the current version space V_i. The position of $\mathbf{w}_i$ in the version space V_i clearly depends on the shape of the region V_i; however, it is often approximately in the center of the version space. Now, we can test each of the unlabeled instances $\mathbf{x}$ in the pool to see how close their corresponding hyperplanes in W come to the centrally placed $\mathbf{w}_i$. The closer a hyperplane in W is to the point $\mathbf{w}_i$, the more centrally it is placed in version space, and the more it bisects version space. Thus we can pick the unlabeled instance in the pool whose hyperplane in W comes closest to the vector $\mathbf{w}_i$. For each unlabeled instance $\mathbf{x}$, the shortest distance between its hyperplane in W and the vector $\mathbf{w}_i$ is simply the distance between the feature vector $\Phi(\mathbf{x})$ and the hyperplane $\mathbf{w}_i$ in F — which is easily computed by $|\mathbf{w}_i \cdot \Phi(\mathbf{x})|$. This results in the natural Simple rule:

- Learn an SVM on the existing labeled data and choose as the next instance to query the pool instance that comes closest to the hyperplane in F.

Figure 3.2(c) presents an illustration. In the stylized picture we have flattened out the surface of the unit weight vector hypersphere that appears in Figure 3.2(a). The white area is version space V_i which is bounded by solid lines corresponding to labeled instances. The five dotted lines represent unlabeled instances in the pool. The circle represents the largest radius hypersphere that can fit in the version space. Note that the edges of the circle do not touch the solid lines — just as the dark sphere in Figure3.2(b) does not meet the hyperplanes on the surface of the larger hypersphere (they meet somewhere under the surface). The instance $\mathbf{b}$ is closest to the SVM $\mathbf{w}_i$ and so we will choose to query $\mathbf{b}$.

Radial basis function kernels are good kernel choices. As noted in Section 3.2, radial basis function kernels have the property that $\| \varPhi(\mathbf{x}_i) \| = \lambda$. The Simple querying method can still be used with other kernels when the training data feature vectors do not have a constant modulus, but the motivating explanation no longer holds since the SVM can no longer be viewed as the center of the largest allowable sphere. However, alternative motivations have recently been proposed by Campbell, Cristianini and Smola [7] that do not require a constraint on the modulus.

3.3.2 Sampling Strategies

For the information retrieval, we have a need for performing multiple pool-queries at the same time. It is not practical to present one data instance at a time for the user to label, because he or she is likely to lose patience after a few rounds. To prevent this from happening, we present the user with multiple data instances (say, h) at each round of pool-querying. Thus, for each round, the active learner has to choose not just one data instance to be labeled but h. Theoretically it would be possible to consider the size of the resulting version spaces for each possible labeling of each possible set of h pool-queries, but clearly this would be impractical. Thus instead, for matters of computational efficiency, SVM_{Active}^{CD} employs heuristic methods for choosing unlabeled instances. We present and examine the following four sampling strategies: *batch-simple*, *speculative*, *angle-diversity*, and *error-reduction*.

3.3.2.1 Batch-Simple Sampling

The *batch-simple* strategy chooses h unlabeled instances closest to the separating hyperplane (between the relevant and the irrelevant instances in the feature space) to solicit user feedback. Figure 3.3 summarizes the algorithm. Based on the labeled pool L, the algorithm first trains a binary classifier f (step 1). The binary classifier f is then applied to the unlabeled pool U to compute each unlabeled instance's distance to the separating hyperplane (step 2). The h unlabeled instances closest to the hyperplane and relatively apart from each other are chosen as the next batch of samples for conducting pool-queries.

3.3.2.2 Speculative Sampling

One can consider a *speculative* procedure, which recursively generates samples by speculating user feedback. The *speculative* procedure is computationally intensive. It can be used as a *yardstick* to measure how well the other active-learning strategies perform. The algorithm starts by finding one most informative sample (the closest unlabeled instance to the hyperplane). It then speculates upon the two possible labels of the sample, and generates two more samples, one based on the positive specula-

Batch-Simple

Input: L, U, h; /* labeled set, unlabeled set, batch size

Output: S; /* Pool-queries or samples

Procedures:

● SVM$_{Train}$() /* SVM training algorithm

● f() /* Classifier learned by SVMs on L

Initialization:

● $S \leftarrow \emptyset$

BEGIN

  1. $f \leftarrow$ SVM$_{Train}$(L);

  2. For each $\mathbf{x}_i \in$ U

    $\mathbf{x}_i.distance \leftarrow |f(\mathbf{x}_i)|$;

  3. While ($|S| < h$)

    $\mathbf{x}_s \leftarrow argmin_{\mathbf{x}_i \in U}(\mathbf{x}_i.distance)$;

    $S \leftarrow S \cup \{\mathbf{x}_s\}; U \leftarrow U - \{\mathbf{x}_s\}$;

  4. return S;

END

Fig. 3.3: Batch-simple sampling algorithma

tion and one based on negative speculation. The algorithm speculates recursively, generating a binary tree of samples. Figure 3.4 presents the *speculative* algorithm. Steps 6 and 8 of the algorithm speculate the pool-query to be positive and negative, respectively, and recursively call the *speculative* procedure to select the next samples. The *speculative* procedure terminates after at least h samples have been generated.

3.3.2.3 Angle-Diversity Sampling

The main idea (described in Step 2 of Figure 3.5) of *angle-diversity* [15] is to select a collection of samples close to the classification hyperplane, and at the same time, maintain their diversity. The diversity of samples is measured by the angles between the samples. Given an example $\mathbf{x}_i$, its normal vector is equal to $\Phi(\mathbf{x}_i)$. The angle between two hyperplanes $\mathbf{h}_i$ and $\mathbf{h}_j$, corresponding to instances $\mathbf{x}_i$ and $\mathbf{x}_j$, can be written in terms of the kernel operator K:

$$|\cos(\angle(\mathbf{h}_i, \mathbf{h}_j))| = \frac{|\Phi(\mathbf{x}_i) \cdot \Phi(\mathbf{x}_j)|}{\|\Phi(\mathbf{x}_i)\| \|\Phi(\mathbf{x}_j)\|} = \frac{|K(\mathbf{x}_i, \mathbf{x}_j)|}{\sqrt{K(\mathbf{x}_i, \mathbf{x}_i)K(\mathbf{x}_j, \mathbf{x}_j)}}.$$

The *angle-diversity* algorithm starts with an initial hyperplane $\mathbf{h}_i$ trained by the given labeled set L. Then, for each unlabeled instance $\mathbf{x}_j$, it computes its distance to the classification hyperplane $\mathbf{h}_i$. The angle between the unlabeled instance $\mathbf{x}_j$ and the current sample set S is defined as the maximal angle from instance $\mathbf{x}_j$ to any instance $\mathbf{x}_s$ in set S. This angle measures how diverse the resulting sample set S would be, if instance $\mathbf{x}_j$ were to be chosen as a sample.

Speculative

Input: L, U, h; /* labeled set, unlabeled set, batch size

Output: S; /* Pool-queries or samples

Procedures:

• SVM$_{Train}$() /* SVM training algorithm

• $f()$ /* Classifier learned by SVMs on L

Initialization:

• $S \leftarrow \emptyset$

BEGIN

  1. if ($h = 0$) then return $\emptyset$;

  2. $f \leftarrow$ SVM$_{Train}$(L);

  3. For each $\mathbf{x}_i \in$ U

    $\mathbf{x}_i.distance \leftarrow | f(\mathbf{x}_i) |$;

  4. $\mathbf{x}_s \leftarrow argmin_{\mathbf{x}_i \in U}(\mathbf{x}_i.distance)$;

  5. U $\leftarrow$ U $- \{\mathbf{x}_s\}$; L $\leftarrow$ L $\cup \{\mathbf{x}_s\}$;

  6. $\mathbf{x}_s.label \leftarrow +$;

  7. $S \leftarrow S \cup Speculative(L, U, (h-1)/2)$;

  8. $\mathbf{x}_s.label \leftarrow -$;

  9. $S \leftarrow S \cup Speculative(L, U, h - (h-1)/2)$;

  10. $S \leftarrow S \cup \{\mathbf{x}_s\}$;

  11. return S;

END

Fig. 3.4: Speculative sampling

Algorithm *angle-diversity* introduces parameter λ to balance two components: the distance to the classification hyperplane and the diversity of angles among samples. Incorporating the trade-off factor, the final score for the unlabeled instance $\mathbf{x}_i$ can be written as

$$\lambda * |f(\mathbf{x}_i)| + (1 - \lambda) * (\max_{\mathbf{x}_j \in S} \frac{|k(\mathbf{x}_i, \mathbf{x}_j)|}{\sqrt{K(\mathbf{x}_i, \mathbf{x}_i)K(\mathbf{x}_j, \mathbf{x}_j)}}), \tag{3.2}$$

where function f computes the distance to the hyperplane, function K is the kernel operator, and S the training set. After that, the algorithm selects the unlabeled instance that enjoys the smallest score in U as the sample. The algorithm repeats the above steps h times to select h samples. In practice, with trade-off parameter λ set at 0.5, [15] shows that the algorithm achieves good performance "on the average." An example in Section 3.4.3.4 shows that λ can be adjusted in a concept-dependent way according to the *diversity* of the target concept.

3.3.2.4 Error-Reduction Sampling

Arriving from another perspective, Roy and McCallum [16] proposed an active learning algorithm that attempts to reduce the expected error on future test exam-

Angle Diversity

Input: L, U, h, λ; /* Training set, unlabeled set, batch size, weighting parameter

Output: S; /* Sample set

Procedures:

• SVM$_{Train}$() /* SVM training algorithm

• $f()$ /* Classifier learned by SVMs on L

• $K()$; /* SVM kernel function

Initialization:

• $S \leftarrow \emptyset$

BEGIN

1. $f \leftarrow$ SVM$_{Train}$(L);

2. While ($|S| < h$)

$\mathbf{x}_s \leftarrow argmin_{\mathbf{x}_i \in U}((\lambda * |f(\mathbf{x}_i)| + (1-\lambda) * (\max_{\mathbf{x}_j \in S} \frac{|k(\mathbf{x}_i, \mathbf{x}_j)|}{\sqrt{K(\mathbf{x}_i, \mathbf{x}_i)K(\mathbf{x}_j, \mathbf{x}_j)}}));$

$S \leftarrow S \cup \{\mathbf{x}_s\};$

3. return S;

END

Fig. 3.5: Angle diversity sampling

ples. In other words, their approach aims to reduce future generalization error. Since the true error on future examples cannot be known in advance, Roy and McCallum proposed a method estimating future error. Suppose we are given a labeled set L and an unlabeled set $U = \{\mathbf{x}_1, \mathbf{x}_2 \ldots \mathbf{x}_n\}$ where each $\mathbf{x}_i$ is a vector. The distribution $P(\mathbf{x})$ of the vectors is assumed to be i.i.d. In addition, each data instance $\mathbf{x}_i$ is associated with a label $y_i \in \{-1, 1\}$ according to some unknown conditional distribution $P(y|\mathbf{x})$. The classifier trained by the labeled set L can estimate an output distribution $\hat{P}_L(y|\mathbf{x})$ for a given input $\mathbf{x}$. Then the expected error of the classifier can be written as

$$E[Error_{\hat{P}_L}] = \int_{\mathbf{x}} Loss(P(y|\mathbf{x}), \hat{P}_L(y|\mathbf{x}))P(\mathbf{x})d\mathbf{x}.$$

The function *Loss* is some loss function employed to measure the difference between the true distribution, $P(y|\mathbf{x})$, and its estimation, $\hat{P}_L(y|\mathbf{x})$. One popular loss function is the log loss function which is defined as follows:

$$Loss(P(y|\mathbf{x}), \hat{P}_L(y|\mathbf{x})) = \sum_{y \in \{-1,1\}} P(y|\mathbf{x}) \log(\hat{P}_L(y|\mathbf{x})).$$

The algorithm proposed by Roy and McCallum selects a query, $\mathbf{x}^*$, that causes minimal error. The algorithm includes $\mathbf{x}^*$ in its sample set if $E[Error_{\hat{P}_{L \cup \{\mathbf{x}^*\}}}]$ is smaller than $E[Error_{\hat{P}_{L \cup \{\mathbf{x}\}}}]$ for any other instance $\mathbf{x}$. Figure 3.6 summarizes the algorithm. Given the training pool L, the algorithm first computes the current classifier's posterior $\hat{P}_L(y|\mathbf{x})$ in step 3. For each unlabeled data instance $\mathbf{x}$ with each possible label y, the algorithm then adds the pair $(\mathbf{x}, y)$ to the training set, re-trains the classifier with the enlarged training set, and computes the expected log loss.

Error Reduction

Input: L, U, h; /* Training set, unlabeled set, batch size

Output: S; /* Sample set

Procedures:

• $\text{SVM}_{Train}()$,

• $f()$, $g()$ /* Classifier learned by SVMs

• $Regression()$ /* Perform logic regression

Variables:

• Γ, Ψ; /* local training and test set

Initialization:

• $S \leftarrow \emptyset$

BEGIN

1. $f \leftarrow \text{SVM}_{Train}(L)$;

2. For each $\mathbf{x}_i \in U$

$\hat{P}_L(y|\mathbf{x}_i) \leftarrow Regression(|f(\mathbf{x}_i)|)$;

$\Psi \leftarrow U - \{\mathbf{x}_i\}$;

$\mathbf{x}_i.loss \leftarrow 0$;

$\Gamma \leftarrow L \cup \{\mathbf{x}_i, 1\}$; /* assign positive label

$g \leftarrow \text{SVM}_{Train}(\Gamma)$;

For each $\mathbf{x}_j \in U$

$\hat{P}_\Gamma(y|\mathbf{x}) \leftarrow Regression(|g(\mathbf{x}_j)|)$;

$\mathbf{x}_i.loss \leftarrow \mathbf{x}_i.loss + \hat{P}_L(y|\mathbf{x}_i) * \sum_{y \in \{-1,1\}} \hat{P}_\Gamma(y|\mathbf{x}) \log(\hat{P}_\Gamma(y|\mathbf{x}))$;

$\Gamma \leftarrow L \cup \{\mathbf{x}_i, -1\}$; /* assign negative label

$g \leftarrow \text{SVM}_{Train}(\Gamma)$;

For each $\mathbf{x}_j \in U$

$\hat{P}_\Gamma(y|\mathbf{x}) \leftarrow Regression(|g(\mathbf{x}_j)|)$;

$\mathbf{x}_i.loss \leftarrow \mathbf{x}_i.loss + (1 - \hat{P}_L(y|\mathbf{x}_i)) * \sum_{y \in \{-1,1\}} \hat{P}_\Gamma(y|\mathbf{x}) \log(\hat{P}_\Gamma(y|\mathbf{x}))$;

3. While ($|S| < h$)

$\mathbf{x}_s \leftarrow argmin_{\mathbf{x}_i \in U}(\mathbf{x}_i.loss)$; $S \leftarrow S \cup \{\mathbf{x}_s\}$;

4. return S;

END

Fig. 3.6: Error reduction sampling

Steps 4 and 5 of the algorithm compute the expected log loss by adding $(\mathbf{x}, y)$ to the training data.

3.4 Concept-Dependent Learning

Ideally, concept learning should be done in a concept-dependent manner. For a simple concept, we can employ, e.g., algorithm *angle-diversity* or *batch-simple* to learn the concept. For a complex concept, we must make proper adjustments in the learning algorithm. We define concept complexity as the level of difficulty in learning a target concept. To model concept complexity, we can use three quantitative measures [17]: *scarcity*, *isolation*, and *diversity*.

The remainder of this section will first define some measures for quantifying query complexity, and then discuss major limitations of current active learning algorithms. We then outline detailed algorithms of the concept-dependent component of $\mathrm{SVM}^{CD}_{Active}$, which uses keywords to guide learning in the feature space, to alleviate the limitations. To illustrate how $\mathrm{SVM}^{CD}_{Active}$ works in this section, we use image retrieval as the example application.

3.4.1 Concept Complexity

Before we can compute the concept complexity, we need to know the user's target concept, which is unavailable beforehand. Fortunately, we often have a rough description of a data instance's semantic content from its keyword-annotation (even though the annotation quality might not be perfect). For instance, most image search engines (e.g., Google) use surrounding texts of an image to provide it some initial (noisy) labels. Query logs can then be used to refine these labels. If an image was clicked often when query "lion" was issued, then that image may contain lion-related semantics in high probability. We treat each label as a concept, and pre-compute *scarcity*, *isolation*, and *diversity* to characterize concept complexity in advance. Similarly, concept complexity of texts, video clips, and music can be obtained in the same fashion.

3.4.1.1 Scarcity

Scarcity measures how well-represented a concept is in the retrieval system. We use *hit-rate*, defined as the percentage of data matching the concept, to indicate scarcity. As we assume that each keyword is equivalent to a concept, the hit-rate of a keyword is the number of images being annotated with that keyword. This parameter is dataset-dependent; while a concept such as *photography* is very general and may produce a high hit-rate, other general concepts such as *raptors* may be scarce simply because the system does not contain many matching images. Similarly, a very specific concept such as *laboratory coat* could have a high hit-rate solely because the system has many such images.

3.4.1.2 Isolation

Isolation characterizes a concept's degree of separation from the other concepts. We measure two types of isolation: *input space* isolation and *keyword* isolation. The input space isolation is considered low (or poor) when the concept is commingled with others in the space formed by extracted features (e.g., colors, shape, and textures of images). When the keywords used to describe a concept has several meanings or senses, the keyword isolation is considered poor; very precise keywords provide good isolation. An example of a poorly-isolated keyword is "feline," which is often used to annotate images of tigers, lions and domestic cats. If the user seed a query with "feline," it is difficult to tell which sense of the word the user has in mind. A well-isolated seed keyword like "Eiffel Tower" has less ambiguity; we can be quite sure that the user is thinking about the famous Parisian landmark. We estimate isolation as follows:

1. *Input Space Isolation.* We characterize the isolation of a concept in the perceptual input space with:

$$I_s(T, \sigma) = \frac{1}{N_T} \sum_{\mathbf{x}_i \in T} \frac{N(T)}{N(\sigma)}. \tag{3.3}$$

For each instance $\mathbf{x}_i$ that belongs to the target concept T, $N(\sigma)$ is the number of instances that are within an L_1 distance of σ from $\mathbf{x}_i$, and $N(T)$ is the number such instances that are from T. N_T is the total number of instances from T in the entire dataset. In essence, $I_s(T, \sigma)$ gives the percentage of nearest neighbors (NNs) within a distance of σ that are from the target concept. The higher the value of I_s, the more isolated is the target concept. A low value indicates that the target concept is commingled with others in the input space. Since I_s is only an estimation, we use σ values ranging from 1 to 9 in increments of one.

2. *Mining Keyword Associations for Keyword Isolation.* To measure keyword isolation, we employ association-rules mining [18] to model the co-occurrences of keywords. Given a query keyword w_q, we find the set of images I that are annotated with w_q. Let W be the set of keywords, besides w_q, that are used to annotate the images in I. For each 2-itemset $\{w_q, w_i\}$, where $w_i \in W$, we compute the confidence (C_{qi}) of the rule $w_q \Rightarrow w_i$. We define isolation of w_q with respect to w_i as

$$I_k(w_q, w_i) = C_{qi} \times (1 - C_{qi}). \tag{3.4}$$

When C_{qi} is close to 0.5, I_k is at its maximum, which implies that the rule is highly ambiguous, and that w_q is poorly isolated from w_i. Let us consider two query keywords "fruit" and "apple." Suppose the confidence value for the rule *fruit* $\Rightarrow$ *apple* is 0.5, and for *apple* $\Rightarrow$ *fruit* it is 0.7. Using Eq.(3.4), we get $I_k = 0.25$ for the first rule, and $I_k = 0.21$ for the second rule. We say that the keyword "fruit" is poorly isolated from "apple", whereas "apple" is well isolated from "fruit." Although "fruit" is a more general word than "apple" according to rules of linguistics, this is not necessarily true in the annotations of the image-set. Therefore, employing a general thesaurus such as Wordnet may be counter-productive for determining keyword isolation. The isolation value must be computed in a dataset-dependent fashion.

The causes of poor keyword isolation could be due that the words are synonyms, or that some images contain semantics or objects described by the words. To quantify isolation, we need not discern the reason of co-occurrences. When some words often appear together with the query keyword in I, it indicates that the query keyword is ambiguous. For instance, the word "apple" in our dataset is

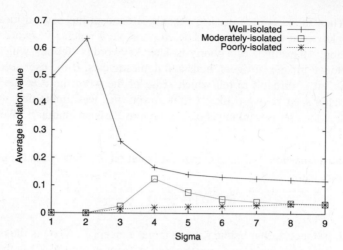

Fig. 3.7: Average isolation values

often associated with either "computer" or "fruit." Therefore, SVM^{CD}_{Active} needs to determine whether it is "computer" or "fruit" that is more relevant to an "apple" query. With association-rules mining, the derived keyword isolation is dataset-dependent. If our dataset only contains images of "fruit apple", the concept "apple" will have good semantic isolation. Thus, we only need to determine the semantic of the word when it appears in more than one 2-itemsets.

Some results of empirical studies are presented here to show that concept isolation can affect retrieval accuracy. The dataset consists of 300K images that are annotated with up to 40K annotation keywords altogether. We chose 6K representative keywords — words that are not too common or too rare — to represent the semantic classes present in this dataset. Figure 3.7 plots the average isolation value (I_s in Eq. 3.3) at $\sigma = 1 \ldots 9$ for the 6K concepts. Based on the I_s values, we divide the concepts into three categories of isolation: well-, moderately- and poorly-isolated. We notice that well-isolated concepts generally have higher I_s values at smaller σ, implying that similar NNs are nearby; poorly-isolated concepts generally have very low I_s values.

Next, we use the *angle-diversity* algorithm with $\lambda = 0.5$ to study the effect of isolation on retrieval accuracy. To circumvent the scarcity problem, we use the keywords to seed each query with a positive and negative example. Relevance feedback is then conducted purely on perceptual feature. We are not able to use keywords to bound the search as we are using keywords to indicate the query concept. This experimental setup is different from that in Section 3.5 where real users interact with the retrieval system to conduct relevance feedback. In Figure 3.8, we plot the average top-20 precision rates using 5 iterations of relevance feedback for the three types of isolation. As expected, the plot shows that concepts with good isolation achieve higher precision rates.

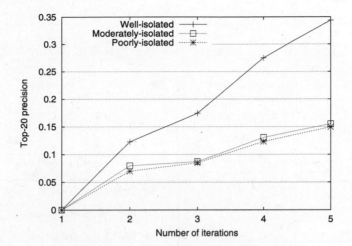

Fig. 3.8: Precision for various input-space isolation types

3.4.1.3 Diversity

The diversity of a concept is characterized by the way relevant images are distributed in the input space; it is an indication on how learnable a concept is. A diverse concept has relevant images scattered all over the input space. For example, the *flowers* concept, which encompasses flowers of different colors and types, is more diverse than the *red roses* concept. Instead of concentrating the sampling effort on a few small subspaces, we want to explore more subregions of the input space and select diverse samples to present to the user. As diversity is related to the spread of the relevant images in the input space, we can characterize concept diversity using the spatial variance of images belonging to a concept in the perceptual input space [19]. Alternatively, we can make use of the TSVQ clusters from the indexer of our dataset [20]. Since clustering will group instances based on how they are scattered in the input space, we characterize diversity using the distance between centroids of the clusters that contain instances from the target concept.

3.4.2 Limitations of Active Learning

When the target concept instances are scarce and not well isolated, active learning can be ineffective for locating relevant images.

   The first limitation is related to scarcity—the availability of images relevant to the concept to be learned. Most active learning algorithms require at least one positive (relevant) and one negative (irrelevant) instance to begin the learning process. A common approach is to randomly sample images from the input space. If none of the samples are labeled as positive, they are all treated as negatives, which will be

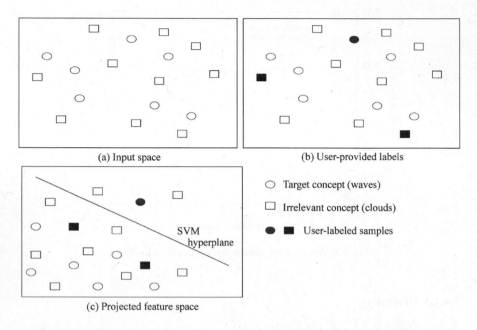

(a) Input space   (b) User-provided labels

○ Target concept (waves)

□ Irrelevant concept (clouds)

● ■ User-labeled samples

(c) Projected feature space

Fig. 3.9: Example illustrating how poor concept isolation can hurt target concept learning

used to prune the sample space by applying an algorithm like [21]. The seed-image sampling continues until we obtain a positive image. While this process is viable, it may not be effective if the matching images are scarce. When the percentage of matching images is low, even if we can winnow irrelevant images from the unlabeled pool, the probability of finding a positive image can still be quite low. For instance, if a concept has 5% matching images, the probability is about 66% that at least one of the 20 randomly selected images is positive. But when the concept is as scarce as 1%, the probability declines to 20%. One can easily find scenarios where the number of concept-matching images is much lower than 1%. Under such conditions, active learning can take many iterations just to find one relevant image.

The second limitation — concept isolation — further complicates the problem. When visually-similar concepts overlap substantially in the input space, the learner might confuse the target concept with an irrelevant one. Figure 3.9 illustrates this problem. The squares and circles represent images from two different concepts, which overlap each other in the input space (Figure 3.9(a)). Suppose three images presented to the user have been labeled as follows: two squares are labeled as "irrelevant" (filled squares), and one circle is labeled "relevant" (filled circle in Figure 3.9(b)). The circles around the filled squared can easily be mistakenly inferred as squares, and the squares around the filled circles as circles. Figure 3.9(c) shows that after using the kernel trick to project the instances into the feature space, several instances lie on the wrong side of the separating hyperplane. This is because the labeled images influence their neighboring images in the input space to be classified into the same category and we have two classes mix together. As a result, it is difficult to separate these two classes apart through any learning algorithm (unless we have substantially larger amount of labeled instances, which defeats the purpose of active learning).

3.4.3 Concept-Dependent Active Learning Algorithms

SVM_{Active}^{CD} consists of a state-transition table and three algorithms, *disambiguate keywords* (Figure 3.11), *disambiguate input-space* (Figure 3.12), and *angle diversity* (Figure 3.5). First, SVM_{Active}^{CD} addresses the scarcity problem by using keywords to seed a query. Thus, we eliminate the need to search the entire dataset for a positive image. The user can type in a keyword (or keywords) to describe the target concept. Images that are annotated with that keyword are added to the initial unlabeled pool U. If the number of images with matching keywords is small, we can perform query expansion[2] using a thesaurus to obtain related words (synonyms) that have matching images in our dataset. For each related semantic, we select matching images and add them to U.

Using U to conduct learning, the *hit-rate* is much improved compared to using the entire dataset. At this juncture, the state of learnability can be in one of the four states depicted in Figure 3.10. The rows show that the specified keyword(s) may enjoy good or suffer from poor isolation. The columns show that the query concept may be well or poorly isolated in the input space formed by the perceptual features. The goal of SVM_{Active}^{CD} is to move the learnability state of the target concept to the ideal state A, where both keywords and perceptual features enjoy good isolation.

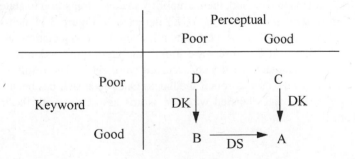

Fig. 3.10: Four learnability states

---

[2] Query expansion is a vital component of any retrieval systems and it remains a challenging research area. This component is not deployed in our system yet and is part of our ongoing research.

Disambiguate Keywords

Input: w, U, n; /* query word, unlabeled pool, sample #

    H; /* keyword association table

Output: U; /* unlabebed pool

Procedures:

• $findAssociations(w, H)$ /* find w's 2-itemsets in H

• $relevanceFeedback(S)$ /* user labels S

Initialization:

• $S \leftarrow \emptyset$

BEGIN

  1. $A \leftarrow findAssociations(w, H)$;

  2. while $(n \neq 0)$

    $\mathbf{s} \leftarrow Random(\mathsf{U})$;

    if $(s.W \cap A \neq \emptyset)$ /* Check annotation set W of $\mathbf{s}$

      $S \leftarrow S \cup \{\mathbf{s}\}$;

      $n \leftarrow n - 1$;

  3. $relevanceFeedback(S)$;

  4. for each $\mathbf{s}^- \in S$

    for each $\mathbf{u} \in \mathsf{U}$

      if $(\mathbf{s}^-.W \cap u.W \neq \emptyset)$

        $\mathsf{U} \leftarrow \mathsf{U} - \{\mathbf{u}\}$;

  5. return U;

END

Fig. 3.11: Disambiguate keywords algorithm

3.4.3.1 State C — Keyword Disambiguation

In state C, the keyword isolation is poor (due to aliasing). Thus, we first need to disambiguate the keywords by presenting images of different semantics to the user. Once the semantic is understood, the learnability makes a transition to state A.

Algorithm *disambiguate keywords* (DK) depicted in Figure 3.11 helps state C to arrive at A. The algorithm first performs a lookup in the association table H for the query keyword w. If 2-itemsets are found, the algorithm randomly selects in U images that are annotated by the co-occurred word set A. The resulting sample set S is presented to the user to solicit feedback. Once feedback has been received, SVM_{Active}^{CD} removes images labeled with the words associated with the irrelevant images from U.

3.4.3.2 State B — Input-Space Disambiguation

State B has poor perceptual isolation but good keyword isolation. If the number of matching images yielded by the good keyword is small, we can perform query expansion to enlarge the unlabeled sample pool. For these poorly-isolated concepts, we attempt to improve the perceptual isolation of the sample pool through input-space

Disambiguate Input-Space

Input: L, U w; /* labeled, unlabeled sample pool, keyword

Output: U;

Procedures:

• *query_expansion*(w);

• *generateWordlist*(L, *type*);

Initialization:

• $P, N \leftarrow \emptyset$

BEGIN

   1. $U \leftarrow U + query\_expansion(w)$;

   2. $P \leftarrow generateWordlist(L, \text{Positive})$;

   3. $N \leftarrow generateWordlist(L, \text{Negative})$;

   4. for each u $\in$ **U**

      if $(|\mathbf{u}.W \cap N| > |\mathbf{u}.W \cap P|)$

         $U \leftarrow U - \{\mathbf{u}\}$;

   5. **return** U;

END

Fig. 3.12: Disambiguate input-space algorithm

disambiguation. SVM^{CD}_{Active} employs *disambiguate input-space* (*DS*) in **Figure 3.12** to achieve this improvement. The algorithm removes from U the instances that can interfere with the learnability of the target concept. Let us revisit Figure 3.9. Rather than learning the labels of both the square and circle classes, *DS* removes the instances of the non-target class from the sample pool.

Given the current labeled pool of images, *DS* first generates two word lists: *P* from positive-labeled images and *N* from negative-labeled images. In the fourth step, *DS* checks the annotation of each image in the unlabeled pool for matches to words in *P* and *N*. If there are more negative keyword matches, SVM^{CD}_{Active} removes that image from U. Once perceptual isolation is enhanced, state *B* makes the transition to *A*.

3.4.3.3 State *D* — Keyword & Space Disambiguation

State *D* suffers from both poor semantic and input-space isolation. To remedy the problem, SVM^{CD}_{Active} first improves the keyword isolation using the same *DK* strategy as state *C*. Thereafter, the state is moved to state *B*, and the perceptual isolation is improved by applying state *B*'s method for disambiguation in the input space. Figure 3.10 shows the path that *D* takes to reach the ideal state *A*.

3.4.3.4 State *A* — Adapt to Diversity

Knowledge of the concept diversity can guide the learner during its exploration of the input space for potential positive samples. For example, if a concept is very diverse, like *flowers*, the learner may need to be more explorative and search for flowers of all colors. For such concepts, we make our learner more explorative by changing the parameter λ used in our classification score function in the *angle diversity* algorithm:

$$\lambda * |f(\mathbf{x}_i)| + (1-\lambda) * (\max_{\mathbf{x}_j \in S} \frac{|k(\mathbf{x}_i, \mathbf{x}_j)|}{\sqrt{K(\mathbf{x}_i, \mathbf{x}_i)K(\mathbf{x}_j, \mathbf{x}_j)}}).$$

If the concept diversity is low, we keep λ at its original value of 0.5. For diverse concepts, we reduce λ, resulting in more weight being assigned to the angle diversity during sample selections.

3.5 Experiments and Discussion

Experiments were conducted to answer five questions:

1. How does active learning perform, compared to passive learning in terms of retrieval accuracy? (Section 3.5.2)
2. Can active learning outperform traditional relevance feedback schemes? (Section 3.5.3)
3. Which sampling strategies, *random*, *simple*, *speculative*, *angle diversity*, or *error reduction*, perform the best for SVM_{Active}^{CD}, and why? (Section 3.5.4)
4. For concepts with differing degrees of isolation, can the multimodal approach of SVM_{Active}^{CD} improve the retrieval accuracy? (Section 3.5.5)
5. Can the *angle diversity* algorithm be tuned to accommodate highly diverse concepts? (Section 3.5.6)

3.5.1 Testbed and Setup

Five datasets were used for empirical evaluation of SVM_{Active}^{CD}: a four-category, a ten-category, a fifteen-category, a 107-category, and a 300K real-world image dataset.

- *Four-category* set. The 602 Corel-CD images in this dataset belong to four categories — *architecture*, *flowers*, *landscape*, and *people*. Each category consists of 100 to 150 images.
- *Ten-category* set. The 1,277 images in this dataset belong to ten categories — *architecture*, *bears*, *clouds*, *flowers*, *landscape*, *people*, *objectionable images*, *tigers*, *tools*, and *waves*. In this set, a few categories were added to increase learning difficulty (i.e., to reduce hit rate and decrease isolation). The tiger category contains images of tigers with landscape and water backgrounds to complicate landscape category. The objectionable (pronographic) images can be confused with people wearing little clothing (beach wear). Clouds and waves have substantial color similarity.

- *Fifteen-category* set. In addition to the ten categories in the above dataset, the total of 1,920 images in this dataset includes *elephants*, *fabrics*, *fireworks*, *food*, and *texture*. We added elephants with landscape and water backgrounds to increase learning difficulty in distinguishing landscape, tigers and elephants. We added colorful fabrics and food to interfere with flowers. Various texture images (e.g., skin, brick, grass, water, etc.) were added to raise learning difficulty for all categories.
- *107-category* set. This set consists of nearly 50,000 images that we collected from Corel Image CDs. The categories are documented in [22].
- *Large* set. A 300K-image dataset with images from a stock-photo company.

Each image is described by 144 perceptual features (108 from color and 36 from texture). (These features have been thoroughly tested and verified to be quite effective. Since the focus of this chapter is not on feature extraction, please consult [1] for details.) In the large image dataset, each image is annotated with up to 50 keywords; 40,000 keywords are used in the entire dataset. For SVM_{Active}^{CD}, we used a Laplacian RBF kernel $K(\mathbf{u}, \mathbf{v}) = (e^{-\gamma \Sigma_i |u_i - v_i|})$, with the γ set at 0.001. For the parameter λ used in the *angle diversity* algorithm, we set its default value at 0.5.

To obtain an objective measure of performance, we assumed that a query concept was an image category. The SVM_{Active}^{CD} learner has no prior knowledge about image categories[3]. The goal of SVM_{Active}^{CD} is to learn a given concept through a relevance feedback process. In this process, at each feedback round SVM_{Active}^{CD} selects sixteen or twenty images to ask the user to label as "relevant" or "not relevant" with respect to the query concept. It then uses the labeled instances to successively refine the concept boundary. After finishing the relevance feedback rounds, SVM_{Active}^{CD} then retrieves the top-k most relevant images from the dataset, based on the final concept it has learned. Accuracy is then computed by looking at the fraction of the k returned result that belongs to the target image category. We note that this computation is equivalent to computing the precision on the top-k images. This measure of performance appears to be the most appropriate for the image retrieval task — particularly since, in most cases, not all of the relevant images can be displayed to the user on one screen. As in the case of web searching, we typically wish the first few screens of returned images to contain a high proportion of relevant images. We are less concerned that not every single instance that satisfies the query concept is displayed.

---

[3] Unlike some recently developed systems [23] that contain a semantic layer between image features and queries to assist query refinement, our system does not have an explicit semantic layer. We argue that having a hard-coded semantic layer can make a retrieval system restrictive. Rather, dynamically learning the semantics of a query concept is more flexible and hence makes the system more useful.

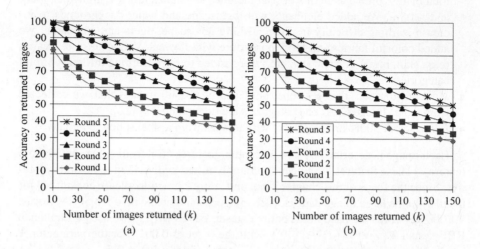

Fig. 3.13: **(a)** Average top-k accuracy over the ten-category dataset. **(b)** Average top-k accuracy over the fifteen-category dataset. Standard error bars are smaller than the curves' symbol size. Legend order reflects order of curves.

3.5.2 Active vs. Passive Learning

In this first experiment, we examined the gain of active learning over passive learning, and did not activate the concept-dependent component of SVM^{CD}_{Active}. Figures 3.13(a-b) show the average top-k accuracy for the 10-category and 15-category datasets. We considered the performance of SVM^{CD}_{Active} after each round of relevance feedback. The graphs indicate that performance clearly increases after each round. Also, the SVM^{CD}_{Active} algorithm's performance degrades gracefully when the concept complexity is increased — for example, after four rounds of relevance feedback, it achieves an average of 95% and 88% accuracy on the top-20 results for the two different sizes of data sets, respectively. It is also interesting to note that SVM^{CD}_{Active} is not only good at retrieving just the top few images with high precision, but it also manages to sustain fairly high accuracy even when asked to return larger numbers of images. For example, after five rounds of querying it attains 84% and 76% accuracy on the top-70 results for the two different sizes of data sets respectively. (For these two datasets, all sampling strategies work equally well. When the concept complexity further increases, we will see shortly in Section 3.5.4 that the *angle-diversity* method outperforms the other.)

We examined the effect that the active querying method had on performance. Figures 3.14(a) and 3.14(b) compare the active querying method with the regular passive method of sampling. The passive method chooses images randomly from the pool to be labeled. This method is typically used with SVMs since it creates a

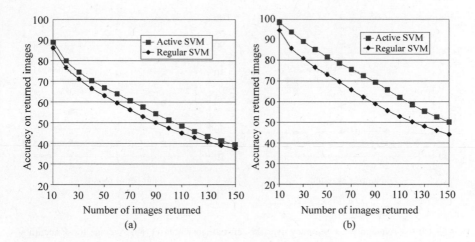

Fig. 3.14: **(a)** Active and regular passive learning on the fifteen-category dataset after three rounds of querying. **(b)** Active and regular passive learning on the fifteen-category dataset after five rounds of querying.

randomly selected dataset. It is clear that the use of active learning is beneficial in the image retrieval domain. We gain a significant increase in performance (a 5% to 10% gain after five iterations) by using the active method.

3.5.3 Against Traditional Relevance Feedback Schemes

We compared SVM_{Active}^{CD} with two traditional query refinement methods: *query point movement* (QPM) and *query expansion* (QEX). In this experiment, each scheme returned the 20 most relevant images after up to five rounds of relevance feedback. To ensure that the comparison to SVM_{Active}^{CD} was fair, we seeded both schemes with one randomly selected relevant image to generate the first round of images. On the ten-category image dataset, Figure 3.15(a) shows that SVM_{Active}^{CD} achieves nearly 90% accuracy on the top-20 results after three rounds of relevance feedback, whereas the accuracies of both QPM and QEX never reach 80%. On the fifteen-image category dataset, Figure 3.15(b) shows that SVM_{Active}^{CD} outperforms the others by even wider margins. SVM_{Active}^{CD} reaches 80% top-20 accuracy after three rounds and 94% after five rounds, whereas QPM and QEX cannot achieve 65% accuracy.

These results hardly surprise us. Traditional information retrieval schemes require a large number of image instances to achieve any substantial refinement. By just refining around current relevant instances, both QPM and QEX tend to be fairly localized in their exploration of the image space and hence rather slow in exploring

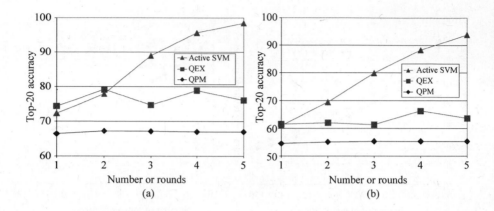

Fig. 3.15: **(a)** Average top-k accuracy over the ten-category dataset. **(b)** Average top-k accuracy over the fifteen-category dataset.

the entire space. In contrast, during the relevance feedback phase SVM_{Active}^{CD} takes both the relevant and irrelevant images into account when choosing the next pool-queries. Furthermore, it chooses to ask the user to label images that it regards as most *informative* for learning the query concept, rather than those that have the most likelihood of being relevant. Thus it tends to explore the feature space more aggressively.

3.5.4 Sampling Method Evaluation

This experiment was conducted to examine the performance of five sampling algorithms: *random, simple active, speculative, angle diversity,* and *error reduction.* As we mentioned in Section 3.5.2, when the concept complexity is low, all sampling methods perform about the same. To conduct this experiment, we used the 107-category dataset.

The first strategy selects random images from the dataset as samples for user feedback. The rest of the sample algorithms have been described in Section 3.3.2. We conducted experiments to compare these five sampling strategies in terms of three factors: (1) the top-k retrieval accuracy, (2) execution times, and (3) database size. In the experiments, we tested the sampling algorithms by fixing the number of sample images per-round at sixteen.

Figures 3.16(a) and 3.16(b) report the results for the 107-category dataset. For this dataset, it is too expensive to scan the entire dataset to perform sample selection and retrieval. Therefore, we conducted sampling and retrieval *approximately*

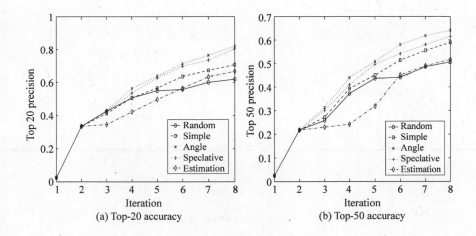

(a) Top-20 accuracy   (b) Top-50 accuracy

Fig. 3.16: Comparison of sampling strategies

through an indexing structure [20]. The figures show that the *angle diversity* algorithm performs the best among all active-learning algorithms. The *angle diversity* algorithm performs as well, and even better in some interactions, as the *speculative* algorithm, which is supposed to achieve nearly optimal performance. This result confirms that the samples should be diverse, as well as semantically uncertain (near the hyperplane).

Next, Figure 3.17 evaluates the execution time (in milliseconds) taken by the five algorithms. The figure shows that the fastest sampling algorithm is the *random* method, followed by *angle diversity*, *simple active*, *speculative*, and then *error-reduction*. Thus, *angle diversity* algorithm is the ideal choice in terms of both effectiveness and efficiency.

3.5.5 Concept-Dependent Learning

This experiment were designed to assess the effectiveness of the concept-dependent approach to image retrieval. The experiments were conducted on a dataset comprising 300K Corbis images.

We evaluated retrieval accuracy (in terms of precision) for concepts belonging to one of the four possible learnability states depicted in Figure 3.10. The representative concepts we queried, based on their learnability states, are shown in Table 3.1. The *angle diversity* (*AD*) sampling strategy is used as the baseline for comparing

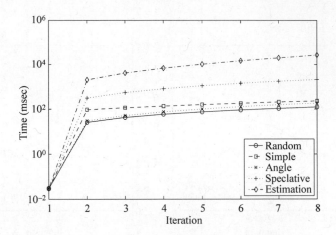

Fig. 3.17: Execution time on large dataset

the performance of the other algorithms used to modify the sample pool, namely *disambiguate input-space* (*DS*) and *disambiguate keywords* (*DK*).

Table 3.1: Queries for various states of learnability

State A	State B	State C	State D
castle	frog	apple	bug
crabs	lion	bird	city
crayfish	Paris	cat	dessert
fireworks	river	china	fruit
outer space	tiger	fish	mountain

3.5.5.1 State B (Figure 3.18(b))

The concepts belonging to state B have good keyword isolation but poor perceptual isolation. We used the *DS* algorithm to weed the sample pool of images annotated by negative keywords. The figure shows the retrieval performance of the *DS* algorithm and the baseline's. In the 5th iteration, *DS* gives a 99% precision rate compared to 90% for the baseline. On average, we can improve the precision rate by 10%. The results show that *DS* is effective for increasing the concept's perceptual isolation.

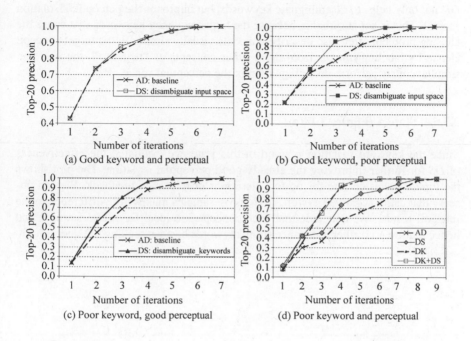

Fig. 3.18: Results for the four states of keyword and perceptual isolation

3.5.5.2 State *C* (Figure 3.18(c))

In order to resolve the aliasing problem for concepts with poor keyword isolation, we applied the *DK* algorithm and compared its retrieval performance with the baseline *AD*'s. We notice that the difference in precision rate after the first round of relevance feedback is 30% for *AD* and 40% for *DK*. The higher precision rate for *DK* persists through subsequent iterations as well. By disambiguating keywords in the initial sample pool, *DK* is able to provide better retrieval results.

3.5.5.3 State *D* (Figure 3.18(d))

Concepts falling into this state are the most problematic. They suffer from both poor keyword and poor perceptual isolation. The figure plots the retrieval results for four algorithms: the baseline, *DS*, *DK*, and *DK* + *DS*. Using only *DS* to improve perceptual isolation, the precision rate is on average 10% higher than the baseline. Using only *DK* to improve the keyword isolation, the average precision increase is

27%. When both *DS* and *DK* are used, the precision rate improves a couple of more percentage points over the case when only *DK* is used. These results suggest that *DK* not only helps to disambiguate keywords, but improves the perceptual isolation as well. This is not surprising, because the high-frequency keywords that cause the aliasing problem often also describe concepts visually similar to the target concept. Hence by labeling images from those other concepts as "negative", the user can simultaneously perform disambiguation in the perceptual space.

3.5.5.4 State A (Figure 3.18(a))

Since the keyword isolation is good in this state, we evaluated the effectiveness of *DS* for further improving the already-good perceptual isolation. The plot shows the top-20 precision rate for up to 7 iterations of relevance feedback. We can observe that using negative keywords to remove samples makes little difference to the retrieval results. We conclude that such concepts are ideal for active learning and hence require no further assistance from keywords besides seeding the query.

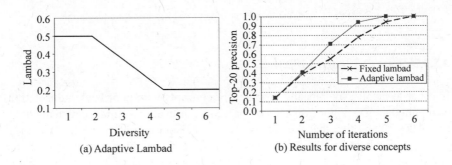

(a) Adaptive Lambad   (b) Results for diverse concepts

Fig. 3.19: Concept diversity

3.5.6 Concept Diversity Evaluation

For highly-diverse concepts, we reduced the value of λ in Eq.(3.2) so more weight is given to the angle diversity during sample selections. Figure 3.19(a) shows how we adapt λ to the concept diversity. For concepts with low diversity values (≤ 1), we set $\lambda = 0.5$. For more diverse concepts, we reduce λ as diversity increases. However, we cannot reduce λ indefinitely, and ignore usefulness of samples near the hyperplane. Therefore, when the diversity value is greater than 3.5, we set $\lambda = 0.2$. We plotted the precision results for queries with diverse concepts in Figure 3.19(b). The dashed curve plots the precision results using λ fixed at 0.5. The solid curve is for the case where λ adapts to the diversity.

At higher iterations, using an adaptive λ improves the precision rate by about 10%. The improvement is not significant for the 2nd iteration since the learner still has insufficient labeled images to construct a good classifier. A qualitative observation reported by users is that when λ is adaptive, the variety of images is significantly better, in both the sample pool and the results set. Thus, our approach of adjusting λ according to the concept's diversity proves to be both simple and effective.

3.5.7 Evaluation Summary

Our experiments have provided us information to answer the five questions posed in the beginning of the section.

1. Active learning outperforms passive learning methods. The gain of active learning becomes more significant when complexity increases.
2. Active learning outperforms traditional relevance feedback schemes by significant margins. This is hardly a surprise, since active learning maximizes information gain when it selects a sample to perform pool-query.
3. The *angle-diversity* sampling method, which strikes a good balance between uncertainty and diversity, works most effectively and efficiently among the sampling methods.
4. When concept complexity increases beyond certain level (e.g., when a target concept is scarce or poorly isolated), the effectiveness of active learning can suffer from severe degradation. We have shown that by taking advantage of keyword profiling (similar to the way that all relational databases perform query optimization based on some profiling techniques), SVM_{Active}^{CD} can perform concept-dependent active learning by disambiguating either keyword semantics or perceptual features. Improving concept learnability assures the algorithm's scalability with respect to concept complexity.
5. Concept-dependent active learning can also adapt its *diversity* parameter to behave exploratively when the target concept is scattered in the feature space.

3.6 Related Reading

Machine learning and relevance feedback techniques have been proposed to learn and to refine query concepts. The problem is that most traditional techniques require a large number of training instances [24, 25, 26, 27, 28], and they require seeding a query with "good" examples [29, 30, 31, 32]. Unfortunately, in many practical scenarios, a learning algorithm must work with a scarcity of training data and a limited amount of training time.

3.6.1 Machine Learning

Ensemble techniques such as *bagging* [33], *arcing* [34], and *boosting* [35, 36, 37] have been proposed to improve classification accuracy for decision trees and neural networks. These ensemble schemes enjoy success in improving classification accuracy through bias or variance reduction, but they do not help reduce the number of samples and time required to learn a query concept. In fact, most ensemble schemes actually increase learning time because they introduce learning redundancy in order to improve prediction accuracy [38, 35, 39, 40].

To reduce the number of required samples, researchers have conducted several studies of active learning [41, 42, 43, 44] for classification. Active learning can be modeled formally as follows: Given a dataset S consisting of an unlabeled subset U and a labeled subset L, an active learner has two components: f and q. The f component is a classifier that is trained on the current set of labeled data L. The second component q is the sampling function that, given a current labeled set L, decides which subset $\mathbf{u}$ in U to select to query the user. The active learner returns a new f after each round of relevance feedback. The sampling techniques employed by the active learner determine the selection of the next batch of unlabeled instances to be labeled by the user.

The *query by committee* (QBC) algorithm [45, 46] is a representative active learning scheme. QBC uses a distribution over all possible classifiers and attempts greedily to reduce the entropy of this distribution. This general purpose algorithm has been applied in a number of domains using classifiers (such as Naive Bayes classifiers [47, 4]) for which specifying and sampling classifiers from a distribution is natural. Probabilistic models such as the Naive Bayes classifier provide interpretable results and principled ways to incorporate prior knowledge. However, they typically do not perform as well as discriminative methods such as SVMs [48, 49], especially when the amount of training data is scarce. For media-data retrieval where a query concept is typical non-linear[4], SVM_{Active}^{CD} with kernel mapping provide more flexible and accurate concept modeling.

---

[4] A query such as "animals", "women", and "European architecture" does not reside contiguously in the space formed by the image features.

Specifically for image retrieval, the PicHunter system [50, 51] uses Bayesian prediction to infer the goal image, based upon users' input. Mathematically, the goal of PicHunter is to find a single goal point in the feature space (e.g., a particular flower image), whereas our goal is to hunt down all points that match a query concept (e.g., the entire flower category, which consists of flowers of different colors, shapes, and textures, and against different backgrounds). Note that the points matching a target concept can be scattered all over the feature space. To find these points quickly with few hints, our learning algorithms must deal with many daunting challenges [52].

3.6.2 Relevance Feedback

Relevance feedback was first proposed by Rocchio in 1971 [53]. The study of [32] puts subsequent relevance feedback techniques proposed by the Information Retrieval (IR) into three categories: *query reweighting*, *query point movement* and *query expansion*.

- *Query reweighting* and *query point movement* [54, 55, 56, 30, 57]. Both query reweighing and query point movement use nearest-neighbor sampling: They return top ranked objects to be examined by the user and then refine the query based on the user's feedback. If the initial query example is good and the query concept is convex in the feature space [54, 31], this nearest-neighbor sampling approach works fine. Unfortunately, most users do not have a good example to start a query, and most image-query concepts are non-convex. Refining a search around bad examples is analogous to trying to find oranges in the middle of an apple orchard by refining one's search to a few rows of apple trees at a time. It will take a long time to find oranges (the desired result).

- *Query expansion* [58, 31]. The *query expansion* approach can be regarded as a multiple-instances sampling approach. The samples of a subsequent round are selected from the neighborhood (not necessarily the nearest ones) of the positive-labeled instances of the previous round. The study of [58] shows that query expansion achieves only a slim margin of improvement (about 10% in precision/recall) over query point movement.

Almost all traditional relevance feedback methods require seeding the methods with "good" positive examples [59, 60, 61, 62, 63, 64], and most methods do not use negative-labeled instances effectively. For instance, sunset images must be supplied as examples in order to search for sunset pictures. However, finding good examples should be the job of a search engine itself. SVM_{Active}^{CD} effectively uses negative-labeled instances to induce more negative instances, and thereby improves the probability of finding positive instances. At the same time, the active-learning approach selects the most informative unlabeled instances to query the user to gather maximum amount of information to disambiguate the user's query concept. Because of the effective use of negative and unlabeled instances, active learning can learn a query concept much faster and more accurately than the traditional relevance-feedback methods.

3.7 Relation to Other Chapters

Chapters 10 and 11 address the scalability issues of SVMs. In addition to learning a hyperplane, the query refinement problem can also consider refining the distance function, which is discussed in Chapter 5. Feature extraction and fusion discussed in Chapters 2 and 6 forms the input space for conduction quer-concept learning.

3.8 Concluding Remarks

We have demonstrated that active learning with support vector machines can provide a powerful tool for searching image databases, outperforming key traditional query refinement schemes. $\mathrm{SVM}_{Active}^{CD}$ not only achieves consistently high accuracy on a wide variety of desired returned results, but also does it quickly and maintains high precision when asked to deliver large quantities of images. Also, unlike recent systems such as SIMPLIcity [23], it does not require an explicit semantic layer to perform well. Our system takes advantage of the intuition that there can be considerable differences between the set of images that we are already confident a user wishes to see, and the set of images that would most informative for the user to label. By decoupling the notions of feedback and retrieval, and by using a powerful classifier with active learning, we have demonstrated that $\mathrm{SVM}_{Active}^{CD}$ can provide considerable gains over other systems.

We have also proposed a multimodal, concept-dependent active learning scheme, which combines keywords with images' perceptual features in a synergistic way to perform image retrieval. In contrast to traditional active learning methods, $\mathrm{SVM}_{Active}^{CD}$ adjusts its learning process based on concept complexity: it meticulously constructs the sample pool in order to ameliorate the query concept's hit-rate and isolation, and enhances the learnability of the query concept by adapting the sampling strategy to the concept's diversity.

References

1. Tong, S., Chang, E. Support vector machine active learning for image retrieval. In *Proceedings of the ACM international conference on Multimedia*, pages 107–118, 2001.
2. Goh, K.S., Chang, E.Y., Lai, W.C. Multimodal concept-dependent active learning for image retrieval. In *Proceedings of the ACM international conference on Multimedia*, pages 564–571, 2004.
3. Lewis, D., Gale, W. A sequential algorithm for training text classifiers. In *Proceedings of the Seventeenth Annual International ACM-SIGIR Conference on Research and Development in Information Retrieval*, pages 3–12. Springer-Verlag, 1994.
4. McCallum, A., Nigam, K. Employing EM in pool-based active learning for text classification. In *Proceedings of the Fifteenth International Conference on Machine Learning*, pages 350–358. Morgan Kaufmann, 1998.

5. Burges, C. A tutorial on support vector machines for pattern recognition. In *Proceedings of ACM KDD*, pages 121–167, 1998.
6. Vapnik, V. *Estimation of Dependences Based on Empirical Data*. Springer Verlag, 1982.
7. Campbell, C., Cristianini, N., Smola, A. Query learning with large margin classifiers. In *Proceedings of the Seventeenth International Conference on Machine Learning*, pages 111–118, 2000.
8. Schohn, G., Cohn, D. Less is more: Active learning with support vector machines. In *Proceedings of the Seventeenth International Conference on Machine Learning*, pages 839–846, 2000.
9. Tong, S., Koller, D. Support vector machine active learning with applications to text classification. In *Proceedings of the 17th International Conference on Machine Learning*, pages 401–412, June 2000.
10. Mendel, M., Poliner, G., Ellis, D. Support vector machine active learning for music retrieval. *Multimedia Systems*, 12(1):3–13, 2006.
11. Mitchell, T. Generalization as search. *Artificial Intelligence*, 28:203–226, 1982.
12. Shawe-Taylor, J., Cristianini, N. Further results on the margin distribution. In *Proceedings of the Twelfth Annual Conference on Computational Learning Theory*, pages 278–285, 1999.
13. Vapnik, V. *Statistical Learning Theory*. Wiley, 1998.
14. Herbrich, R., Graepel, T., Campbell, C. Bayes point machines: Estimating the bayes point in kernel space. In *International Joint Conference on Artificial Intelligence Workshop on Support Vector Machines*, pages 23–27, 1999.
15. Brinker, K. Incorporating diversity in active learning with support vector machines. In *Proceedings of the Twentieth International Conference on Machine Learning (ICML)*, pages 59–66, August 2003.
16. Roy, N., McCallum, A. Toward optimal active learning through sampling estimation of error reduction. In *Proceedings of the Eighteenth International Conference on Machine Learning (ICML)*, pages 441–448, August 2001.
17. Lai, W.C., Goh, K., Chang, E.Y. On scalability of active learning for formulating query concepts. In *Proceedings of Workshop on Computer Vision Meets Databases (CVDB) in cooperation with ACM International Conference on Management of Data (SIGMOD)*, pages 11–18, 2004.
18. Agrawal, R. Fast algorithms for mining association rules in large databases. In *Proceedings of VLDB*, pages 487–499, 1994.
19. Duda, R., Hart, P., Stork, D.G. *Pattern Classification*. Wiley, New York, 2 edition, 2001.
20. Li, C., Chang, E., Garcia-Molina, H., Wilderhold, G. Clindex: Approximate similarity queries in high-dimensional spaces. *IEEE Transactions on Knowledge and Data Engineering (TKDE)*, 14(4):792–808, July 2002.
21. Chang, E., Li, B. MEGA: The maximizing expected generalization algorithm for learning complex query concepts. *ACM Transactions on Information Systems*, 21(4):347–382, December 2003.
22. Chang, E., Goh, K., Sychay, G., Wu, G. Content-based soft annotation for multimodal image retrieval using bayes point machines. *IEEE Transactions on Circuits and Systems for Video Technology Special Issue on Conceptual and Dynamical Aspects of Multimedia Content Description*, 13(1):26–38, 2003.
23. Wang, J., Li, J., Wiederhold, G. Simplicity: Semantics-sensitive integrated matching for picture libraries. In *Proceedings of ACM Multimedia Conference*, pages 483–484, 2000.
24. Bishop, C. *Neural Networks for Pattern Recognition*. Oxford, 1998.
25. Kearns, M., Vazirani, U. *An Introduction to Computational Learning Theory*. MIT Press, 1994.
26. Mitchell, T.M. *Machine Learning*. McGraw-Hill, 1997.
27. Zhou, X.S., Huang, T.S. Comparing discriminating transformations and svm for learning during multimedia retrieval. In *Proc. of ACM Conf. on Multimedia*, pages 137–146, 2001.
28. Zhou, X.S., Huang, T.S. Relevance feedback for image retrieval: a comprehensive review. *ACM Multimedia Systems Journal, special issue on CBIR*, 8:536–544, 2003.

29. Jones, K.S., (Editors), P.W. *Readings in Information Retrieval.* Morgan Kaufman, July 1997.
30. Porkaew, K., Chakrabarti, K., Mehrotra, S. Query refinement for multimedia similarity retrieval in mars. In *Proceedings of ACM International Conference on Multimedia*, pages 235–238, 1999.
31. Wu, L., Faloutsos, C., Sycara, K., Payne, T.R. Falcon: Feedback adaptive loop for content-based retrieval. In *Proceedings of the 26th VLDB Conference*, pages 279–306, September 2000.
32. Ortega-Binderberger, M., Mehrotra, S. Relevance feedback techniques in the MARS image retrieval system. *Multimedia Systems*, 9(6):535–547, 2004.
33. Breiman, L. Bagging predicators. *Machine Learning*, pages 123–140, 1996.
34. Breiman, L. Arcing classifiers. *The Annals of Statistics*, pages 801–849, 1998.
35. Grove, A., Schuurmans, D. Boosting in the limit: Maximizing the margin of learned ensembles. In *Proc. 15th National Conference on Artificial Intelligence (AAAI)*, pages 692–699, 1998.
36. Schapire, R., Freund, Y., Bartlett, P., Lee, W. Boosting the margin: A new explanation for the effectiveness of voting methods. In *Proceeding of the Fourteenth International Conference on Machine Learning*, pages 322–330. Morgan Kaufmann, 1997.
37. Wu, H., Lu, H., Ma, S. A practical svm-based algorithm for ordinal regression in image retri eval. In *Proceedings of ACM International Conference on Multimedia*, pages 612–621, 2003.
38. Dietterich, T., Bakiri, G. Solving multiclass learning problems via error-correcting output codes. *Journal of Artifical Intelligence Research*, 2:263–286, 1995.
39. James, G., Hastie, T. Error coding and substitution PaCTs. In *Proceedings of NIPS*, 1997.
40. Moreira, M., Mayoraz, E. Improved pairwise coupling classification with error correcting classifiers. In *Proceedings of ECML*, pages 160–171, April 1998.
41. Cohn, D., Ghahramani, Z., Jordan, M. Active learning with statistical models. *Journal of Artificial Intelligence Research*, 4:129–145, 1996.
42. Cesa-Bianchi, N., Freund, Y., Haussler, D., Helmbold, D.P., Schapire, R.E., Warmuth, M.K. How to use expert advice. *Journal of ACM*, 44(3):427–485, 1997.
43. Jaakkola, T., Siegelmann, H. Active information retrieval. In *Proceedings of NIPS*, pages 777–784, 2001.
44. Tong, S., Chang, E. Support vector machine active learning for image retrieval. In *Proceedings of ACM International Conference on Multimedia*, pages 107–118, October 2001.
45. Freund, Y., Seung, H., Shamir, E., Tishby, N. Selective sampling using the Query by Committee algorithm. *Machine Learning*, 28:133–168, 1997.
46. Seung, H., Opper, M., Sompolinsky, H. Query by committee. In *Proceedings of the Fifth Workshop on Computational Learning Theory*, pages 287–294. Morgan Kaufmann, 1992.
47. Dagan, I., Engelson, S. Committee-based sampling for training probabilistic classifiers. In *Proceedings of the Twelfth International Conference on Machine Learning*, pages 150–157. Morgan Kaufmann, 1995.
48. Joachims, T. Text categorization with support vector machines. In *Proceedings of ECML*, pages 137–142. Springer-Verlag, 1998.
49. Dumais, S.T., Platt, J., Heckerman, D., Sahami, M. Inductive learning algorithms and representations for text categorization. In *Proceedings of the Seventh International Conference on Information and Knowledge Management*, pages 148–155. ACM Press, 1998.
50. Cox, I.J., Miller, M.L., Omohundo, S.M., Yianilos, P.N. Pichunter: Bayesian relevance feedback for image retrieval. In *Proceedings of International Conference on Pattern Recognition*, pages 361–369, August 1996.
51. Cox, I.J., Miller, M.L., Minka, T.P., Papathomas, T.V., Yianilos, P.N. The bayesian image retrieval system, pichunter: Theory, implementation and psychological experiments. *IEEE Transactions on Image Processing*, 9(1):20–37, 2000.
52. Chang, E.Y., Li, B., Wu, G., Goh, K.S. Statistical learning for effective visual information retrieval (invited paper). In *Proceedings of IEEE International Conference on Image Processing (ICIP)*, pages 609–612, 2003.

53. Rocchio, J.J. Relevance feedback in information retrievalIn, editor, *The SMART Retrieval System — Experiments in Automatic Document Processing*, pages 313–323. Prentice Hall, 1971.
54. Ishikawa, Y., Subramanya, R., Faloutsos, C. Mindreader: Querying databases through multiple examples. In *Proceedings of VLDB*, pages 218–227, 1998.
55. Ortega, M., Rui, Y., Chakrabarti, K., Warshavsky, A., Mehrotra, S., Huang, T.S. Supporting ranked boolean similarity queries in mars. *IEEE Transactions on Knowledge and Data Engineering*, 10(6):905–925, December 1999.
56. Ortega, M., Rui, Y., Chakrabarti, K., Mehrotra, S., Huang, T.S. Supporting similarity queries in mars. In *Proceedings of ACM International Conference on Multimedia*, pages 403–413, 1997.
57. Rui, Y., Huang, T.S., Ortega, M., Mehrotra, S. Relevance feedback: A power tool in interactive content-based image retrieval. *IEEE Transactions on Circuits and Systems for Video Technology*, 8(5):644–655, 1998.
58. Porkaew, K., Mehrota, S., Ortega, M. Query reformulation for content based multimedia retrieval in mars. In *Proceedings of ICMCS*, pages 747–751, 1999.
59. Flickner, M., Sawhney, H., Ashley, J., Huang, Q., Dom, B., Gorkani, M., Hafner, J., Lee, D., Petkovic, D., Steele, D., Yanker, P. Query by image and video content: the QBIC system. *IEEE Computer*, 28(9):23–32, 1995.
60. Gupta, A., Jain, R. Visual information retrieval. *Communications of the ACM*, 40(5):69–79, 1997.
61. Hua, K.A., Vu, K., Oh, J.H. Sammatch: A flexible and efficient sampling-based image retrieval technique for image databases. In *Proceedings of ACM Multimedia*, pages 225–234, 1999.
62. Manjunath, B.S., Ma, W.Y. Texture features for browsing and retrieval of image data. *IEEE Transactions on Pattern Analysis and Machine Intelligence*, 18(8):837–842, August 1996.
63. Smith, J.R., Chang, S.F. VisualSEEk: A fully automated content-based image query system. In *Proceedings of ACM Multimedia*, pages 87–98, 1996.
64. Wang, J.Z., Wiederhold, G., Firschein, O., Wei, S.X. Wavelet-based image indexing techniques with partial sketch retrieval capability. In *Proceedings of the ADL*, pages 13–24, May 1997.

Chapter 4
Similarity

Abstract How to account for similarity between two data instances is fundamental for any data management, retrieval, and analysis tasks. This chapter[†] shows that traditional distance functions such as the Minkowski metric and weighted Minkowski are not effective in accounting similarity. Through mining a large set of visual data, we discovered a perceptual distance function, which works much more effectively for finding similar images than the Minkowski family. We call the discovered function *dynamic partial function* (DPF). We demonstrate the effectiveness of DPF through empirical studies and explain why it works better by cognitive theories. **Keywords**: Cognitive theory, distance function, DPF, perceptual similarity.

4.1 Introduction

To achieve effective management, retrieval, and analysis, an image/video system must be able to accurately characterize and quantify perceptual similarity. However, a fundamental challenge — how to measure perceptual similarity — remains largely unanswered. Various distance functions, such as the Minkowski metric [2], earth mover distance [3], histogram Cosine distance [4], and fuzzy logic [5], have been used to measure similarity between feature vectors representing images (and hence video frames). Unfortunately, our experiments show that they frequently overlook obviously similar objects and hence are not adequate for measuring perceptual similarity.

Quantifying perceptual similarity is a difficult problem. Indeed, we may be decades away from fully understanding how human perception works (as we have discussed in Chapter 2). In this chapter, we show how we employed a data-driven approach to analyze the characteristics of similar data instances, and how that led to our formulation of a new distance function. Our mining hypothesis is this: Sup-

---

[†] ©Springer, 2003. This chapter is a minor revision of the author's work with Beitao Li and Yi-Leh Wu [1] published in ACM Multimedia Systems'03. Permission to publish this chapter is granted under copyright license #2591350681815.

pose most of the similar data instances can be clustered in a feature space. We can then claim with high confidence that (1) the feature space can adequately capture the characteristics of those data instances, and (2) the distance function used for clustering data instances in that feature space can accurately model similarity. Our target task was to formulate a distance function that can keep similar data instances in the same cluster, while keeping dissimilar ones away.

We performed our *discovery through mining* operation in two stages. In the first stage, we isolate the distance function factor (we used the Euclidean distance) to find a reasonable feature set. In the second stage, we froze the features to discover a perceptual distance function that could better cluster similar data instances in the feature space. We call the discovered function *dynamic partial distance function* (DPF). When we empirically compare DPF to Minkowski-type distance functions in image retrieval, video shot-transition detection, and new-article near-duplicate detection, DPF performs significantly better.

Similarity is one of the central theoretical constructs in psychology [6, 7], probably related to human survival instincts. We believe that being able to quantify similarity accurately must also hold a central place in theories of information management and retrieval. Our excitement in discovering DPF does not arise merely from the practical effectiveness we found in three applications. More importantly, we find that DPF has roots in cognitive psychology. While we will discuss the links between DPF and some *similarity theories* in cognitive psychology in Section 4.5, let us use an example to explain both the *dynamic* and *partial* aspects. Suppose we are asked to name two places that are similar to England. Among several possibilities, Scotland and New England could be two reasonable answers. However, the respects England is similar to Scotland differ from those in which England is similar to New England. If we use the shared attributes of England and Scotland to compare England and New England, the latter pair might not be similar, and vice versa. Objects can be similar to the query object in different respects. A distance function using a fixed set of respects cannot capture objects that are similar in different sets of respects. A distance function for measuring a pair of objects is formulated only after the objects are compared, not before the comparison is made. The respects for the comparison are activated in this formulation process. The activated respects are more likely to be those that can support coherence between the compared objects.

The rest of his chapter is organized as follows:

1. We first show our data mining process to determine a reasonable feature space. In that feature space, we find distinct patterns of similar and dissimilar images, which lead to the discovery of DPF.
2. We derive DPF based on the observed patterns, and we provide methods for finding the optimal settings for the function's parameters.
3. Through case studies, we demonstrate that DPF is very effective in finding images that have been transformed by rotation, scaling, downsampling, and cropping, as well as images that are perceptually similar to the query image. Applying DPF to video shot-transition detection and new-article near-duplicate detection, we show that DPF is also more effective than the Minkowski metric.

4.2 Mining Image Feature Set

This section depicts how the mining dataset was constructed in three steps: testbed setup (Section 4.2.1), feature extraction (Section 4.2.2), and feature selection (Section 4.2.3).

4.2.1 Image Testbed Setup

To ensure that sound inferences can be drawn from our mining results, we carefully construct the dataset. First, we prepare for a dataset that is comprehensive enough to cover a diversified set of images. To achieve this goal, we collect 60,000 JPEG images from Corel CDs and from the Internet. Second, we define "similarity" in a slightly restrictive way so that individuals' subjectivity can be excluded.[1] For each image in the 60,000-image set, we perform 24 transformations (described shortly), and hence form 60,000 similar-image sets. The total number of images in the testbed is 1.5 million.

The 24 image transformations we perform include the following:

1. Scaling.
   - Scale up then down. We scale each image up by 4 and 16 times, respectively, and then scale it back to the original size.
   - Scale down then up. We scale each image down by factors of 2, 4, and 8, respectively, then scale it back to the original size.

2. Downsampling. We downsample each image by seven different percentages: 10%, 20%, 30%, 40%, 50%, 70%, and 90%.

3. Cropping. We evenly remove the outer borders to reduce each image by 5%, 10%, 20%, 30%, 40%, 50%, 60%, and 70%, respectively, and then scale it back up to the original size.

4. Rotation. We rotate each image by 90, 180, and 270 degrees.

5. Format transformation. We obtain the GIF version of each JPEG image.

---

[1] We have considered adding images taken under different lighting conditions or with different camera parameters. We decided not to include them because they cannot be automatically generated from an image. Nevertheless, our experimental results (see Section 4.4) show that the perceptual distance function discovered during the mining process can be used effectively to find other perceptually similar images. In other words, our testbed consists of a good representation of similar images, and the mining results (i.e., training results) can be generalized to testing data consisting of perceptually similar images produced by other methods.

4.2.2 Feature Extraction

To describe images, we must find a set of features that can represent those images adequately. Finding a universal representative feature set can be very challenging, since different imaging applications may require different feature sets. For instance, the feature set that is suitable for finding tumors may not be effective for finding landscape images, and vice versa. However, we believe that by carefully separating perception from intelligence (i.e., domain knowledge), we can identify meaningful perceptual features. Chapter 2 shows both model-based and data-driven approaches for extracting features. We used a data-driven approach in this study to find useful features from a large set of feature candidates.

Psychologists and physiologists divide the human visual system into two parts: the *perceiving part*, and the *inference part* [8]. The perceiving part receives photons, converts electrical signals into neuro-chemical signals, and delivers the signals to our brains. The inference part then analyzes the perceived data based on our knowledge and experience. A baby and an adult have equal capability for perceiving, but differing capability for understanding what is perceived. Among adults, specially trained ones can interpret an X-ray film, but the untrained cannot. In short, the perceiving part of our visual system is task-independent, so it can be characterized in a domain-independent manner.

We extract features such as color, shape, and texture from images. In the color channel, we characterize color in multiple resolutions. We first divide color into 12 color bins including 11 bins for culture colors and one bin for outliers [9]. At the coarsest resolution, we characterize color using a color mask of 12 bits. To record color information at finer resolutions, we record nine additional features for each color. These nine features are color histograms, color means (in H, S and V channels), color variances (in H, S and V channels), and two shape characteristics: elongation and spreadness. Color elongation characterizes the shape of a color, and spreadness characterizes how that color scatters within the image [10]. Table 4.1 summarizes color features in coarse, medium and fine resolutions.

Table 4.1: Multi-resolution color features

Filter Name	Resolution	Representation
Masks	Coarse	Appearance of culture colors
Spread	Coarse	Spatial concentration of a color
Elongation	Coarse	Shape of a color
Histograms	Medium	Distribution of colors
Average	Medium	Similarity comparison within the same culture color
Variance	Fine	Similarity comparison within the same culture color

Texture is an important characteristic for image analysis. Studies [11, 12, 13, 14] have shown that characterizing texture features in terms of structuredness, orientation, and scale (coarseness) fits well with models of human perception. From the wide variety of texture analysis methods proposed in the past, we choose a discrete wavelet transformation (DWT) using quadrature mirror filters [13] because of its computational efficiency.

Each wavelet decomposition on a 2-D image yields four subimages: a $\frac{1}{2} \times \frac{1}{2}$ scaled-down image of the input image and its wavelets in three orientations: horizontal, vertical and diagonal. Decomposing the scaled-down image further, we obtain the tree-structured or wavelet packet decomposition. The wavelet image decomposition provides a representation that is easy to interpret. Every subimage contains information of a specific scale and orientation and also retains spatial information. We obtain nine texture combinations from subimages of three scales and three orientations. Since each subimage retains the spatial information of texture, we also compute elongation and spreadness for each texture channel. Figure 4.1 summarizes texture features.

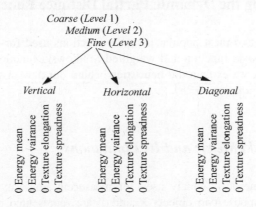

Fig. 4.1: Multi-resolution texture features

4.2.3 Feature Selection

Once the testbed is set up and relevant features extracted, we fix the distance function to examine various feature combinations. For the time being, we employ the Euclidean distance function to quantify the similarity between two feature vectors. We use the Euclidean function because it is commonly used, and it achieves acceptable results. (However, we will offer a replacement distance function for the Euclidean distance in Section 4.3.)

Using different feature combinations, we employ the Euclidean function to find the distance rankings of the 24 images that are similar to the original image (i.e., the query image). If a feature set can adequately capture the characteristics of images, the 24 similar images should be among those closest to the query image. (In an ideal case, the 24 similar images should be the 24 images closest to the query image.)

Our experiments reveal that when only individual features (e.g., color histograms, color elongation, and color spreadness) are employed, the distance function cannot easily capture the similar images even among the top-100 nearest neighbors. For a top-100 query, all individual features suffer from a dismal recall lower than 30%. When we combine all color features, the top-100 recall improves slightly, to 45%. When both color and texture features are used, the recall improves to 60%.

At this stage, we can go in either of two directions to improve recall. One, we can add more features, and two, we can replace the Euclidean distance function. We will consider adding additional features in our future work. In this chapter, we focus on finding a perceptual distance function that improves upon the Euclidean Function.

4.3 Discovering the Dynamic Partial Distance Function

We first examine two most popular distance functions used for measuring image similarity: Minkowski function and weighted Minkowski function. Building upon those foundations, we explain the heuristics behind our new distance function — *Dynamic Partial Function (DPF)*.

4.3.1 Minkowski Metric and Its Limitations

The Minkowski metric is widely used for measuring similarity between objects (e.g., images). Suppose two objects X and Y are represented by two p dimensional vectors $(x_1, x_2, \cdots, x_p)$ and $(y_1, y_2, \cdots, y_p)$, respectively. The Minkowski metric $d(X, Y)$ is defined as

$$d(X,Y) = \left(\sum_{i=1}^{p} |x_i - y_i|^r\right)^{\frac{1}{r}}, \qquad (4.1)$$

where r is the Minkowski factor for the norm. Particularly, when r is set as 2, it is the well known Euclidean distance; when r is 1, it is the Manhattan distance (or L_1 distance). An object located a smaller distance from a query object is deemed more similar to the query object. Measuring similarity by the Minkowski metric is based on one assumption: that similar objects should be similar to the query object in all dimensions. This assumption is true for abstract points in mathematical space. However, for multimedia objects (e.g., images), this assumption may not hold. Human perception of similarity may not strictly follow the rules of mathematical space [7].

A variant of the Minkowski function, the weighted Minkowski distance function, has also been applied to measure image similarity. The basic idea is to introduce weighting to identify important features. Assigning each feature a weighting coefficient w_i $(i = 1, 2, \cdots p)$, the weighted Minkowski distance function is defined as:

$$d_w(X,Y) = (\sum_{i=1}^{p} w_i |x_i - y_i|^r)^{\frac{1}{r}}. \tag{4.2}$$

By applying a static weighting vector for measuring similarity, the weighted Minkowski distance function assumes that similar images resemble the query images in the same features. For example, when the function weights color features high and ignores texture features, this same weighting is applied to all pair-wise distance computation with the query image. We will show shortly that this *fixed* weighting method is restrictive in finding similar objects of different kinds.

We can summarize the assumptions of the traditional distance functions as follows:

- Minkowski function: All similar images must be similar in all features.
- Weighted Minkowski function: All similar images are similar in the same way (e.g., in the same set of features).

We questioned the above assumptions upon observing how similar objects are located in the feature space. For this purpose, we carried out extensive data mining work on a 1.5M-image dataset introduced in Section 4.2. To better discuss our findings, we introduce a term we have found useful in our data mining work. We define the *feature distance* on the ith feature as

$$\delta_i = |x_i - y_i|. \qquad (i = 1, 2, \cdots, p)$$

The expressions of Eq.(4.1) and Eq.(4.2) can be simplified into

$$d(X,Y) = (\sum_{i=1}^{p} \delta_i^r)^{\frac{1}{r}} \text{ and } d_w(X,Y) = (\sum_{i=1}^{p} w_i \delta_i^r)^{\frac{1}{r}}.$$

In our mining work, we first tallied the feature distances between similar images (denoted as δ^+), and also those between dissimilar images (denoted as δ^-). Since we normalized feature values to be between zero and one, the ranges of both δ^+ and δ^- are between zero and one. Figure 4.2 presents the distributions of δ^+ and δ^-. The x-axis shows the possible value of δ, from zero to one. The y-axis (in logarithmic scale) shows the percentage of the features at different δ values.

The figure shows that δ^+ and δ^- have different distribution patterns. The distribution of δ^+ is much skewed toward small values (Figure 4.2(a)), whereas the distribution of δ^- is more evenly distributed (Figure 4.2(b)). We can also see from Figure 4.2(a) that a moderate portion of δ^+ is in the high value range (≥ 0.5), which indicates that similar images may be quite dissimilar in some features. From this

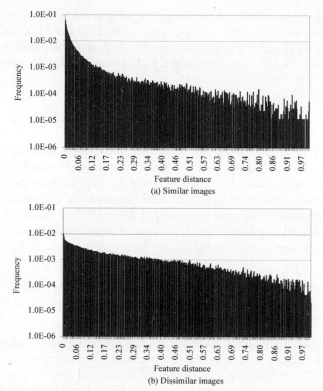

Feature distance
(a) Similar images

Feature distance
(b) Dissimilar images

Fig. 4.2: The distributions of feature distances

observation, we infer that the assumption of the Minkowski metric is inaccurate. Similar images are not necessarily similar in all features.

Furthermore, we examined whether similar images resemble the query images in the same way. We tallied the *distance* (δ^+) of the 144 features for different kinds of image transformations. Figure 4.3 presents four representative transformations: GIF, cropped, rotated, and scaled. The x-axis of the figure depicts the feature numbers, from 1 to 144. The first 108 features are various color features, and the last 36 are texture features. The figure shows that various similar images can resemble the query images in very different ways. GIF images have larger δ^+ in color features (the first 108 features) than in texture features (the last 36 features). In contrast, cropped images have larger δ^+ in texture features than in color features. For rotated images, the δ^+ in colors comes close to zero, although its texture feature distance is much greater. A similar pattern appears in the scaled and the rotated images. However, the magnitude of the δ^+ of scaled images is very different from that of rotated images.

Our observations show that the assumptions made by the Minkowski and weighted Minkowski function are questionable.

1. Similar images do not resemble the query images in all features. Figure 4.2 shows that similar images different from a query image in many respects.

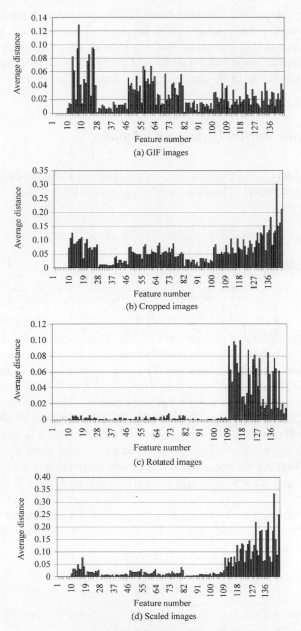

Fig. 4.3: The average feature distances

2. Images similar to the query images can be similar in differing features. Figure 4.3 shows that some images resemble the query image in texture, others in color.

The above observations not only refute the assumptions of Minkowski-type distance functions, but also provide hints as to how a good distance function would work. The first point is that a distance function does not need to consider all features equally, since similar images may match only some features of the query images. The second point is that a distance function should weight features dynamically, since various similar images may resemble the query image in differing ways. These points lead to the design of the *dynamic partial* distance function.

4.3.2 *Dynamic Partial Distance Function*

Based on the observations explained above, we designed a distance function to better represent the perceptual similarity. Let $\delta_i = |x_i - y_i|$, for $i = 1, 2, \cdots, p$. We first define sets Δ_m as

$$\Delta_m = \{ \text{ The smallest } m \ \delta's \text{ of } (\delta_1, ..., \delta_p) \}.$$

Then we define the *Dynamic Partial Function* (DPF) as

$$d(m, r) = (\sum_{\delta_i \in \Delta_m} \delta_i^r)^{\frac{1}{r}}. \tag{4.3}$$

DPF has two adjustable parameters: m and r. Parameter m can range from 1 to p. When $m = p$, it degenerates to the Minkowski metric. When $m < p$, it counts only the smallest m feature distances between two objects, and the influence of the $(p - m)$ largest feature distances is eliminated. Note that DPF dynamically selects features to be considered for different pairs of objects. This is achieved by the introduction of Δ_m, which changes dynamically for different pairs of objects. In Section 4.4, we will show that if a proper value of m is chosen, it is possible to make similar images aggregate more compactly and locate closer to the query images, simultaneously keeping the dissimilar images away from the query images. In other words, similar and dissimilar images are better separated by DPF than by earlier methods.

The idea employed by DPF can also be generalized to improve the weighted Minkowski distance function. We modify the weighted Minkowski distance by defining the weighted DPF as

$$d_w(m, r) = (\sum_{\delta_i \in \Delta_m} w_i \delta_i^r)^{\frac{1}{r}}. \tag{4.4}$$

In Section 4.4, we will show that DPF can also improve the retrieval performance of the weighted Minkowski distance function.

4.3.3 Psychological Interpretation of Dynamic Partial Distance Function

The *Just Noticeable Difference* (JND) is the smallest difference between two stimuli that a person can detect. B. Goldstein [15] uses the following example to illustrate the JND: A person can detect the difference between a 100-gram weight and a 105-gram weight but cannot detect a smaller difference, so the JND for this person is 5 grams. For our purpose, we introduce a new term. The term is *Just Not the Same* (JNS). Using the same weight example, we may say that a 100-gram weight is just not the same as a weight that is more than 120 grams. So the JNS is 20 grams. When the weight is between 105 and 120 grams, we say that the weight is similar to a 100-gram weight (to a degree).

Now, let us apply JND and JNS to our color perception. We can hardly tell the difference between *deep sky blue* (whose RGB is 0,191,255) and *dodger blue* (whose RGB is 30,144,255). The perceptual difference between these two colors is below JND. On the other hand, we can tell that blue is different from green, and yellow is different from red. In both cases, the colors are perceived as JNS.

For an image search engine, JND and JNS indicate that using Euclidean distance for measuring color difference may not be appropriate. First, JND reveals that when the difference between two colors is insignificant, the two colors are perceived as the same. Second, JNS reveals that when the difference is significant, we say two colors are not the same, and it may not be meaningful to account the full magnitude of difference. (E.g., saying that blue is more different from red than from green is meaningless for our purpose.)

The JND and JNS values for each feature can be obtained only through extensive psychological experiments. Moreover, different people may have different subjective values of JND and JNS. Being aware of the practical difficulty of obtaining exact values of JND and JNS for each feature, DPF addresses this issue reasoning as follows:

- JND is not vital for designing a perceptual distance function, since a feature distance below JND usually is very small and has little effect on the aggregated distance. It does not make much difference to consider it as zero or as a small value.
- JNS is vital for designing a perceptual distance function. A feature distance greater than JNS can introduce significant noise on the aggregated distance.

Though it is difficult to obtain the exact value of JNS for each feature, DPF circumvents this difficulty by taking a probabilistic view: The largest $(p - m)$ feature distances are likely to exceed their JNS values. Removing the $(p - m)$ largest feature distances from the final aggregated distance between objects can reduce the noise above JNS. First, the distances of the $(p - m)$ features are all scaled back to their respective JNS. Second, removing these JNS from the aggregated distance does not affect the relative distance between objects.

In short, DPF considers only the m smallest feature distances and does not count the $(p - m)$ largest feature distances. In this sense, DPF provides a good approximation to consider JND and JNS.

4.4 Empirical Study

We conducted an empirical study to examine the effectiveness of DPF. Our experiments consisted of three parts.

1. We compared DPF with the Euclidean distance function and L_1 distance function, the most widely used similarity functions in image retrieval. We also compared DPF with the histogram Cosine[2] distance function, which is also commonly used in information retrieval [16, 4] (Section 4.4.1).
2. We tested whether DPF can be generalized to video shot-transition detection, the foundation of video analysis and retrieval applications (Section 4.4.2).
3. We experimented DPF with a set of news articles to identify near-duplicates.
4. In addition to the unweighted versions, we also examined whether the weighted DPF is effective for enhancing the performance of the weighted Minkowski distance function (Section 4.4.4).

4.4.1 Image Retrieval

Our empirical study of image retrieval consisted of two parts: training and testing. In the training part, we used the 1.5M-image dataset to predict the optimal m value for DPF. In the testing part, we set DPF with the optimal m value, and tested it on an independently constructed 50K-image dataset to examine its effectiveness.

4.4.1.1 Predicting m Through Training

The design goal of DPF is to better separate similar images from dissimilar ones. To meet this design goal, we must judiciously select parameter m. (We take the Euclidean distance function as the baseline, thus we set $r = 2$ for both DPF and the Minkowski distance function.) Alternatively, we can set a JND threshold for selecting features to be considered by DPF. If we find enough number of features

---

[2] The Cosine metric computes the direction difference between two feature vectors. Specifically, given two feature vectors $\mathbf{x}$ and $\mathbf{y}$, the Cosine metric is given as

$$D = 1 - \frac{\mathbf{x}^T \mathbf{y}}{|\mathbf{x}||\mathbf{y}|}.$$

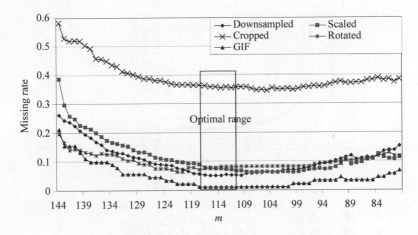

Fig. 4.4: Training for the optimal m value

between two images having a difference below JND, we can say the pair to be similar. One advantage of the threshold method is that the value of m is also pairwise dependent. Please see [17] for this threshold method.

To find the optimal m value, we used the 60,000 original images to perform queries. we applied DPF of different m values to the 1.5M-image dataset. The 24 images with the shortest distance from each query image were retrieved. For each of the five similar-image categories (i.e., GIF, cropped, downsampled, rotated, or scaled), we observed how many of them failed to appear in the top-24 results. Figure 4.4 presents the average rate of missed images for each similar-image category. The figure shows that when m is reduced from 144 to between 110 and 118, the rates of missing are near their minimum for all five similar-image categories. (Note that when $m = 144$, DPF degenerates into the Euclidean function.) DPF outperforms the Euclidean distance function by significant margins for all similar-image categories.

To investigate why DPF works effectively when m is reduced, we tallied the distances from these 60,000 queries to their similar images and their dissimilar images, respectively. We then computed the average and the standard deviation of these distances. We denote the average distance of the similar images to their queries as μ_d^+, of the dissimilar images as μ_d^-. We denote the standard deviation of the similar images' distances as σ_d^+, of the dissimilar images as σ_d^-.

Figure 4.5 depicts the effect of m (in the x-axis) on μ_d^+, μ_d^-, σ_d^+, and σ_d^-. Figure 4.5(a) shows that as m becomes smaller, both μ_d^+ and μ_d^- decrease. The average distance of similar images (μ_d^+), however, decreases at a faster pace than that of dissimilar images (μ_d^-). For instance, when we decrease m from 144 to 130, μ_d^+ decreases from 1.0 to about 0.3, a 70% decrease, whereas μ_d^- decreases from 3.2 to about 2.0, a 38% decrease. This gap indicates μ_d^+ is more sensitive to the m value

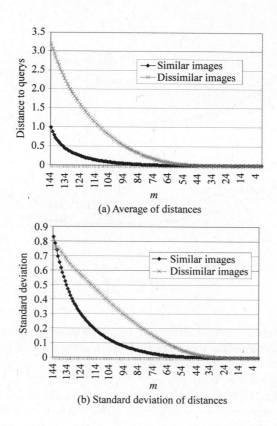

(a) Average of distances

(b) Standard deviation of distances

Fig. 4.5: The effect of DPF

than μ_d^-. Figure 4.5(b) shows that the standard deviations σ_d^+ and σ_d^- observe the same trend as the average distances do. When m decreases, similar images become more compact in the feature space at a faster pace than dissimilar images do.

To provide more detailed information, Figure 4.6 depicts the distance distributions at four different m values. Figure 4.6(a) shows that when $m = 144$, a significant overlap occurs between the distance distributions of similar and dissimilar images to the query images. (When $m = 144$, DPF degenerates to the Euclidean function.) In other words, many similar images and dissimilar images may reside about the same distance from their query image, which causes degraded search performance. When we decrease m to 124, Figure 4.6(b) shows that both distributions shift toward the left. The distribution of similar images becomes more compact, and this leads to a better separation from dissimilar images. Further decreasing the m value moves both distributions leftward (as shown in Figures 4.6(c) and 4.6(d)). When little room is left for the distance distribution of similar images to move leftward, the overlap can eventually increase. Our observations from these figures confirm

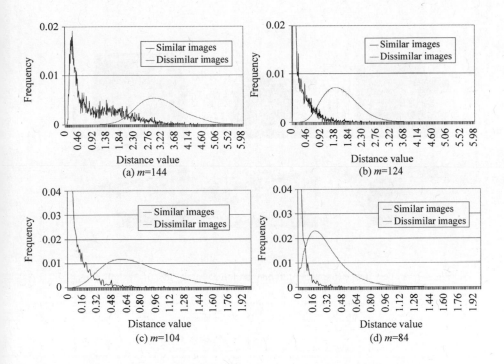

Fig. 4.6: Distance distributions vs. m

that we need to find the optimal m value to achieve best separation for similar and dissimilar images.

4.4.1.2 Testing DPF

We tested our distance functions on a dataset that was independently constructed from the 1.5M-image dataset used for conducting mining and parameter training.

The test dataset consisted of 50K randomly collected World Wide Web images. Among these images we identified 100 images as query images. For each query image, we generated 24 similar images using the transformation methods described in Section 4.2. We also visually identified 5 similar images for each query image. (See Figure 4.7 for examples of visually-identified similar images).

We conducted 100 queries using the 100 query images. For each query, we recorded the distance ranks of its similar images. For DPF, we fixed m value as 114 based on the training results in Section 4.4.1.1. Figure 4.8 depicts the experimental results. The precision-recall curves in the figure show that the search performance of DPF is significantly better than the other traditional distance functions. For in-

Fig. 4.7: A query image (the upper left one) and five visually identified similar images

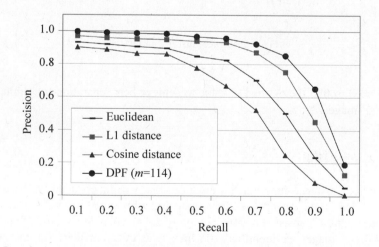

Fig. 4.8: Precision/recall for similar images

stance, to achieve a recall of 80%, the retrieval precision of DPF is 84%, whereas the precision of the L_1 distance, the Euclidean distance, and the histogram Cosine distance is 70%, 50%, and 25%, respectively.

We were particularly interested in the retrieval performance of the visually identified similar images, which were not included into the training-image dataset. Figure 4.9 compares the retrieval performance of DPF and traditional distances for the visually identified similar images. The precision-recall curves indicate that, even though the visually identified similar images were not included in the training-image dataset, DPF could still find them effectively in the testing phase. This indicates that the trained DPF parameters can be generalized to find similar images produced by methods other than those for producing the training dataset.

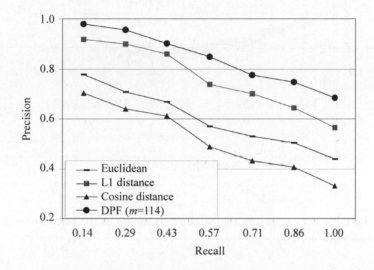

Fig. 4.9: Precision/recall for visually identified similar images

4.4.2 Video Shot-Transition Detection

To further examine the generality of the DPF, we experimented DPF in another application — video shot-transition detection.

Our video dataset consisted of 150 video clips which contained thousands of shots. The videos covered the following subjects:

- Cartoon: 30 clips, each clip lasting for 50 seconds (from commercial CDs).
- Comedy: 50 clips, each lasting for up to 30 seconds.
- Documentary: 70 clips, each lasting for two to five minutes [18].

For characterizing a frame, we extracted the same set of 144 features for each frame, since these features can represent images to a reasonable extent. Our experiments had two goals. The first was to find the optimal parameter m settings for DPF (Section 4.4.2.1). The second was to compare the shot detection accuracy between

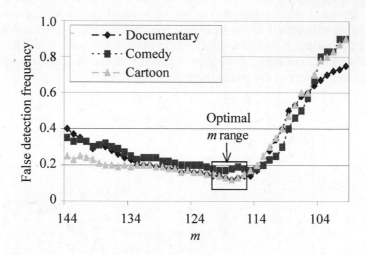

Fig. 4.10: Optimal m

employing DPF and employing the Minkowski metric as the inter-frame distance function (Section 4.4.2.2).

4.4.2.1 Parameter m

We fixed $r = 2$ in our empirical study. Then we took a machine learning approach to train the value of m. We sampled 40% of the video clips as the training data to discover a good m. We then used the remaining 60% of video clips as testing data to examine the effectiveness of the learned m.

In the training phase, we labeled the accurate positions of shot boundaries. We then experimented with different values of m on three video datasets (cartoon, comedy, and documentary). Figure 4.10 shows that for all three video types, the false detection rates are reduced to a minimum as m is reduced from 144 to between 115 and 120. (Recall that when $m = 144$, DPF degenerates into the Minkowski distance function.) It is evident that the Minkowski distance function is not the best choice for our purpose.

4.4.2.2 DPF vs. Minkowski

We next compared two inter-frame distance functions, DPF and Euclidean, on the testing data. For DPF, we set $m = 117$ based on the training results in Section 4.4.2.1. Table 4.2 shows that DPF improves the detection accuracy over the Euclidean distance function on both precision and recall for all video categories. The average improvement as shown in Figure 4.11 is about 7% in both recall and precision. In other words, for every 100 shot transitions to be detected, DPF makes 7 fewer detection errors, a marked improvement.

Table 4.2: Precision and recall

Distance Functions	Video Type	Comedy	Cartoon	Documentary
	# of Shot Boundaries	425	167	793
Euclidean	*# of false*	93	39	192
	# of miss	97	37	183
	Precision	78.1%	76.6%	75.8%
	Recall	77.2%	77.8%	76.9%
DPF	*# of false*	61	26	140
	# of miss	67	25	129
	Precision	85.6%	84.4%	82.3%
	Recall	84.2%	85.0%	83.7%

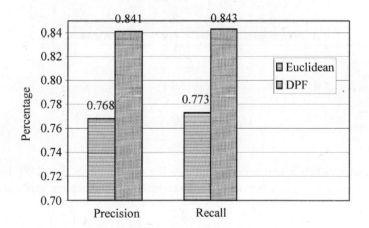

Fig. 4.11: Overall precision and recall comparison

Figure 4.12 illustrates why DPF can better detect shot boundaries than Euclidean distance, from the signal/noise ratio perspective. The x-axis of the figure depicts the frame number; the y-axis depicts the inter-frame distance between the ith and the $(i+1)^{th}$ frames. We mark each real shot boundary with a circle and a false detection with a cross. Figure 4.12(a) shows that the Euclidean distance function identified four shot boundaries, in which the left-most one was a false positive. Figure 4.12(b) shows that DPF separates the distances between shot boundaries and

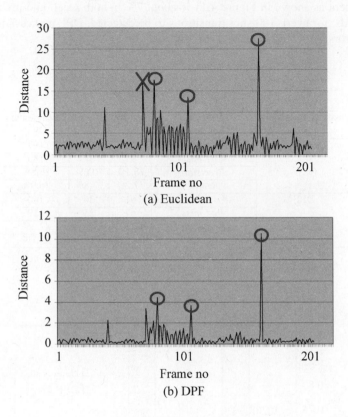

Fig. 4.12: Euclidean vs. DPF

non-boundaries better, and hence eliminates the one mis-detection. DPF improves the signal/noise ratio, and therefore, it is more effective in detecting shot transitions.

4.4.3 Near Duplicated Articles

A piece of news is often quoted or even included by several articles. For instance, a piece of new released by the Reuters may be included in some Blogger posts. A search engine would like to cluster all near-duplicated articles and present them together to avoid information redundancy.

We compared two distance functions on Google News in 2006. Between DPF and a hashing algorithm very similar to LSH, DPF outperforms the hash algorithm by about 10% in both precision and recall. However, since the computation complexity of hashing is linear but DPF quadratic. When the number of candidate articles is very large, DPF encounters scalability problem. To deal with this practical deployment challenge, the work of Dyndex [19] proposes an approximate indexing method to speed up similar-instance lookup. The basic idea is to ignore the non-metric nature of DPF, or using the full Euclidean space to perform indexing. A lookup is performed in the Euclidean space. Though precision/recall may be degraded, this approximation compromises slightly degraded accuracy for speedup. For details, please consult reference GohLC02.

4.4.4 Weighted DPF vs. Weighted Euclidean

We were also interested in applying weighted DPF to improve the weighted Minkowski distance function, which has been used extensively to personalize similarity measures. For weighted Minkowski distance, a weighting vector is learned for each query. Usually, the weight of a feature is set as the inverse of the variance of its values among similar images. Here, we allowed the weighted Euclidean distance function to work under the ideal condition — that is, it knows all similar images a priori and can compute the ideal weighting vector for each query. Figure 4.13 shows that the weighted Euclidean function outperforms its unweighted counterpart. This result confirms that the weighted version [20, 21] is indeed a better choice than the unweighted version (provided that the appropriate weighting can be learned). However, there is still much room for improvement. When we applied weighted DPF using the same weighting vector, its retrieval performance was better than that of the weighted Euclidean distance function. For instance, at 80% recall rate, the retrieval precision of the weighted Euclidean distance is about 68%, whereas the weighted DPF could achieve a precision of above 85%. Again, our empirical study shows that the generalized form of DPF, weighted DPF, can be used to markedly enhance the weighted Minkowski distance for measuring image similarity.

4.4.5 Observations

We summarize the results of our experiments as follows:

1. DPF is more effective than some most representative distance functions used in the CBIR community (e.g., Minkowski-like and histogram Cosine distance functions) for measuring image similarity and for detecting shot transitions.
2. The weighted version of DPF outperforms the weighted version of the Euclidean distance function.
3. We believe that DPF can be generalized to find similar images of some other ways, and that DPF can be effective when a different set of low-level features are employed. Our belief is partially supported by our empirical results, and partially justified by similar theories in cognitive science, which we discuss next.

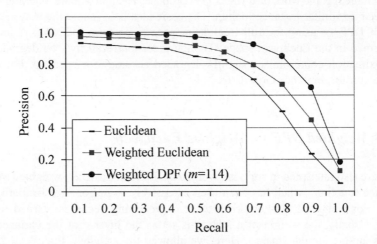

Fig. 4.13: Comparison of weighted functions

4.5 Related Reading

Similarity is one of the most central theoretical constructs in psychology [6, 7]. Its also plays a central role in information categorization and retrieval. Here we summarize related work in similarity distance functions. Using our experimental results, together with theories and examples in cognitive psychology, we explain why DPF works effectively as we discuss the progress of the following three similarity paradigms in cognitive psychology.

1. *Similarity is a measure of all respects.* As we discussed in Section 4.3, a Minkowski-like metric accounts for all respects (i.e., all features) when it is employed to measure similarity between two objects. Our mining result shown in Figure 4.2 is just one of a large number of counter-examples demonstrating that the assumption of the Minkowski-like metric is questionable. The psychology studies of [6, 7] present examples showing that the Minkowski model appears to violate human similarity judgements.
2. *Similarity is a measure of a fixed set of respects.* Substantial work on similarity has been carried out by cognitive psychologists. The most influential work is perhaps that of Tversky [7], who suggests that similarity is determined by matching features of compared objects, and integrating these features by the formula

$$S(A,B) = \theta f(A \cap B) - \alpha f(A - B) - \beta f(B - A). \tag{4.5}$$

The similarity of A to B, $S(A,B)$, is expressed as a linear combination of the common and distinct features. The term $(A \cap B)$ represents the common features of A and B. $(A - B)$ represents the features that A has but B does not; $(B - A)$ represents the features that B has but A does not. The terms θ, α, and β reflect the weights given to the common and distinctive components, and function f is often assumed to be additive [6]. The weighted Minkowski function [22] and the quadratic-form distances [23, 24] are the two representative distance functions that match the spirit of Eq.(4.5). The weights of the distance functions can be learned via techniques such as relevance feedback [20, 22], principal component analysis, and discriminative analysis [25]. Given some similar and some dissimilar objects, the weights can be adjusted so that similar objects can be better distinguished from other objects.

However, the assumption made by these distance functions, that all similar objects are similar in the same respects [25], is questionable. As we have shown in Figure 4.3, GIF, cropped, rotated, and scaled images are all similar to the original images, but in differing features. For this reason, DPF allows objects to be matched to the query in different respects dynamically.

3. *Similarity is a process that provides respects for measuring similarity.* Murphy and Medin [26] provide early insights into how similarity works in human perception: "The explanatory work is on the level of determining which attributes will be selected, with similarity being at least as much a consequence as a cause of a concept coherence." Goldstone [27] explains that similarity is the process that determines the respects for measuring similarity. In other words, a distance function for measuring a pair of objects is formulated only after the objects are compared, not before the comparison is made. The respects for the comparison are activated in this formulation process. The activated respects are more likely to be those that can support coherence between the compared objects.

With those paradigms in mind, let us re-examine how DPF works. DPF activates different features for different object pairs. The activated features are those with minimum differences — those which provide coherence between the objects. If coherence can be maintained (because sufficient a number of features are similar), then the objects paired are perceived as similar. Cognitive psychology seems able to explain much of the effectiveness of DPF.

4.6 Concluding Remarks

We have presented DPF, its formulation via data mining and its explanation in cognitive theories. There are several avenues to improve DPF. First, the activation of respects is believed to be context-sensitive [28, 29, 30]. Also, certain respects may be more salient than others, and hence additional weighting factors should be considered. In Chapter 5 we discuss how weights can be learned from user feedback via some supervised approach. As we discussed in the chapter, the parameters of DPF can be learned using a threshold method, and the quadratic nature of DPF can be alleviated through an approximate indexing scheme. For details, please consult [17, 19].

References

1. Li, B., Chang, E.Y., Wu, Y.L. Discovery of a perceptual distance function for measuring image similarity. *Multimedia Syst.*, 8(6):512–522, 2003.
2. Richardson, M.W. Multidimensional psychophysics. *Psychological Bulletin*, 35:659–660, 1938.
3. Rubner, Y., Tomasi, C., Guibas, L. Adaptive color-image embedding for database navigation. In *Proceedings of the the Asian Conference on Computer Vision*, pages 104–111, January 1998.
4. Witten, I., Moffat, A., Bell, T. *Managing Gigabytes: Compressing and Indexing Documents and Images*. Van Nostrand Reinhold, New York, NY, 1994.
5. Li, J., Wang, J.Z., Wiederhold, G. Irm: Integrated region matching for image retrieval. In *Proceedings of ACM Multimedia*, pages 147–156, October 2000.
6. Medin, D.L., Goldstone, R.L., Gentner, D. Respects for similarity. *Psychological Review*, 100(2):254–278, 1993.
7. Tversky, A. Feature of similarity. *Psychological Review*, 84:327–352, 1977.
8. Wandell, B. *Foundations of Vision*. Sinauer, 1995.
9. Hua, K.A., Vu, K., Oh, J.H. Sammatch: A flexible and efficient sampling-based image retrieval technique for image databases. In *Proceedings of ACM Multimedia*, pages 225–234, 1999.
10. Leu, J.G. Computing a shape's moments from its boundary. *Pattern Recognition*, pages Vol.24, No.10,pp.949–957, 1991.
11. Ma, W.Y., Zhang, H. Benchmarking of image features for content-based retrieval. In *Proceedings of Asilomar Conference on Signal, Systems & Computers*, 1998.
12. Manjunath, B., Wu, P., Newsam, S., Shin, H. A texture descriptor for browsing and similarity retrieval. *Signal Processing Image Communication*, 2001.
13. Smith, J., Chang, S.F. Automated image retrieval using color and texture. *IEEE Transactions on Pattern Analysis and Machine Intelligence*, November 1996.
14. Tamura, H., Mori, S., Yamawaki, T. Texture features corresponding to visual perception. *IEEE Transactions on Systems Man Cybernet (SMC)*, 1978.
15. Goldstein, E., Fink, S. Selective attention in vision. *Journal of Experimental Psychology*, 7:954–967, 1981.
16. Smith, J.R. *Integrated Spatial and Feature Image Systems: Retrieval, Analysis and Compression*. PhD Thesis, Columbia University, 1997.
17. Qamra, A., Meng, Y., Chang, E.Y. Enhanced perceptual distance functions and indexing for image replica recognition. *IEEE Trans. Pattern Anal. Mach. Intell.*, 27(3):379–391, 2005.
18. www-nlpir.nist.gov/projects/t01v/t01v.html.
19. Goh, K., Li, B., Chang, E.Y. Dyndex: a dynamic and non-metric space indexer. In *Proceedings of ACM International Conference on Multimedia*, pages 466–475, 2002.
20. Porkaew, K., Mehrota, S., Ortega, M. Query reformulation for content based multimedia retrieval in mars. In *Proceedings of ICMCS*, pages 747–751, 1999.
21. Ortega, M., Rui, Y., Chakrabarti, K., Mehrotra, S., Huang, T.S. Supporting similarity queries in mars. In *Proceedings of ACM International Conference on Multimedia*, pages 403–413, 1997.
22. Rocchio, J. Relevance feedback in information retrievalIn, editor, *The SMART retrival system: Experiments in automatic document processing*. Prentice-Hall, 1971.
23. Flickner, M., Sawhney, H., Ashley, J., Huang, Q., Dom, B., Gorkani, M., Hafner, J., Lee, D., Petkovic, D., Steele, D., Yanker, P. Query by image and video content: the QBIC system. *IEEE Computer*, 28(9):23–32, 1995.
24. Ishikawa, Y., Subramanya, R., Faloutsos, C. Mindreader: Querying databases through multiple examples. In *Proceedings of VLDB*, pages 218–227, 1998.
25. Zhou, X.S., Huang, T.S. Comparing discriminating transformations and svm for learning during multimedia retrieval. In *Proc. of ACM Conf. on Multimedia*, pages 137–146, 2001.
26. Murphy, G., Medin, D. The role of theories in conceptual coherence. *Psychological Review*, 92:289–316, 1985.

27. Goldstone, R.L. Similarity, interactive activation, and mapping. *Journal of Experimental Psychology: Learning, Memory, and Cognition*, 20:3–28, 1994.
28. Jurisica, I., Glasgow, J.I. Improving performance of case-based classification using context-based relevance. *IJAIT*, 6(4):511–536, 1997.
29. Schyns, P.G., Goldstone, R.L., Thibaut, J.P. The development of features in object concepts. *Behavioral and Brain Science*, 21:1–54, 1998.
30. Tong, S., Chang, E. Support vector machine active learning for image retrieval. In *Proceedings of ACM International Conference on Multimedia*, pages 107–118, October 2001.

Chapter 5
Formulating Distance Functions

Abstract Tasks of data mining and information retrieval depend on a good distance function for measuring similarity between data instances. The most effective distance function must be formulated in a context-dependent (also application-, data- and user-dependent) way. In this chapter[†], we present a method, which learns a distance function by capturing the nonlinear relationships among contextual information provided by the application, data, or user. We show that through a process called the "kernel trick," such nonlinear relationships can be learned efficiently in a projected space. Theoretically, we substantiate that our method is both sound and optimal. Empirically, using several datasets and applications, we demonstrate that our method is effective and useful.

Keywords: Distance function, kernel methods, kernel trick, similarity.

5.1 Introduction

At the heart of data-mining and information-retrieval tasks is a distance function that measures *similarity* between data instances. As mentioned in Chapter 4, to date, most applications employ a variant of the *Euclidean distance* for measuring similarity. We show in Chapter 4 that DPF is more effective than the Minkowsky metric to account for similarity. In this chapter, we discuss how to incorporate the idiosyncrasies of the application, data, and user (which we refer as contextual information) into distance-function formulation.

How do we consider contextual information in formulating a good distance function? One extension of DPF is to weight the data attributes (features) based on their importance for a target task [2, 3, 4]. For example, for answering a *sunset* image-query, color features should be weighted higher. For answering an *architecture* image-query, shape and texture features may be more important. Weighting these

---

[†] ©ACM, 2005. This chapter is a minor revision of the author's work with Gang Wu and Navneet Panda [1] published in KDD'05. Permission to publish this chapter is granted under copyright license #2587641486368.

features is equivalent to performing a *linear* transformation in the space formed by the features. Although linear models enjoy the twin advantages of simplicity of description and efficiency of computation, this same simplicity is insufficient to model similarity for many real-world datasets. For example, it has been widely acknowledged in the image/video retrieval domain that a query concept is typically a nonlinear combination of perceptual features (color, texture, and shape) [5, 6]. In this chapter we perform a *nonlinear* transformation on the feature space to gain greater flexibility for mapping features to semantics.

We name our method *distance-function alignment* (DFA for short). The inputs to DFA are a prior distance function (e.g., a DPF with a learned parameter m), and *contextual information*. Contextual information can be conveyed in the form of *training data* (discussed in detail in Section 5.2). For instance, in the information-retrieval domain, Web users can convey information via relevance feedback showing which documents/images are relevant to their queries. In the biomedical domain, physicians can indicate which pairs of proteins may have similar functions. DFA transforms the prior function to capture the nonlinear relationships among the *contextual information*. The similarity scores of unseen data-pairs can then be measured by the transformed function to better reflect the idiosyncrasies of the application, data, and user.

At first it might seem that capturing nonlinear relationships among contextual information can suffer from high computational complexity. DFA avoids this concern by employing the *kernel trick*[1]. The kernel trick lets us generalize distance-based algorithms to operate in the *projected space* (defined next), usually nonlinearly related to the *input space*. The *input space* (denoted as $\mathscr{I}$) is the original space in which data vectors are located (e.g., in Figure 5.2(a)), and the *projected space* (denoted as $\mathscr{P}$) is that space to which the data vectors are projected, linearly or nonlinearly, (e.g., in Figure 5.2(b)). The advantage of using the *kernel trick* is that, instead of explicitly determining the coordinates of the data vectors in the projected space, the distance computation in $\mathscr{P}$ can be efficiently performed in $\mathscr{I}$ through a kernel function. Specifically, given two vectors $\mathbf{x}_i$ and $\mathbf{x}_j$, kernel function $K(\mathbf{x}_i, \mathbf{x}_j)$ is defined as the inner product of $\phi(\mathbf{x}_i)$ and $\phi(\mathbf{x}_j)$, where ϕ is a basis function that maps the vectors $\mathbf{x}_i$ and $\mathbf{x}_j$ from $\mathscr{I}$ to $\mathscr{P}$. The inner product between two vectors can be thought of as a measure of their similarity. Therefore, $K(\mathbf{x}_i, \mathbf{x}_j)$ returns the similarity of $\mathbf{x}_i$ and $\mathbf{x}_j$ in $\mathscr{P}$. The distance between $\mathbf{x}_i$ and $\mathbf{x}_j$ in terms of the kernel is defined as

$$
\begin{aligned}
d(\mathbf{x}_i, \mathbf{x}_j) \\
= \|\phi(\mathbf{x}_i) - \phi(\mathbf{x}_j)\|_2 \\
= \sqrt{K(\mathbf{x}_i, \mathbf{x}_i) + K(\mathbf{x}_j, \mathbf{x}_j) - 2K(\mathbf{x}_i, \mathbf{x}_j)}.
\end{aligned}
\tag{5.1}
$$

Since a kernel function can be either linear or nonlinear, the traditional feature-weighting approach (e.g., [2, 4]) is just a special case of DFA.

---

[1] The *kernel trick* was first published in 1964 in the paper of M. Aizerman, E. Braverman, and L. Rozonoer's [7]. The kernel trick has been applied to several algorithms in statistics, including Support Vector Machines and kernel PCA.

How does DFA work? Given a distance function and a dataset, DFA first uses the prior function to compute pairwise similarity for every data pair. It then selects a subset of the data to query the user's subjective similarity scores. (The feedback provided by the user can be in a qualitative or quantitative form, which we will discuss in Section 5.2.2.) The difference between the similarity scores computed by the prior function and the scores provided by the user is the gap that DFA aims to bridge. Our goal is to transform the function in such a way that it can produce similarity scores in better agreement with the user's perceptions. More specifically, given a prior function (for example, a polynomial function or a Gaussian function) that produces default similarity scores between data items, DFA performs a linear transformation on the prior function in $\mathscr{P}$, based on the contextual information provided by the user. Effectively, our DFA procedure ensures that the distances between similar vectors are decreased, and the distances between dissimilar vectors are increased. Since performing a linear transformation in $\mathscr{P}$ can result in a nonlinear transformation in $\mathscr{I}$, DFA achieves both model flexibility and computational efficiency. Theoretically, we prove that DFA achieves optimal alignment to the *ideal* kernel defined by [8]. Empirically, using several datasets and applications, we show that our experimental results back up our theoretical analysis.

Illustrative Examples

We use two examples to illustrate the effectiveness of DFA. The first example demonstrates why contextual information is needed, and the second one demonstrates why nonlinear transformation is necessary to gain adequate model complexity.

Figure 5.1 presents different groupings of the same raw data. Figure 5.1(a) depicts the raw data of different locations, colors, and sizes. Suppose we use the *k*-means algorithm to organize the data into two clusters. Without knowing the desired grouping rule, we do not know which of the rules is the choice of the application or user: *group by proximity*, *group by color*, or *group by size*. The only way to know the intent is to query the user for contextual information. The boxes in the figure depict the similar sets conveyed to DFA through training data. For each grouping rule, DFA uses the contextual information to formulate the distance function. The *k*-means algorithm can then use the function to achieve the desired grouping.

Notice that to achieve *group by proximity* and *group by color* in this example, performing a linear transformation in the feature space will suffice. For proximity grouping, we can weight the vertical dimension higher than the horizontal dimension; and for color grouping, we can weight the horizontal dimension higher than the vertical. This linear transformation can be achieved by the traditional IR methods (e.g., [2, 4]). For *group by size*, however, a nonlinear transformation is required, and this is what DFA can achieve and what the traditional methods cannot.

Figure 5.2 shows how DFA learns a nonlinear distance function using projection. Figure 5.2(a) shows two clusters of data, one in circles and the other in

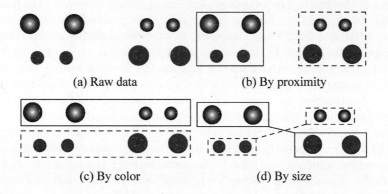

(a) Raw data   (b) By proximity

(c) By color   (d) By size

Fig. 5.1: Clustering by different functions (See color insert)

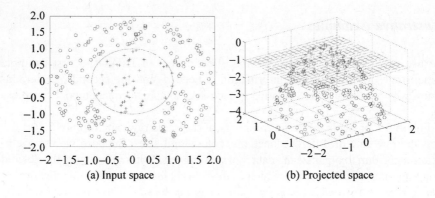

(a) Input space   (b) Projected space

Fig. 5.2: Clustering via the kernel trick (See color insert)

crosses. These two clusters of data obviously are not linearly separable in the two-dimensional input space $\mathscr{I}$. After we have used the "kernel trick" to implicitly project the data onto a three-dimensional projected space $\mathscr{P}$ (shown in Figure 5.2(b)), the two clusters can be separated by a linear hyperplane in the projected space. What DFA accomplishes is to learn the distance function in the projected space based on contextual information. From the perspective of the input space, $\mathscr{I}$, the learned distance function captures the nonlinear relationships among the training data.

In summary, we address in this chapter a core problem of data mining and information retrieval: formulating a context-based distance function to improve the accuracy of similarity measurement. In Section 5.2, we propose DFA, an efficient method for adapting a similarity measure to contextual information, and also provide the proof of optimality of the DFA algorithm. We empirically demonstrate the effectiveness of DFA on clustering and classification in Section 5.3. Finally, we offer our concluding remarks and suggestions for future work in Section 5.5.

5.2 DFA Algorithm

Given the prior kernel function K and contextual information, DFA transforms K. Kernel function $K(\mathbf{x}_i, \mathbf{x}_j)$, as discussed in Section 5.4.2, can be considered as a similarity measure between instances $\mathbf{x}_i$ and $\mathbf{x}_j$. We assume $0 \leq K(\mathbf{x}_i, \mathbf{x}_j) \leq 1$. A value of 1 indicates that the instances are identical while a value of 0 means that they are completely dissimilar. Commonly used kernels like the Gaussian and the Laplacian are normalized to produce a similarity measure between 0 and 1. The polynomial kernel, though not necessarily normalized, can easily be normalized by using

$$K(\mathbf{x}_i, \mathbf{x}_j) = \frac{K(\mathbf{x}_i, \mathbf{x}_j)}{\sqrt{K(\mathbf{x}_i, \mathbf{x}_i) K(\mathbf{x}_j, \mathbf{x}_j)}}. \tag{5.2}$$

The contextual information is represented by sets $\mathscr{S}$ and $\mathscr{D}$, where $\mathscr{S}$ denotes the set of similar pairs of instances, and $\mathscr{D}$ the set of dissimilar pairs of instances. Sets $\mathscr{S}$ and $\mathscr{D}$ can be constructed either directly or indirectly. Directly, users can return the information about whether two instances $\mathbf{x}_i$ and $\mathbf{x}_j$ are similar or dissimilar. In such cases, the similar set $\mathscr{S}$ can be written as $\{(\mathbf{x}_i, \mathbf{x}_j) | \mathbf{x}_i \sim \mathbf{x}_j\}$, and the dissimilar set $\mathscr{D}$ as $\{(\mathbf{x}_i, \mathbf{x}_j) | \mathbf{x}_i \nsim \mathbf{x}_j\}$. Indirectly, we may know only the class-label of instance $\mathbf{x}_i$ as y_i. In this case, we can consider $\mathbf{x}_i$ and $\mathbf{x}_j$ to be a similar pair if $y_i = y_j$, and a dissimilar pair otherwise.

In the remainder of this section, we first propose a transformation model to formulate the contextual information in terms of the prior kernel K (Section 5.2.1). Next, we discuss methods to generalize the model to compute the distance between unseen instances (Section 5.2.2).

5.2.1 Transformation Model

The goal of our transformation is to increase the kernel value for the similar pairs, but decrease the kernel value for the dissimilar pairs. DFA performs transformation in $\mathscr{P}$, to modify the kernel from K to $\tilde{K}$. Let β_1 and β_2 denote the slopes of transformation curves for dissimilar and similar pairs, respectively. For a given $\mathscr{S}$ and $\mathscr{D}$, the kernel matrix $\mathbf{K}$, corresponding to the kernel K, is then modified as follows

$$\tilde{k}_{ij} = \begin{cases} \beta_1 k_{ij}, & \text{if } (\mathbf{x}_i, \mathbf{x}_j) \in \mathscr{D}, \\ \beta_2 k_{ij} + (1 - \beta_2), & \text{if } (\mathbf{x}_i, \mathbf{x}_j) \in \mathscr{S}, \end{cases} \tag{5.3}$$

where $0 \leq \beta_1, \beta_2 \leq 1$ and $\tilde{k}_{ij}$ is the ij^{th} component of the new kernel matrix $\tilde{\mathbf{K}}$.

In what follows, we prove two important theorems. Theorem 5.1 demonstrates that under some constraints on β_1 and β_2, our proposed similarity transformation model in Eq.(5.3) ensures a valid kernel. Theorem 5.2 mathematically demonstrates that under the constraints from Theorem 5.1, the transformed $\tilde{\mathbf{K}}$ in Eq.(5.3) guarantees a better alignment to the ideal $\mathbf{K}^*$ in Eq.(5.30) than the $\mathbf{K}$ to the $\mathbf{K}^*$.

Theorem 5.1. *Under the assumption that* $0 \leq \beta_1 \leq \beta_2 \leq 1$, *the transformed kernel* $\tilde{K}$ *is positive (semi-) definite if the prior kernel* K *is positive (semi-) definite.* (The assumption $\beta_1 \leq \beta_2$ means that we place more emphasis on the decreasing similarity value (K) for dissimilar instance-pairs.)

Proof. The transformed kernel $\tilde{K}$ can be written as follows:

$$\tilde{K} = \beta_1 K + (\beta_2 - \beta_1) K \otimes K^* + (1 - \beta_2) K^*, \tag{5.4}$$

which corresponds to the kernel matrix $\tilde{\mathbf{K}}$, associated with the training set $\mathscr{X}$, in Eq.(5.3). If the prior K is positive (semi-) definite, using the fact that the ideal kernel K^* is positive (semi-) definite [8], we can derive that $\tilde{K}$ in Eq.(5.4) is also positive (semi-) definite if $0 \leq \beta_1 \leq \beta_2 \leq 1$. Here, we use the closure properties of kernels, namely that the product and summation of valid kernels also give a valid kernel [9].

Theorem 5.2. *The kernel matrix* $\tilde{\mathbf{K}}$ *of the transformed kernel* $\tilde{K}$ *obtains a better alignment than the prior kernel matrix* $\mathbf{K}$ *to the ideal kernel matrix* $\mathbf{K}^*$, *if* $0 \leq \beta_1 \leq \beta_2 \leq 1$. *Moreover, a smaller* β_1 *or* β_2 *would induce a higher alignment score.*

Proof. In [10], it has been proven that a kernel with the following form has a higher alignment score with the ideal kernel K^* than the original kernel K,

$$\overline{K} = K + \gamma K^*, \quad \gamma > 0, \tag{5.5}$$

where we use $\overline{K}$ to distinguish with our $\tilde{K}$ defined in Eq.(5.3). According to the definition of kernel target alignment [8], we have

$$\left[\frac{< \mathbf{K}, \mathbf{K}^* >}{\sqrt{< \mathbf{K}, \mathbf{K} >}} \right]^2 < \left[\frac{< \mathbf{K} + \gamma \mathbf{K}^*, \mathbf{K}^* >}{\sqrt{< \mathbf{K} + \gamma \mathbf{K}^*, \mathbf{K} + \gamma \mathbf{K}^* >}} \right]^2, \tag{5.6}$$

where the common item $\sqrt{< \mathbf{K}^*, \mathbf{K}^* >}$ is omitted at both sides of inequality, and the Frobenius norm of two matrices, say $\mathbf{M} = [m_{ij}]$ and $\mathbf{N} = [n_{ij}]$, is defined as $< \mathbf{M}, \mathbf{N} > = \sum_{i,j} m_{ij} n_{ij}$.

Cristianini et al. [8] proposed the notion of "ideal kernel" (K^*). Suppose $y(\mathbf{x}_i) \in \{1, -1\}$ is the class label of $\mathbf{x}_i$. K^* is defined as

$$K^*(\mathbf{x}_i, \mathbf{x}_j) = \begin{cases} 1, & \text{if } y(\mathbf{x}_i) = y(\mathbf{x}_j), \\ 0, & \text{if } y(\mathbf{x}_i) \neq y(\mathbf{x}_j), \end{cases} \tag{5.7}$$

which is the target kernel that a given kernel is supposed to align with. Employing Eqs.(5.5) and (5.30), we expand Eq.(5.6) as follows

$$\frac{(\sum_{\mathscr{S}} k_{ij})^2}{\sum_{\mathscr{S}} k_{ij}^2 + \sum_{\mathscr{D}} k_{ij}^2} < \frac{[\sum_{\mathscr{S}} (k_{ij} + \gamma)]^2}{\sum_{\mathscr{S}} (k_{ij} + \gamma)^2 + \sum_{\mathscr{D}} k_{ij}^2}. \tag{5.8}$$

Defining $\gamma = \frac{1-\beta_2}{\beta_2} > 0$, where β_2 is the parameter in Eq.(5.3), we then rewrite the right side in (Eq.(5.8)) as follows

$$\frac{\left[\sum_{\mathscr{S}} (k_{ij} + \frac{1-\beta_2}{\beta_2})\right]^2}{\sum_{\mathscr{S}} (k_{ij} + \frac{1-\beta_2}{\beta_2})^2 + \sum_{\mathscr{D}} k_{ij}^2}$$

$$= \frac{\beta_2^2 \left[\sum_{\mathscr{S}} (k_{ij} + \frac{1-\beta_2}{\beta_2})\right]^2}{\beta_2^2 \sum_{\mathscr{S}} (k_{ij} + \frac{1-\beta_2}{\beta_2})^2 + \beta_2^2 \sum_{\mathscr{D}} k_{ij}^2}$$

$$= \frac{[\sum_{\mathscr{S}} (\beta_2 k_{ij} + (1 - \beta_2))]^2}{\sum_{\mathscr{S}} (\beta_2 k_{ij} + (1 - \beta_2))^2 + \beta_2^2 \sum_{\mathscr{D}} k_{ij}^2} \tag{5.9}$$

$$\leq \frac{[\sum_{\mathscr{S}} (\beta_2 k_{ij} + (1 - \beta_2))]^2}{\sum_{\mathscr{S}} (\beta_2 k_{ij} + (1 - \beta_2))^2 + \beta_1^2 \sum_{\mathscr{D}} k_{ij}^2} \tag{5.10}$$

$$= \left[\frac{<\tilde{\mathbf{K}}, \mathbf{K}^*>}{\sqrt{<\tilde{\mathbf{K}}, \tilde{\mathbf{K}}>}}\right]^2, \tag{5.11}$$

where we apply the assumption of $\beta_1 \leq \beta_2$ from (5.9) to (5.10), and employ Eq.(5.3) in the last step (5.11). Combining Eqs.(5.6) and (5.11), we obtain

$$\frac{<\mathbf{K}, \mathbf{K}^*>}{\sqrt{<\mathbf{K}, \mathbf{K}><\mathbf{K}^*, \mathbf{K}^*>}} < \frac{<\tilde{\mathbf{K}}, \mathbf{K}^*>}{\sqrt{<\tilde{\mathbf{K}}, \tilde{\mathbf{K}}><\mathbf{K}^*, \mathbf{K}^*>}}.$$

Therefore, the transformed kernel $\tilde{\mathbf{K}}$ in Eq.(5.3) can achieve a better alignment than the original kernel K under the assumption of $0 \leq \beta_1 \leq \beta_2 \leq 1$. Moreover, a greater γ in Eq.(5.5) will have a higher alignment score [10]. Recall that $\beta_2 = \frac{1}{1+\gamma}$. Hence, a smaller β_2 will have a higher alignment. On the other hand, from Eqs.(5.10) and (5.11), we can see that the alignment score of $\tilde{K}$ is a decreasing function w.r.t. β_1. Therefore, a smaller β_1 or β_2 will result in a higher alignment score. ∎

For a prior kernel K, the inner product of two instances $\mathbf{x}_i$ and $\mathbf{x}_j$ is defined as $\phi(\mathbf{x}_i)^T \phi(\mathbf{x}_j)$ in $\mathscr{P}$. For simplicity, we denote $\phi(\mathbf{x}_i)$ as ϕ_i. The distance between $\mathbf{x}_i$ and $\mathbf{x}_j$ in $\mathscr{P}$ can thus be computed as $d_{ij}^2 = k_{ii} + k_{jj} - 2k_{ij}$, where $k_{ij} = \phi_i^T \phi_j$. Therefore, for the transformed kernel $\tilde{K}$ in Eq.(5.3), the corresponding distance $\tilde{d}_{ij}^2 = \tilde{k}_{ii} + \tilde{k}_{jj} - 2\tilde{k}_{ij}$ in $\mathscr{P}$ can be written in terms of d_{ij}^2 as

$$\tilde{d}_{ij}^2 = \begin{cases} \beta_2 d_{ij}^2, & \text{if } (\mathbf{x}_i, \mathbf{x}_j) \in \mathscr{S}, \\ \beta_1 d_{ij}^2 + 2(1 - \beta_2) + (\beta_2 - \beta_1)(k_{ii} + k_{jj}), & \text{if } (\mathbf{x}_i, \mathbf{x}_j) \in \mathscr{D}. \end{cases} \tag{5.12}$$

Since K has been normalized as in Eq.(5.2), we have the distance $d_{ij}^2 = 2 - 2k_{ij}$. Eq.(5.12) can thus be rewritten as

$$\tilde{d}_{ij}^2 = \begin{cases} \beta_2 d_{ij}^2, & \text{if } (\mathbf{x}_i, \mathbf{x}_j) \in \mathcal{S}, \\ \beta_1 d_{ij}^2 + 2(1 - \beta_1), & \text{if } (\mathbf{x}_i, \mathbf{x}_j) \in \mathcal{D}. \end{cases} \tag{5.13}$$

Using the property $0 \le d_{ij} \le 2$ and the condition $0 \le \beta_1 \le \beta_2 \le 1$, we can see that $\tilde{d}_{ij}^2 \le d_{ij}^2$ if the two instances are similar, and $\tilde{d}_{ij}^2 \ge d_{ij}^2$ if the two instances are dissimilar. In other words, the transformed distance metric in Eq.(5.13) decreases the intra-class pairwise distance in $\mathcal{P}$, and increases the inter-class pairwise distance in $\mathcal{P}$. The developed distance metric (Eq.(5.13)) is a valid distance metric (non-negativity, symmetry, and triangle inequality) since the transformed kernel in Eq.(5.3) is a valid kernel, according to Theorem 5.1.

Sometimes users may not return the class label y_i for each instance $\mathbf{x}_i$ in the feedback contextual information. Instead, they may only return the information about whether two instances $\mathbf{x}_i$ and $\mathbf{x}_j$ are similar or dissimilar. Denoting the set of pairs of similar instances in the contextual information $\mathcal{X}$ as $\mathcal{S} = \{(\mathbf{x}_i, \mathbf{x}_j) | \mathbf{x}_i \sim \mathbf{x}_j\}$, and the set of pairs of dissimilar instances as $\mathcal{D} = \{(\mathbf{x}_i, \mathbf{x}_j) | \mathbf{x}_i \nsim \mathbf{x}_j\}$, we can extend Eq.(5.13) to the case where only $\mathcal{S}$ and $\mathcal{D}$ information is available for classification or clustering

Since the class label information y_i in Eq.(5.13) can be also considered as a kind of (dis)similarity information by thinking $y_i = y_j$ as $\mathbf{x}_i \sim \mathbf{x}_j$ and $y_i \ne y_j$ as $\mathbf{x}_i \nsim \mathbf{x}_j$, for both cases, we use the notations $\mathcal{S}$ and $\mathcal{D}$ in the rest of the paper as the contextual information from users.

$$\tilde{d}_{ij}^2 = \begin{cases} \beta_2 d_{ij}^2, & \text{if } (\mathbf{x}_i, \mathbf{x}_j) \in \mathcal{S}, \\ \beta_1 d_{ij}^2 + 2(1 - \beta_1), & \text{if } (\mathbf{x}_i, \mathbf{x}_j) \in \mathcal{D}. \end{cases} \tag{5.14}$$

5.2.2 Distance Metric Learning

In this subsection, we show how to generalize the model in Eq.(5.13) to unseen instances without overfitting the contextual information. We use the feature-weighting method [11] by modifying the inner product $k_{ij} = \phi_i^T \phi_j$ in $\mathcal{P}$ as $\phi_i^T \mathbf{A} \mathbf{A}^T \phi_j$. Here, $\mathbf{A}_{m' \times m'}$ is the weighting matrix and m' is the dimension of $\mathcal{P}$. Based on the idea of feature reduction [11], we aim to achieve a small rank of $\mathbf{A}$, which means that the dimensionality of feature vector ϕ is kept small in projected space $\mathcal{P}$. The corresponding distance function thus becomes

$$\tilde{d}_{ij}^2 = (\phi_i - \phi_j)^T \mathbf{A} \mathbf{A}^T (\phi_i - \phi_j). \tag{5.15}$$

To solve for $\mathbf{A} \mathbf{A}^T$ so that the distance metric in Eq.(5.15) equals the distance in Eq.(5.13), we formulate the problem as a convex optimization problem whose objective function is to minimize the rank of the weighting matrix $\mathbf{A}$. However, it induces an NP-Hard problem by directly minimizing rank($\mathbf{A}$), the so-called zero-norm problem in [12]. Since minimizing the trace of a matrix tends to give a low-rank solution when the matrix is symmetric [13], in this chapter, we approximate the rank of a matrix by its trace. Moreover, since $\text{rank}(\mathbf{AA}^T) = \text{rank}(\mathbf{AA}^T\mathbf{AA}^T) \approx \text{trace}(\mathbf{AA}^T\mathbf{AA}^T) = \|\mathbf{AA}^T\|_F^2$, we approximate the problem of minimizing the rank of $\mathbf{A}$ by minimizing its Frobenius norm $\|\mathbf{AA}^T\|_F^2$. The corresponding primal problem is formulated as

$$\min_{\mathbf{AA}^T,\beta_1,\beta_2} \frac{1}{2}\|\mathbf{AA}^T\|_F^2 + C_{\mathscr{D}}\beta_1 + C_{\mathscr{S}}\beta_2, \tag{5.16}$$

$$\begin{aligned}
\text{s.t.} \quad & \beta_2 d_{ij}^2 = \bar{d}_{ij}^2, & (\mathbf{x}_i,\mathbf{x}_j) \in \mathscr{S} \\
& \bar{d}_{ij}^2 = \beta_1 d_{ij}^2 + 2(1-\beta_1), & (\mathbf{x}_i,\mathbf{x}_j) \in \mathscr{D} \\
& \beta_2 \geq \beta_1, & \tag{5.17} \\
& \beta_1 \geq 0, \\
& 1 - \beta_2 \geq 0,
\end{aligned}$$

where $C_{\mathscr{S}}$ and $C_{\mathscr{D}}$ are two non-negative hyper-parameters. Theorem 5.2 shows that a large β_1 or β_2 will induce a lower alignment score. However, on the contrary, $\beta_1 = \beta_2 = 0$ would overfit the training dataset. We hence add two penalty terms $C_{\mathscr{D}}\beta_1$ and $C_{\mathscr{S}}\beta_2$ to control the alignment degree. This strategy is similar to that used in Support Vector Machines [14], which limits the length of weight vector $\|\mathbf{w}\|^2$ in projected space $\mathscr{P}$ to combat the overfitting problem.

The constrained optimization problem above can be solved by considering the corresponding Lagrangian formulation

$$\begin{aligned}
&\mathscr{L}(\mathbf{AA}^T, \beta_1, \beta_2, \eta, \mu, \gamma, \pi) \tag{5.18}\\
&= \frac{1}{2}\|\mathbf{AA}^T\|_F^2 + C_{\mathscr{D}}\beta_1 + C_{\mathscr{S}}\beta_2 \\
&\quad - \sum_{(\mathbf{x}_i,\mathbf{x}_j)\in\mathscr{S}} \alpha_{ij}\left[(\phi_i-\phi_j)^T(\beta_2\mathbf{I}-\mathbf{AA}^T)(\phi_i-\phi_j)\right] \\
&\quad - \sum_{(\mathbf{x}_i,\mathbf{x}_j)\in\mathscr{D}} \alpha_{ij}\left[(\phi_i-\phi_j)^T(\mathbf{AA}^T-\beta_1\mathbf{I})(\phi_i-\phi_j)-2(1-\beta_1)\right] \\
&\quad - \mu\beta_1 - \gamma(\beta_2-\beta_1) - \pi(1-\beta_2),
\end{aligned}$$

where the Lagrangian multipliers (dual variables) are $\alpha_i, \mu, \gamma, \pi \geq 0$. This function has to be minimized w.r.t. the primal variables $\mathbf{AA}^T$, β_1, β_2, and maximized w.r.t. the dual variables $\alpha, \eta, \mu, \gamma, \pi$. To eliminate the primal variables, we set the corresponding partial derivatives to be zero, obtaining the following conditions:

$$\mathbf{A}\mathbf{A}^T = \sum_{(\mathbf{x}_i, \mathbf{x}_j) \in \mathscr{D}} \alpha_{ij} (\phi_i - \phi_j)(\phi_i - \phi_j)^T$$

$$- \sum_{(\mathbf{x}_i, \mathbf{x}_j) \in \mathscr{S}} \alpha_{ij} (\phi_i - \phi_j)(\phi_i - \phi_j)^T, \tag{5.19}$$

$$C_{\mathscr{S}} + \pi - \gamma = \sum_{(\mathbf{x}_i, \mathbf{x}_j) \in \mathscr{S}} \alpha_{ij} (\phi_i - \phi_j)^T (\phi_i - \phi_j), \tag{5.20}$$

$$C_{\mathscr{D}} + \gamma - \mu = \sum_{(\mathbf{x}_i, \mathbf{x}_j) \in \mathscr{D}} \alpha_{ij} \left[2 - (\phi_i - \phi_j)^T (\phi_i - \phi_j) \right]. \tag{5.21}$$

Substituting the conditions of (5.20) and (5.21) into (5.18), we obtain the following dual formulation

$$\mathscr{W}(\alpha, \pi) = \frac{1}{2} \|\mathbf{A}\mathbf{A}^T\|_F^2 + 2 \sum_{(\mathbf{x}_i, \mathbf{x}_j) \in \mathscr{D}} \alpha_{ij}$$

$$- \sum_{(\mathbf{x}_i, \mathbf{x}_j) \in \mathscr{D}} \alpha_{ij} (\phi_i - \phi_j)^T \mathbf{A}\mathbf{A}^T (\phi_i - \phi_j)$$

$$+ \sum_{(\mathbf{x}_i, \mathbf{x}_j) \in \mathscr{S}} \alpha_{ij} (\phi_i - \phi_j)^T \mathbf{A}\mathbf{A}^T (\phi_i - \phi_j)$$

$$- \pi, \tag{5.22}$$

which has to be maximized w.r.t. α_{ij}'s and $\pi \geq 0$. Actually, π can be removed from the dual function (5.22), since $\frac{\partial \mathscr{W}(\alpha, \pi)}{\partial \pi} = -1 < 0$, which means that $\mathscr{W}(\alpha, \pi)$ is a decreasing function w.r.t π. Hence, $\mathscr{W}(\alpha, \pi)$ is maximal at $\pi = 0$.

Now, the dual formulation (5.22) becomes a convex quadratic function w.r.t. only α_{ij}. Next, we examine the constraints of dual formulation on α_{ij}. According to the KKT theorem [15], we have the following conditions

$$\alpha_{ij} \left[\beta_2 d_{ij}^2 - \tilde{d}_{ij}^2 \right] = 0, \quad (\mathbf{x}_i, \mathbf{x}_j) \in \mathscr{S} \tag{5.23}$$

$$\alpha_{ij} \left[\tilde{d}_{ij}^2 - \beta_1 d_{ij}^2 - 2(1 - \beta_1) \right] = 0, \quad (\mathbf{x}_i, \mathbf{x}_j) \in \mathscr{D} \tag{5.24}$$

$$\gamma(\beta_2 - \beta_1) = 0, \tag{5.25}$$

$$\mu \beta_1 = 0, \tag{5.26}$$

$$\pi(1 - \beta_2) = 0. \tag{5.27}$$

The constraint (5.17) requires $\beta_2 \geq \beta_1$. In the case of $\beta_1 = \beta_2 = 0$, the training dataset would be overfitted. In addition, in the case of $\beta_1 = \beta_2 = 1$, $\tilde{d}_{ij}^2$ is exactly equal to the original distance metric d_{ij}^2 but we do not get any improvement. To avoid these cases, we then change (5.17) to be a strict inequality constraint of $\beta_2 > \beta_1$. Therefore, we have $\gamma = 0$ from (5.25). Using the properties of $\gamma = 0$ and $\pi = 0$, we can then change the dual formulation (5.22) and its constraints of (5.20) and (5.21) by substituting the expansion form of $\mathbf{A}\mathbf{A}^T$ in (5.19) as follows:

$$\max_{\alpha} \quad \mathscr{W}(\alpha) \tag{5.28}$$

$$= 2 \sum_{(\mathbf{x}_i, \mathbf{x}_j) \in \mathscr{D}} \alpha_{ij}$$

$$- \frac{1}{2} \sum_{(\mathbf{x}_i, \mathbf{x}_j) \in \mathscr{S}} \sum_{(\mathbf{x}_k, \mathbf{x}_l) \in \mathscr{S}} \alpha_{ij} \alpha_{kl} \left((\phi_i - \phi_j)^T (\phi_k - \phi_l) \right)^2$$

$$- \frac{1}{2} \sum_{(\mathbf{x}_i, \mathbf{x}_j) \in \mathscr{D}} \sum_{(\mathbf{x}_k, \mathbf{x}_l) \in \mathscr{D}} \alpha_{ij} \alpha_{kl} \left((\phi_i - \phi_j)^T (\phi_k - \phi_l) \right)^2$$

$$+ \sum_{(\mathbf{x}_i, \mathbf{x}_j) \in \mathscr{S}} \sum_{(\mathbf{x}_k, \mathbf{x}_l) \in \mathscr{D}} \alpha_{ij} \alpha_{kl} \left((\phi_i - \phi_j)^T (\phi_k - \phi_l) \right)^2,$$

$$\text{s.t.} \quad \begin{cases} C_{\mathscr{S}} = \sum_{(\mathbf{x}_i, \mathbf{x}_j) \in \mathscr{S}} \alpha_{ij} (\phi_i - \phi_j)^T (\phi_i - \phi_j), \\ C_{\mathscr{D}} \geq \sum_{(\mathbf{x}_i, \mathbf{x}_j) \in \mathscr{D}} \alpha_{ij} \left[2 - (\phi_i - \phi_j)^T (\phi_i - \phi_j) \right], \\ 0 \quad \leq \alpha_{ij}, \end{cases}$$

where $\phi_i^T \phi_j = K(\mathbf{x}_i, \mathbf{x}_j)$ from the kernel trick. The dual formulation (5.28) is very similar to that of C-SVMs [14]. It is a standard convex quadratic programming, which can result in a global optimal solution without any local minima.

After solving the convex quadratic optimization problem in (5.28), we can then generalize our distance-metric model defined in Eq.(5.19) to the unseen test instances. Suppose $\mathbf{x}$ and $\mathbf{x}'$ are two test instances with unknown class labels. Their pairwise distance $\tilde{d}^2(\mathbf{x}, \mathbf{x}')$ after feature weighting is computed as

$$\tilde{d}^2(\mathbf{x}, \mathbf{x}') = \left(\phi(\mathbf{x}) - \phi(\mathbf{x}') \right)^T \mathbf{A} \mathbf{A}^T \left(\phi(\mathbf{x}) - \phi(\mathbf{x}') \right).$$

Substituting (5.19) into the above equation, we obtain

$$\tilde{d}^2(\mathbf{x}, \mathbf{x}') \tag{5.29}$$

$$= \left(\phi(\mathbf{x}) - \phi(\mathbf{x}') \right)^T \left(\sum_{(\mathbf{x}_i, \mathbf{x}_j) \in \mathscr{D}} \alpha_{ij} (\phi_i - \phi_j)(\phi_i - \phi_j)^T \right.$$

$$\left. - \sum_{(\mathbf{x}_i, \mathbf{x}_j) \in \mathscr{S}} \alpha_{ij} (\phi_i - \phi_j)(\phi_i - \phi_j)^T \right) \left(\phi(\mathbf{x}) - \phi(\mathbf{x}') \right)$$

$$= \sum_{(\mathbf{x}_i, \mathbf{x}_j) \in \mathscr{D}} \alpha_{ij} (K_{\mathbf{x}\mathbf{x}_i} - K_{\mathbf{x}\mathbf{x}_j} - K_{\mathbf{x}_i\mathbf{x}'} + K_{\mathbf{x}_j\mathbf{x}'})^2$$

$$- \sum_{(\mathbf{x}_i, \mathbf{x}_j) \in \mathscr{S}} \alpha_{ij} (K_{\mathbf{x}\mathbf{x}_i} - K_{\mathbf{x}\mathbf{x}_j} - K_{\mathbf{x}_i\mathbf{x}'} + K_{\mathbf{x}_j\mathbf{x}'})^2.$$

Remark 5.1. We note that here our learned distance function $\tilde{d}^2(\mathbf{x}, \mathbf{x}')$ is expressed in terms of the prior kernel K which has been chosen before we apply the algorithm. For example, such a prior kernel could be chosen as a Gaussian RBF function

Table 5.1: Datasets description

dataset	# dim	# class	# instance
toy	11	2	100
soybean	35	4	47
wine	12	3	178
glass	10	6	214
seg	19	2	210
image	144	14	1,897

$\exp\left(-\frac{\|\mathbf{x}-\mathbf{x}'\|_2^2}{2\sigma^2}\right)$. $K_{\mathbf{x}\mathbf{x}_i}$ and $K_{\mathbf{x}\mathbf{x}_j}$ in Eq.(5.29) can thus be computed as $\exp\left(-\frac{\|\mathbf{x}-\mathbf{x}_i\|_2^2}{2\sigma^2}\right)$ and $\exp\left(-\frac{\|\mathbf{x}-\mathbf{x}_j\|_2^2}{2\sigma^2}\right)$, and so are $K_{\mathbf{x}_i\mathbf{x}'}$ and $K_{\mathbf{x}_j\mathbf{x}'}$. Therefore, even in the case where both $\mathbf{x}$ and $\mathbf{x}'$ are unseen test instances, their pairwise distance can still be calculated from Eq.(5.29). Moreover, when a linear kernel, $K(\mathbf{x}_i, \mathbf{x}_j) = <\mathbf{x}_i, \mathbf{x}_j>$, is employed, the projected space $\mathscr{P}$ is exactly equivalent to the original input space $\mathscr{I}$. *DFA* then becomes a distance-function-learning in $\mathscr{I}$. Therefore, DFA is a general algorithm which can learn a distance function in both $\mathscr{P}$ and $\mathscr{I}$. ∎

5.3 Experimental Evaluation

We conducted an extensive empirical study to examine the effectiveness of our context-based distance-function learning algorithm in two aspects.

1. *Contextual information.* We compared the quality of our learned distance function when given quantitatively and qualitatively different contextual information for learning.
2. *Learning effectiveness.* We compared our DFA algorithm with the regular Euclidean metric, Kwok et al. [10], and Xing et al.'s [16] on classification and clustering applications.

To conduct our experiments, we used six datasets: one toy dataset, four UCI benchmark datasets, and one 2K image dataset, which are described as follows:

One toy dataset The *toy* dataset was first introduced in [10]. We used it to compare the effectiveness of our method to other methods. The toy dataset has two classes and eleven features. The first feature of the first class was generated by using the Gaussian distribution $N(3, 1)$, and the first feature of the second class by $N(-3, 1)$; the other ten features were generated by $N(0, 25)$. Each feature was generated independently. The first row of Table 5.1 provides the detailed composition of the *toy* dataset.

Four UCI benchmark datasets The four UCI datasets we experimented with are *soybean*, *wine*, *glass*, and *segmentation* (abbreviated as *seg*). The first three UCI datasets are multi-dimensional. The last dataset, *seg*, is processed as a binary-class

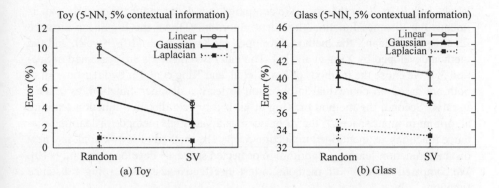

Fig. 5.3: Quality of contextual information vs. classification error

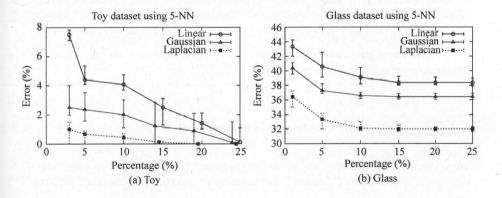

Fig. 5.4: Percentage of contextual information vs. classification error

dataset by choosing its first class as the target class, and all the other classes as the non-target class. Table 5.1 presents the detailed description of these four UCI datasets.

2K-image dataset The image dataset was collected from the Corel Image CDs. Corel images have been widely used by the computer vision and image-processing research communities for conducting various experiments. This set contains 2K representative images from fourteen categories: *architecture, bears, clouds, elephants, fabrics, fireworks, flowers, food, landscape, people, textures, tigers, tools,* and *waves*. Each image is represented by a vector of 144 dimensions including color, texture, and shape features [6].

The contextual-information sets $\mathscr{S}$ (the similar set) and $\mathscr{D}$ (the dissimilar set) were constructed by defining two instances as *similar* if they had the same class label[2], and *dissimilar* otherwise. We compared four distance functions in the experiments: our distance-function-alignment algorithm (DFA), the Euclidean distance function (Euclidean), the method developed by Kwok et al. [10] (Kwok), and the method developed by Xing et al. [16]. The latter two methods are presented in Section 5.4. We chose the methods of Kwok et al. and Xing et al. for two reasons. First, both are based on contextual information to learn a distance function, as is used in DFA. Second, the method of Xing et al. is a typical distance-function-learning algorithm in input space $\mathscr{I}$ that been seen as a yardstick method for learning distance functions, as mentioned in Section 5.4.1. The method of Kwok et al. is a new distance-function-learning algorithm in projected space $\mathscr{P}$ developed recently [10]. We compared DFA to both methods to test its effectiveness on learning a distance function.

Our evaluation procedure was as follows: First, we chose a prior kernel and derived a distance function via the kernel trick. We then ran DFA, Kwok's, and Xing's[3] algorithms, respectively, to learn a new distance function. Finally, we ran k-NN and k-means using the prior and the learned distance functions, and compared their results. Three prior kernels we used in the experiments are linear ($\mathbf{x}^T \mathbf{x}'$), Gaussian ($\exp\left(-\frac{\|\mathbf{x}-\mathbf{x}'\|_2^2}{2\sigma^2}\right)$), and Laplacian ($\exp(-\gamma\|\mathbf{x}-\mathbf{x}'\|_1)$). For each dataset, we carefully tuned kernel parameters including σ, γ, $C_{\mathscr{S}}$ and $C_{\mathscr{D}}$ via cross-validation. All measurements were averaged over 20 runs.

5.3.1 Evaluation on Contextual Information

We chose two datasets, *toy* and *glass*, to examine the performance of our learned distance function when given a different quality or quantity of contextual information.

5.3.1.1 Quality of Contextual Information

We examined two different schemes for choosing contextual information for the k-NN classifier. The first scheme, denoted as *random*, randomly samples a subset of contextual information from $\mathscr{S}$ and $\mathscr{D}$ sets. The second scheme chooses the most uncertain boundary instances as contextual information. Those instances are the hardest to classify as similar or dissimilar. Without any prior knowledge or any

---

[2] In this chapter, we only consider the case where $\mathscr{S}$ and $\mathscr{D}$ are obtained from the class-label information. How to construct $\mathscr{S}$ and $\mathscr{D}$ has been explained in the beginning of Section 5.2.

[3] Xing's algorithm cannot be run when the dimensionality of $\mathscr{I}$ is very high or when nonlinear kernels, such as Gaussian and Laplacian, are employed. This is because its computational time does not scale well with the high dimensionality of input space and non-linear kernels [10]. We thus did not report the corresponding results in these cases.

help provided by the user, we can consider those boundary instances to be the most informative. One way to achieve such uncertain instances is to run Support Vector Machines (SVMs) to identify the instances along the class boundary (support vectors), and samples these boundary instances to construct contextual information. Some other strategies can also be employed to help select the boundary instances. In this chapter, we denote such a scheme as *SV*. We chose SVMs because it is easy to identify the boundary instances (support vectors).

We tested both contextual information selection schemes using our distance-learning algorithm on three prior kernels — linear, Gaussian, and Laplacian. For each scheme, we sampled 5% contextual information from $\mathscr{S}$ and $\mathscr{D}$. Figures 5.3(a) and (b) present the classification errors (y-axis) of k-NN using the both schemes (x-axis). Figure 5.3(a) shows the result on the *toy* dataset, and Figure 5.3(b) shows on the *glass* dataset. We can see that scheme *SV* yielded lower error rates than scheme *random* on both datasets and on all three prior kernels. This shows that choosing the informative contextual information is very useful for learning a good distance function. In the remaining experiments, we employed the *SV* scheme.

5.3.1.2 Quantity of Contextual Information

We tested the performance of our learned distance function using different amounts of contextual information. Figures 5.4 (a) and (b) show the classification error rates (y-axis) with different amounts of contextual information (x-axis) available for learning. We ran the k-NN algorithm using the distance metric learned from our algorithm based on different prior kernels. We see that for both *toy* and *glass* datasets, more contextual information is always helpful. However, the improvement on the *glass* dataset did level off after more than about 5% to 10% contextual information was used. Therefore, in the rest of our experiments, we used 5% contextual information.

5.3.2 Evaluation on Effectiveness

We used the k-means and k-NN algorithms to examine the effectiveness of our distance-learning algorithm on clustering and classification problems. In the following, we first report k-means results, then k-NN results.

5.3.2.1 Clustering Results

For clustering experiment, we used the *toy* dataset and the four UCI datasets. The size of the contextual information was chosen as roughly 5% of the total number of all possible pairs. DFA uses the contextual information to modify three prior distance functions: linear, Gaussian, and Laplacain. The value of k for k-means was

Table 5.2: Clustering error rates on non-image datasets. (No results reported for Xing on Gaussian and Laplacian kernels since this algorithm can only work in input space $\mathscr{I}$)

Dataset	Kernel	Euclidean	Learned		
			DFA	Kwok	Xing
toy	Linear	50.5	**48.2**	48.5	48.9
	Gaussian	50.5	**43.2**	47.1	-
	Laplacian	48.5	**33.5**	38.9	-
soybean	Linear	33.2	**23.0**	24.2	25.8
	Gaussian	32.6	27.0	**22.4**	-
	Laplacian	32.1	**22.8**	34.8	-
wine	Linear	37.1	36.0	**35.8**	36.3
	Gaussian	36.3	**35.7**	35.9	-
	Laplacian	36.3	**26.8**	32.0	-
glass	Linear	31.5	**31.3**	37.0	35.5
	Gaussian	**31.0**	33.6	40.8	-
	Laplacian	31.5	**30.3**	33.3	-
seg	Linear	43.3	**24.2**	25.3	33.0
	Gaussian	40.5	**19.6**	34.4	-
	Laplacian	36.2	20.3	**14.1**	-

set to be the number of classes for each dataset. To measure the quality of clustering, we used the clustering error rate defined in [16] as follows:

$$\sum_{i>j} \frac{1\{1\{c_i = c_j\} \neq 1\{\hat{c}_i = \hat{c}_j\}\}}{0.5n(n-1)},$$

where $\{c_i\}_{i=1}^n$ denotes the true cluster to which $\mathbf{x}_i$ belongs, $\{\hat{c}_i\}_{i=1}^n$ denotes the cluster predicted by a clustering algorithm, n is the number of instances in the dataset, and $1\{\cdot\}$ the indicator function ($1\{\text{true}\} = 1$, and $1\{\text{false}\} = 0$).

Table 5.2, we report the k-means clustering results for the five datasets. From Table 5.2, we can see that DFA achieves the best clustering results in almost all testing scenarios except for three combinations of the *soybean* and Gaussian, the *wine* and Linear, and the *glass* and Gaussian. DFA performs better than Xing in all cases where the linear kernel is used.

5.3.2.2 Classification Results

When performing classification, we randomly extracted a fraction of the data as training data, and used the remaining data as testing data. For each dataset, the training/testing ratio was empirically chosen via cross-validation so that the classifier using the regular Euclidean metric performs best for a fair comparison. The number of training instances for each dataset is reported in the last column For clas-

Table 5.3: Classification Error Rates on Non-image Datasets. (No results reported for Xing on Gaussian and Laplacian kernels since this algorithm can only work in input space $\mathcal{I}$.)

Dataset	Kernel	Euclidean	Learned		
			DFA	Kwok	Xing
toy	Linear	10.00	**4.40**	4.80	5.32
	Gaussian	10.00	**2.22**	2.61	-
	Laplacian	10.00	**0.68**	1.10	-
soybean	Linear	2.50	**0.00**	0.12	0.39
	Gaussian	2.50	**1.00**	1.34	-
	Laplacian	2.50	**0.30**	1.19	-
wine	Linear	5.00	**4.86**	4.94	4.98
	Gaussian	5.00	**3.30**	3.50	-
	Laplacian	6.67	**1.65**	1.68	-
glass	Linear	40.58	**39.67**	42.03	40.34
	Gaussian	40.58	**37.28**	47.83	-
	Laplacian	37.68	**33.10**	33.33	-
seg	Linear	2.94	**1.47**	1.51	1.86
	Gaussian	2.94	**0.10**	0.31	-
	Laplacian	**0.00**	**0.00**	2.94	-

sification experiment, we used the *toy* dataset, the four UCI datasets, and the 2K image dataset. The size of the contextual information chosen was again about 5% of the total number of all possible pairs.

When performing classification, we employed different distance functions: linear, Gaussian, and Laplacian. We compared the performance of using these distance functions before and after applying DFA, and competing methods. For k-NN, we randomly extracted 80% of the dataset as training data, and used the remaining data 20% as testing data. (Notice that the 80% training data here is for training k-NN, not for modifying distance function.) Such a training/testing ratio was empirically chosen via cross-validation so that the classifier using the regular Euclidean metric performed best for a fair comparison. We set k in the k-NN algorithm to be 5 for non-image datasets and to be 1 for the 2K-image dataset. Our setting is empirically validated as the optimal one for the classifier using the regular Euclidean metric.

We first report the k-NN classification results using five non-image datasets — *toy*, *soybean*, *wine*, *glass*, and *seg*. Table 5.3.2.2 reports all classification error rates (in percentages). We compared our *DFA* with three other metrics: Euclidean, Kwok, and Xing. For each dataset, we used three different prior kernels — linear, Gaussian, and Laplacian — and then experimented with the four metric candidates. In the third column of the table we report the error rates using the Euclidean distance. The last three columns report the results of using *DFA*, Kwok, and Xing, respectively. The best results achieved are shown in bold. First, compared to Euclidean, *DFA* performs better on almost all datasets, with only one exception being tying on *seg* with Laplacian (both achieved zero classification error rate). Second, DFA outperforms Kwok on all datasets, improving the classification error rates by an average of 0.40%, 0.45%, 0.12%, 5.38%, and 1.06% on the five datasets, respectively. Finally, DFA obtains better results than Xing's in all testing scenarios.

Table 5.4: Image-dataset prediction error rates

Category	Euclidean	DFA	Kwok
Architecture	29.3	**28.0**	31.7
Bear	64.0	68.0	**63.0**
Cloud	**38.3**	42.1	39.3
Elephant	63.3	**43.9**	60.2
Fabric	**59.0**	67.0	61.0
Firework	14.3	13.3	**12.2**
Flower	45.0	**36.0**	**36.0**
Food	32.5	**29.4**	33.1
Landscape	59.3	**59.2**	58.7
People	52.5	**46.9**	47.0
Texture	38.0	**35.0**	43.0
Tiger	46.7	**37.8**	52.2
Tool	15.0	**12.0**	15.0
Wave	44.3	**39.6**	48.1
total	41.6	**38.9**	41.4

Then, we report the prediction error-rates using k-NN on the 2K-image dataset in Table 5.4. We empirically chose $k = 1$ and prior kernel as Gaussian. Since Xing's algorithm can run only in input space $\mathscr{I}$, not in projected space $\mathscr{P}$, we only compared our distance metric with Kwok et al.'s and the regular Euclidean metric and in $\mathscr{P}$. Among fourteen categories, the distance metric learned by *DFA* beats the Euclidean in twelve categories and Kwok in eleven. It also improves the total prediction error rate by 2.7 and 2.5 percentile points, respectively.

5.3.3 Observations

From the experimental results, we make the following observations:

1. *Quality of contextual information counts.* Choosing contextual information with good quality can be very useful for learning a good distance metric, as shown in Figure 5.3. The best contextual information is that which can provide maximal information to depict the context. As illustrated in our experiment, the boundary instances tend to be most useful, since when they are disambiguated through additional contextual information, we can achieve the best context-based alignment on the boundary (on the function).
2. *Quantity matters little.* Choosing more contextual information can be useful for learning a good distance metric. However, as shown in Figure 5.4, the cost of

increasing quantity can outweigh the benefit, once the quality of information starts to deteriorate.

3. *DFA helps*. DFA improves upon prior functions in almost all of our test scenarios. Occasionally, DFA performs slightly worse than the prior function. We conjecture that this may be caused by overfitting in certain circumstances: specifically, when some chosen prior kernels may not be the best model for the datasets, further aligning these kernels could be counter-productive. We will further investigate related issues in our future work.

5.4 Related Reading

Distance-function learning approaches can be divided broadly into metric-learning and kernel-learning approaches. In the rest of this section we discuss representative work using these two approaches.

5.4.1 Metric Learning

Metric learning attempts to find the optimal linear transformation for the given set of data vectors to better characterize the similarity between vectors. The transformation by itself is linear, but the data vectors may first be *explicitly* mapped to a new set of vectors using a nonlinear function $\phi(\mathbf{x})$. The transformation of the data vectors is equivalent to assigning weights to the features of the vectors; therefore, metric learning is often called *feature weighting*. The metric learning approach is given a set of data vectors $\mathscr{X} = \{\mathbf{x}\}_{i=1}^{m}$ in $\mathfrak{R}^n$ and similarity information in the form of $(\mathbf{x}_i, \mathbf{x}_j) \in S$ (a similar set), if $\mathbf{x}_i$ and $\mathbf{x}_j$ are similar. Metric learning aims to learn a distance metric $d_\mathbf{M}(\mathbf{x}_i, \mathbf{x}_j)$ between data vectors $\mathbf{x}_i$ and $\mathbf{x}_j$ that respects the similarity information. Mathematically the distance metric can be represented as

$$d_\mathbf{M}(\mathbf{x}_i, \mathbf{x}_j) = \sqrt{(\phi(\mathbf{x}_i) - \phi(\mathbf{x}_j))^T \mathbf{M}(\phi(\mathbf{x}_i) - \phi(\mathbf{x}_j))},$$

where $\mathbf{M}$ needs to be positive (semi-) definite so as to satisfy metric properties — non-negativity and triangle inequality. More generally, $\mathbf{M}$ parameterizes a family of Mahalanobis distances over $\mathfrak{R}^n$. The choice of the basis function ϕ and the scaling matrix $\mathbf{M}$ will differentiate the various metric learning algorithms.

Wettschereck et al. [17] provide a review of the performance of feature-weighting algorithms with emphasis on their performance for the k-nearest neighbor classifier. Here, we discuss only a few representative algorithms. (For the other algorithms, please refer to [17].) A number of papers address the problem of learning distance metrics using contextual information[4] in the form of groups of similar

---

[4] Contextual information is also called side information in some papers such as [18, 16].

vectors [18, 16]. Contextual information can be user-provided information on the similarity characteristics of a subset of data. Based on this information, the work of [18] uses Relevant Component Analysis (RCA) to efficiently learn a full rank Mahalanobis metric [19]. The authors use equivalence relations for the contextual information. They compute

$$\hat{C} = \frac{1}{p} \sum_{j=1}^{|G|} \sum_{i=1}^{|S_j|} (\mathbf{x}_{ji} - \hat{\mathbf{m}}_j)(\mathbf{x}_{ji} - \hat{\mathbf{m}}_j)^T,$$

where $\hat{\mathbf{m}}_j$ is the mean of the j-th group of vectors and $|G|$ and $|S_j|$ denote the number of groups and the number of samples in the j-th group, respectively. The matrix $W = \hat{C}^{-\frac{1}{2}}$ is used for transformation and the inverse of $\hat{C}$ as the Mahalanobis matrix. Xing et al. [16] treat the same problem as a convex optimization problem, hence producing local-optima-free algorithms. They present techniques for learning the weighting matrix both for the diagonal and for the full matrix case. The major difference between the two approaches is that RCA uses closed-form expressions, whereas [16] uses iterative methods that can be sensitive to parameter tuning and that are computationally expensive.

C. Aggarwal [2] discusses a systematic framework for designing distance functions sensitive to particular characteristics of the data. The models used are the parametric Minkowski model

$$D(\mathbf{x}_i, \mathbf{x}_j, \lambda) = \left(\sum_{r=1}^{m} \lambda_r |x_{ir} - x_{jr}|^p \right)^{\frac{1}{p}}$$

and the the parametric cosine model

$$\cos(\mathbf{x}_i, \mathbf{x}_j, \lambda) = \sum_{r=1}^{m} \frac{\lambda_r . x_{ir} . x_{jr}}{\sqrt{\sum_{r=1}^{m} \lambda_r^2 x_{ir}^2 \sum_{r=1}^{m} \lambda_r^2 x_{jr}^2}}.$$

Both these models attempt to minimize the error with respect to each λ_r. The parametric Minkowski model can be thought of as feature weighting in the input space $\mathscr{I}$. Similarly, the parametric cosine model can be thought of as the inner product in $\mathscr{I}$.

In summary, metric learning aims to learn a good distance function by computing the optimal feature weighings in $\mathscr{I}$. Clearly, this linear transformation is restrictive in terms of modeling complex semantics. Although one can perform a non-linear transformation on the features via a basis function ϕ in $\mathscr{I}$, such a transformation is *explicit* and the resulting computational complexity renders this approach impractical. The kernel learning approach, which we discuss next, successfully addresses the concern about computational complexity.

5.4.2 Kernel Learning

Kernel-based methods attempt to *implicitly* map a set of data vectors $\mathscr{X} = \{\mathbf{x}\}_{i=1}^{m}$ in $\mathscr{I}$ to some other high-dimensional (possibly infinite) projected space $\mathscr{P}$, using a basis function (usually nonlinear) ϕ, where the mapped data can be separated by applying a linear procedure [14]. Kernel function K is defined as an inner product between two vectors in projected space $\mathscr{P}$, $\phi(\mathbf{x}_i)$ and $\phi(\mathbf{x}_j)$, as

$$K(\mathbf{x}_i, \mathbf{x}_j) = < \phi(\mathbf{x}_i), \phi(\mathbf{x}_j) > .$$

Kernel-based methods use these inner products (K) as a similarity measure. (Theoretical justifications are presented in [9].) The kernel K provides an elegant way of dealing with nonlinear algorithms by reducing them to linear ones in $\mathscr{P}$. Any algorithm that can be expressed in terms of inner products can be made nonlinear by substituting kernel values for the inner products. A typical example is Support Vector Machines [14].

The requirement for choosing a valid K is that it should be positive (semi-) definite and symmetric. Three popular kernels employed are the polynomial kernel ($< \mathbf{x}_i, \mathbf{x}_j > +1)^p$, the Gaussian radial basis function (RBF) kernel $\exp(-\frac{\|\mathbf{x}_i - \mathbf{x}_j\|_2^2}{2\sigma^2})$, and Laplacian RBF kernel $\exp(-\gamma \| \mathbf{x}_i - \mathbf{x}_j \|_1)$. In $\mathscr{P}$, the distance between $\mathbf{x}_i$ and $\mathbf{x}_j$ can then be computed via the kernel trick

$$d(\mathbf{x}_i, \mathbf{x}_j) = \sqrt{K(\mathbf{x}_i, \mathbf{x}_i) + K(\mathbf{x}_j, \mathbf{x}_j) - 2K(\mathbf{x}_i, \mathbf{x}_j)},$$

where the validity of the kernel K ensures that the resulting distance function $d(\mathbf{x}_i, \mathbf{x}_j)$ will be a valid metric.

An important advantage of using kernels lies in the ease of computing the inner product (similarity measure) in $\mathscr{P}$ without actually having to know ϕ. There has been a lot of work [20, 9, 14] in classification, clustering, and regression methods, using the kernel $K(\mathbf{x}_i, \mathbf{x}_j)$ for indirect computations of similarity measures (and hence the distance $d(\mathbf{x}_i, \mathbf{x}_j)$) in $\mathscr{P}$.

Due to the central role of the kernel, a poor kernel function can lead to significantly poor performance using kernel methods. Instead of choosing a pre-defined kernel for training, many recent efforts aim to learn a kernel from the training data [8, 10]. All these papers are based on the notion of *kernel alignment* proposed by Cristianini et al. [8] to measure the similarity between two kernel functions. Geometrically, when given a set of training instances $\mathscr{X}$, the *alignment* score is defined as the cosine of the angle between the two kernel matrices[5], after flattening the two matrices into vectors. They also proposed the notion of "ideal kernel" (K^*). Suppose $y(\mathbf{x}_i) \in \{1, -1\}$ is the class label of $\mathbf{x}_i$. K^* is defined as

---

[5] Given a kernel function K and a set of instances $\mathscr{X}$, the kernel matrix (Gram matrix) is the matrix of inner-products of all possible pairs from $\mathscr{X} \times X$, $\mathbf{K} = [k_{ij}]$, where $k_{ij} = K(\mathbf{x}_i, \mathbf{x}_j)$.

$$K^*(\mathbf{x}_i, \mathbf{x}_j) = \begin{cases} 1, & \text{if } y(\mathbf{x}_i) = y(\mathbf{x}_j), \\ 0, & \text{if } y(\mathbf{x}_i) \neq y(\mathbf{x}_j), \end{cases} \tag{5.30}$$

which is the target kernel that a given kernel is supposed to align with. Kernel-alignment calculates the alignment score of a given kernel K to the ideal kernel K^* to indicate the degree to which the kernel matches the training data. Cristianini et al. [8] prove the connection between the alignment and the generalization performance of the resulting classifier. Basically, their work shows that with a high alignment on the training set, we can expect a good generalization performance of the resulting distance-based classifier.

Since a kernel K defines a pairwise distance from Eq.(5.1), kernel learning has been recently applied to distance metric learning. Zhang [21] proposed to idealize a given kernel K using $\tilde{K} = K + K^*$, to achieve a good distance metric which is Euclideanly and Fisherly separable. Then, the achieved distance metric was embedded into a new Euclidean space via Multi-Dimensional Scaling (MDS) [22]. However, this iterative embedding procedure is computationally expensive, and its idealization model might not be optimal. Kwok et al. [10] considered the same problem of metric learning as a convex optimization problem. The approach works in both the input and the kernel-induced projected spaces. They modify the prior kernel using $\tilde{K} = K + \frac{\gamma}{2}K^*$ and derive distance functions using this modified kernel. Since the number of parameters is related to the number of patterns but not to the dimensionality of the patterns, the approach allows feature weighting to be done efficiently in the possibly infinite dimensional projected spaces. However, their learned distance metric cannot be guaranteed to be positive (semi-) definite. Hence, the induced distance function might not be a valid one. Moreover, their kernel transformation model, $\tilde{K} = K + \frac{\gamma}{2}K^*$, is not an optimal one for kernel idealization because $\tilde{K}(\mathbf{x}_i, \mathbf{x}_j) = K(\mathbf{x}_i, \mathbf{x}_j)$ when two instances $\mathbf{x}_i$ and $\mathbf{x}_j$ are similar ($K^*(\mathbf{x}_i, \mathbf{x}_j) = 1$), which means the intra-class similarity scores remain unchanged using this transformation model.

Our *DFA* algorithm transforms a prior kernel function[6] in a projected space $\mathscr{P}$ when given a set of training data. We theoretically prove that our method leads to a valid distance function. More importantly, we also show that since the optimization is performed in a convex space, the solution obtained is globally optimal. Consequently, given a prior kernel and contextual information, the alignment needs to be performed just once. Our empirical results show that our proposed method outperforms competing methods on a variety of testbeds.

---

[6] The *kernel trick* in Eq.(5.1) uniquely links the kernel functions with the distance function. The former (K) provides the pairwise-similarity measurement between two instances, whereas the latter (d) provides the pairwise-dissimilarity measurement between two instances. Therefore, when we say a transformation on a prior kernel function, it also means a transformation on a prior distance function, vise versa.

5.5 Concluding Remarks

In this chapter, we have reported our study of an important database issue — formulating a context-based distance function for measuring similarity. We show that *DFA* learns a distance function by considering the nonlinear relationships among the contextual information, without incurring high computational cost. Theoretically, we substantiated that our method achieves optimal alignment toward the ideal kernel. Empirically, we demonstrated that *DFA* improves similarity measures and hence leads to improved performance in clustering and classification applications.

Considering the contextual information in information retrieval has been identified as an important research area. Our work provides a mechanism that is more effective than the traditional ones because it achieves both model flexibility and computational efficiency. One of our future research tasks is to theoretically analyze the generalization ability of *DFA*. Cristianini et al. [8] give a proof on the connection between the "ideal" kernel and the generalization ability of the classifier. This might inspire us to formulate a method to analyze *DFA*. Other future research projects might be to investigate how one might best model context (for an application or for a user), and how one might effectively formulate and gather contextual information to take advantage of our proposed mechanisms.

References

1. Wu, G., Chang, E.Y., Panda, N. Formulating distance functions via the kernel trick. In *Proceedings of ACM SIGKDD*, pages 703–709, 2005.
2. Aggarwal, C.C. Towards systematic design of distance functions for data mining applications. In *Proceedings of ACM SIGKDD*, pages 9–18, 2003.
3. Fagin, R., Kumar, R., Sivakumar, D. Efficient similarity search and classification via rank aggregation. In *Proceedings of ACM SIGMOD Conference on Management of Data*, pages 301–312, June 2003.
4. Wang, T., Rui, Y., Hu, S.M., Sun, J.Q. Adaptive tree similarity learning for image retrieval. *Multimedia Systems*, 9(2):131–143, 2003.
5. Rui, Y., Huang, T. Optimizing learning in image retrieval. In *Proceedings of IEEE CVPR*, pages 236–245, June 2000.
6. Tong, S., Chang, E. Support vector machine active learning for image retrieval. In *Proceedings of ACM International Conference on Multimedia*, pages 107–118, 2001.
7. Aizerman, M.A., Braverman, E.M., Rozonoer, L.I. Theoretical foundations of the potential function method in pattern recognition learning. *Automation and Remote Control*, 25:821–837, 1964.
8. Cristianini, N., Shawe-Taylor, J., Elisseeff, A., Kandola, J. On kernel target alignment. pages 367–373, 2001.
9. Schölkopf, B., Smola, A. *Learning with Kernels: Support Vector Machines, Regularization, Optimization, and Beyond*. MIT Press, Cambridge, MA, 2002.
10. Kwok, J.T., Tsang, I.W. Learning with idealized kernels. In *Proceedings of the Twentieth International Conference on Machine Learning*, pages 400–407, Washington DC, August 2003.
11. Grandvalet, Y., Canu, S. Adaptive scaling for feature selection in SVMs. In *Proceedings of NIPS*, pages 553–560, 2002.
12. Amaldi, E., Kann, V. On the approximability of minimizing non-zero variables or unsatisfied relations in linear systems. *Theoretical Computer Science*, 209:237–260, 1998.

13. Fazel, M. Matrix rank minimization with applications. *Ph.D. Thesis, Electrical Engineering Dept, Stanford University*, March 2002.
14. Vapnik, V. *Statistical Learning Theroy*. John Wiley and Sons, 1998.
15. Kuhn, H.W., Tucker, A.W. Nonlinear programming. In *Proceedings of the Second Berkeley Symposium on Mathematical Statistics and Probabilistics*, pages 481–492, Berkeley, 1951. University of California Press.
16. Xing, E., Ng, A., Jordan, M., Russell, S. Distance metric learning, with application to clustering with side-information. In *Proceedings of NIPS*, pages 505–512, 2002.
17. Wettschereck, D., Aha, D., Mohri, T. A review and empirical evaluation of feature weighting methods for a class of lazy learning algorithms. *Artificial Intelligence Review*, 11:273–314, 1997.
18. Bar-hillel, A., Hertz, T., Shental, N., Weinshall, D. Learning distance functions using equivalence relations. In *Proceedings of International Conference on Machine Learning (ICML)*, pages 11–18, Washington DC, August 2003.
19. Nadler, M., Smith, E.P. *Pattern Recognition Engineering*. New York: Wiley, 1993.
20. Ben-Hur, A., Horn, D., Siegelmann, H.T., Vapnik, V. Support vector clustering. *Journal of Machine Learning Research*, 2:125–137, March 2002.
21. Zhang, Z. Learning metrics via discriminant kernels and multidimensional scaling: Towards expected euclidean representation. In *Proceedings of International Conference on Machine Learning (ICML)*, pages 872–879, August 2003.
22. Webb, A.R. Multidimensional scaling by iterative majorization using radial basis functions. *Pattern Recognition*, 28(5):753–759, 1995.

Chapter 6
Multimodal Fusion

Abstract Multimedia data instances consist of metadata from multiple sources. Given a set of features extracted from these sources (e.g., features extracted from the visual, audio, and caption track of videos), how do we determine the best modalities? Once a set of modalities has been identified, how do we best fuse them to map to semantics? This chapter[†] presents a two-step approach. The first step finds *statistically independent modalities* from raw features. In the second step, we use *super-kernel fusion* to determine the optimal combination of individual modalities. We carefully analyze the tradeoffs between three design factors that affect fusion performance: *modality independence*, *curse of dimensionality*, and *fusion-model complexity*. Through analytical and empirical studies, we demonstrate that the two-step approach, which achieves a careful balance of the three design factors, can improve class-prediction accuracy over traditional techniques.
Keywords: Feature combination, multimodal fusion, PCA, ICA, super kernel.

6.1 Introduction

Multimedia data such as images and videos are represented by features from multiple media sources. Traditionally, images are represented by keywords and perceptual features such as color, texture, and shape [2, 3]. Videos are represented by features embedded in the tracks of visual, audio, caption text, etc. [4]. Besides, contextual information associated with a data instance, such as camera parameters, user profile, social interactions, and search logs, can also be considered for analyzing multimedia data. These features are extracted and then fused in a complementary way for understanding the semantics of multimedia data.

Traditional work on multimodal integration has largely been heuristic-based. It lacks theories to answer two fundamental questions: (1) what are the *best* modal-

---

[†] ©ACM, 2004. This chapter is a minor revision of the author's work with Yi Wu, Kevin Chang, and John R. Smith [1] published in MULTIMEDIA'04. Permission to publish this chapter is granted under copyright license #2587660035739.

ities? and (2) how can we optimally fuse information from multiple modalities? Suppose we extract l, m, n features from the visual, audio, and caption tracks of videos. At one extreme, we could treat all these features as one modality and form a feature vector of $l + m + n$ dimensions. At the other extreme, we could treat each of the $l + m + n$ features as one modality. We could also regard the extracted features from each media-source as one modality, formulating a visual, audio, and caption modality with l, m, and n features, respectively. Almost all prior multimodal-fusion work in the multimedia community employs one of these three approaches [5, 6]. But, can any of these feature compositions yield the optimal result?

Statistical methods such as principle component analysis (PCA) and independent component analysis (ICA) have been shown to be useful for feature transformation and selection. PCA is useful for denoising data, and ICA aims to transform data to a space of independent axises (components). Despite their best attempt under some error-minimization criteria, PCA and ICA do not guarantee to produce independent components. In addition, the created feature space may be of very high dimensions and thus be susceptible to the *curse of dimensionality*[1]. In the first part of this chapter, we present an *independent modality analysis* scheme, which identifies independent modalities, and at the same time, avoids the curse-of-dimensionality challenge.

Once a good set of modalities has been identified, the second research challenge is to fuse these modalities in an optimal way to perform data analysis (e.g., classification). Suppose we can yield truly independent modalities, and each modality can derive accurate posterior probability for class prediction. We can simply use the *product-combination* rule to multiply the probabilities for predicting class membership. Unfortunately, the above two conditions do not hold in general for a multimedia data-analysis task (see Section 6.2 for detailed discussion). Using the product-combination rule to fuse information is thus inappropriate. Another popular fusion method is the *weighted-sum* rule, which performs a linear combination on the modalities. The weighted-sum rule enjoys the advantage of simplicity, but its linear constraint forbids high model complexity; hence it cannot adequately explore the inter-dependencies left unresolved by PCA and ICA. In this chapter, we present a discriminative approach (whereas in Chapter 8 we present a generative approach) to address multimodal fusion. Our discriminative approach employs the *super-kernel fusion* scheme to fuse individual modalities in a non-linear way (linear fusion is a special case of our method). The *super-kernel fusion* scheme finds the best combination of modalities through supervised training.

Let us use a simple example to explain the shortcomings of some traditional multimodal integration schemes that invite further research. Figure 6.1 shows the existence of feature dependencies in a real image dataset, before and after performing PCA/ICA. This figure plots the normalized correlation matrix in absolute value derived from a 2K-image dataset of 14 classes. (Detailed description for this image dataset is given in Section 6.5.) A total of 144 features are considered: the first 108

---

[1] The work of [7] shows that, when data dimension is high, the distances between pairs of objects in the space become increasingly similar to each other due to the *central limit theory*. This phenomenon is called the *dimensionality curse* [8], because it can severely hamper the effectiveness of data analysis.

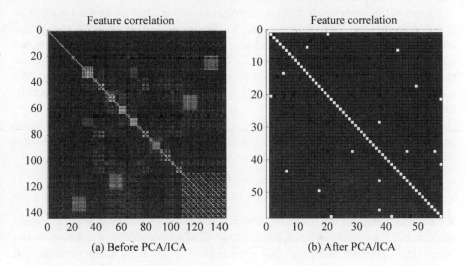

(a) Before PCA/ICA   (b) After PCA/ICA

Fig. 6.1: Feature correlation matrix

are color features; the other 36 are texture features. Correlation between features within the same media source and across different media sources is measured by computing the covariance matrix:

$$C = \frac{1}{N} \sum_{\mathbf{x}_i \in X} (\mathbf{x}_i - \bar{\mathbf{x}})(\mathbf{x}_i - \bar{\mathbf{x}})^T \text{ with } \bar{\mathbf{x}} = \frac{1}{N} \sum_{\mathbf{x}_i \in X} \mathbf{x}_i, \qquad (6.1)$$

where N is the total number of sample data, $\mathbf{x}_i$ is a feature vector to represent the ith sample, and X is the set of feature vectors for N samples. Normalized correlation between features i and j is defined by

$$\hat{C}(i,j) = \frac{C(i,j)}{\sqrt{C(i,i) \times C(j,j)}}. \qquad (6.2)$$

In the figure, both the x- and y-axis depict the 144 features. The light-colored areas in the figure indicate high correlation between features, and the dark-colored areas indicate low correlation. If any feature correlates only with itself, only the diagonal elements will be light-colored. The off-diagonal light-colored areas in Figure 6.1(a) indicate that this image dataset exhibits not only a high correlation of features within the same media source, but also between certain features from different media sources (e.g., color and texture). Color and texture are traditionally treated as orthogonal modalities, but this example shows otherwise. These correlated and even noisy "raw" features may affect the learning algorithm by obscuring the distributions of truly relevant and representative features. (The weighted-sum fusion rule cannot deal with these inter-dependencies.)

Figure 6.1(b) presents the feature correlation matrix after we applied both PCA and ICA to the data. The process yields 58 "improved" components. Although the components exhibit better independence, inter-dependencies between components still exist. This chapter first deals with grouping components like these 58 into a smaller number of independent modalities to avoid the *dimensionality curse*. We then explore non-linear combinations of the modalities to improve the effective multimodal fusion.

As the main contribution of this work, we propose a discriminative fusion scheme for multimedia data analysis. Given a list of features extracted from multiple media-sources, we tackle two core issues:

- Formulating independent feature modalities (Section 6.3).
- Fusing multiple modalities optimally (Section 6.4).

We carefully analyze the tradeoffs between three design factors that affect fusion performance: *modality independence*, *curse of dimensionality*, and *fusion-model complexity*. Through analytical and empirical studies on an image dataset and TREC-Video 2003 benchmarks, we show that a careful balance of the three design factors consistently leads to superior performance for multimodal fusion.

6.2 Related Reading

We discuss related work in *modality identification* and *modality fusion*.

Table 6.1: Related work summarization (m—No. of media sources; k—No. of independent components; D—No. of independent modalities)

No. of Modality	Fusion Methods	Evaluation
1	No	No need to do fusion; curse of dimensionality
m	Any	Loss of inter-dependency relationship between features
k	Any	High model complexity; no perfect independent components
	Product	Very sensitive to the accuracy of individual classifiers
D	Linear	Not suitable for independent feature spaces
	Super-kernel	Suitable

6.2.1 Modality Identification

Let D denote the number of modalities. Given $d_1, d_2, \cdots d_m$ features extracted from m media sources, respectively, prior modality identification work can be divided into two representative categories.

1. $D = 1$, or treating all features as one modality. This approach does not require the fusion step. Goh et al. [9] used the raw color and texture features to form a high-dimensional feature vector for each image. Recently, statistical methods such as PCA and ICA have been widely used in the Computer Vision, Machine Learning, Signal Processing communities to denoise data and to identify independent information sources (e.g., [10, 11, 12, 13]). In the multimedia community, the work of [14, 15] observed that audio and visual data of a video stream exhibit some statistical regularity, and that regularity can be explored for joint processing. Smaragdis et al. [16] proposed to operate on a fused set of audio/visual features and to look for combined subspace components amenable to interpretation. Vinokourov et al. [17] found a common latent/semantic space from multi-language documents using independent component analysis for cross-language document retrieval. The major shortcoming of these works is that the curse of dimensionality arises, causing ineffective feature-to-semantics mapping and inefficient indexing [2]. (Please refer to [7, 18, 19] for the discussion of dimensionality-curse and why dimension reduction can greatly enhance the effectiveness of statistical analysis and the efficiency of query processing.)
2. $D = m$, or treating each source as one modality. This approach treats the features as m modalities, with d_i features in the ith modality ($i = 1, 2, \cdots, m$). Most work in image and video retrieval analysis (e.g., [4, 20, 21, 22, 23]) employs this approach. For example, the QBIC system [20] supported image queries based on combining distances from the color and texture modalities. Velivelli et al. [23] separated video features into audio and visual modalities. Adams et al. [4] also regarded each media track (visual, audio, textual, etc.) as one modality. For each modality, these works trained a separate classification model, and then used the weighted-sum rule to fuse a class-prediction decision. This modality-decomposition method can alleviate the "curse of dimensionality." However, since media sources are treated separately, the inter-dependencies between sources are left unexplored.

Our method is to apply independent component analysis on the raw feature sets to identify k "independent" components. Thereafter, we group these components into D modalities to (1) minimize the dependencies between modalities, and (2) mitigate the dimensionality-curse problem.

6.2.2 Modality Fusion

Given that we have obtained D modalities, we need to fuse D classifiers, one for each modality, for interpreting data.

PCA and ICA cannot perfectly identify independent components for at least two reasons. First, like the way that the k-mean algorithm works, all well-known ICA algorithms (fixed-point algorithm [24], Infomax [25, 26], kernel canonical analysis [17], and kernel independent analysis [27]) need a good estimate of the number of independent components k to find them effectively. Second, as we discussed in Section 6.1, ICA only performs the best attempt under some error-minimization criteria to find k independent components. But the resulting components, as shown in Figure 6.1(b), may still exhibit inter-dependencies.

Now, given D modalities, not entirely independent of each other, we need an effective fusion strategy. Various fusion strategies for multimodal information have been presented and were discussed in [28], including *product combination*, *weighted-sum*, *voting*, and *min-max aggregation*. Among them, *product combination* and *weighted-sum* are by far the most popular fusion methods.

1. *Product combination.* Supposing that D modalities are independent of each other, and we can estimate posterior probability for each modality accurately, the product-combination rule is the optimal fusion model from the Bayesian perspective. However, in addition to the fact that we will not have D truly independent modalities, we generally cannot estimate posterior probability with high accuracy. The work of [29] concluded that the product-combination rule works well only when the posterior probability of individual classifiers can be accurately estimated. In a multimedia data-understanding task, we often assert similarity between data based on our beliefs. (E.g., one can "believe" two videos to be 87% similar or 90% similar. This estimate does not come from classical probability experiments, so the sum of beliefs may not be equal to one.) Because of this subjective process, and because the product-combination rule is highly sensitive to noise, this strategy is not appropriate.

2. *Weighted-sum.* The weighted-sum strategy is more tolerant to noise because *sum* does not magnify noise as severely as *product*. Weighted-sum (e.g., [30]) is a linear model, not equipped to explore the inter-dependencies between modalities. Recently, Yan and Hauptmann [31] presented a theoretical framework for bounding the average precision of a linear combination function in video retrieval. Concluding that the linear combination functions have limitations, they suggested that non-linearity and cross-media relationships should be introduced to achieve better performance.

In this chapter, we depict a super-kernel scheme, which can fuse multimodal information non-linearly to explore the cross-modality relationship. Chapters 7 and 8 present two generative schemes. Both discriminative and generative models enjoy their pros and cons, which we will discuss throughout these three chapters.

6.3 Independent Modality Analysis

In this section, we present our approach to transform m raw features to D modalities. Given input in the form of an $m \times n$ matrix X (n denotes the number of training instances), our independent modality analysis procedure produces $M_1, M_2, \cdots, M_D$ modalities. The procedure consists of the following three steps:

1. Run principal component analysis (PCA) on X to remove noise and reduce the feature dimensionality. Let U denote the matrix containing the first k eigenvectors. The PCA representation of zero-mean feature vectors X is defined as $U^T X$.
2. Run independent component analysis (ICA) on the PCA output $U^T X$ to obtain estimates of independent feature components S and an estimate of a mixing matrix W. We can recover the independent components by computing $S = WU^T X$.
3. Run independent modality grouping (IMG) on S to form independent modalities $M_1, M_2, \cdots, M_D$.

6.3.1 PCA

PCA has been frequently used as a technique for removing noises and redundancies between feature dimensions [32]. PCA projects the original data to a lower dimensionality space such that the variance of the data is best maintained. Let's assume that we have n samples, $\{x_1, x_2, \cdots, x_n\}$, and each x_i is an m-dimensional vector. We can represent the n samples as a matrix $X_{m \times n}$. It is known in linear algebra that any such matrix can be decomposed in the following form (known as singular value decomposition or SVD):

$$X = UDV^T,$$

where matrices $U_{m \times p}$ and $V_{n \times p}$ represent orthonormal basis vectors matrices (eigenvectors of the symmetric matrix XX^T and $X^T X$), with p as the number of largest principal components. The $D_{p \times p}$ matrix is a diagonal matrix, and the diagonal elements of D are the eigenvalues of XX^T and $X^T X$. Consider the projection onto the subspace spanned by the p largest principal components (PC's), i.e., $U^T X$.

6.3.2 ICA

Compared to PCA, the spirit of ICA is to find statistically independent hidden sources from a given set of mixture signals. Both ICA and PCA project data matrices into components in different spaces. However, the goals of the two methods are different. PCA finds the uncorrelated components of maximum variance. It is ideal for compressing data into a lower-dimensional space by removing the least significant components. ICA finds the statistically independent components. ICA is

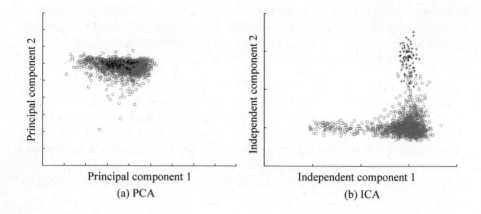

Fig. 6.2: Scatter plots of the 2K image dataset

the ideal choice for separating mixed signals and finding the most representative components.

To formalize an ICA problem, we assume that there are k unknown independent components $S = \{s_1, s_2, \cdots, s_k\}$. What we observe is a set of m-dimensional samples $\{x_1, x_2, \cdots, x_n\}$, which are mixture signals coming from k independent components, $k \leq m$. We can represent all the observation data as a matrix $X_{m \times n}$. A linear mixture model can be formulated as:

$$X = AS,$$

where $A_{m \times k}$ is a mixing matrix. Our goal is to find $W = A^{-1}$; therefore, given training set X, we can recover the independent components (IC's) through the transformation of $S = WX$.

ICA establishes a common latent space for the media, which can be viewed as a method for learning the inter-relations between the involved media [33, 16]. For multimedia data, observation data x_i usually contains features coming from more than one medium. The different independent components $\{s_1, s_2, \cdots, s_k\}$ provide a meaningful segmentation of the feature space. The kth column of W^{-1} constitutes the original multiple features associated with the kth independent component. These independent components can provide a better interpretation for multimedia data. Figures 6.3.1(a) and (b) show the scatter plots of the 2K image dataset, projected to a two-dimensional subspace identified by the first two principal components and the first two independent components. Dark points correspond to the class of *tools* (one of the 14 classes), and green (light) points correspond to the other 13 classes. Compared with PC's in Figure 6.3.1(a), IC's found from ICA in Figure 6.3.1(b) can better separate data from different semantic classes. Figure 6.3.1(b) strongly suggests an ICA interpretation to differentiate semantics. The main attraction of ICA is that it provides unsupervised groupings of data that have been shown to be well aligned with manual grouping in different media [11]. The representative and non-redundant feature representations form a solid base for later processing.

Lacking any prior information about the number of independent components, ICA algorithms usually assume that the number of independent components is the same as the dimension of observed mixtures, that is, $k = m$. PCA technique can be used as preprocessing to ICA to reduce noise in the data and control the number of independent components [34]. Then ICA is performed on the main eigenvectors of PCA representations ($k = p$, where p is the number of PC's) to determine which PC's actually are independent and which should be grouped together as parts of a multidimensional component. Finally, the independent components are recovered by computing $S = WU^T X$.

6.3.3 IMG

As discussed in Sections 6.1, though ICA makes a best attempt to find independent components, the resulting k components might not be independent, and the number of components can be too large to face the challenge of "dimensionality curse" during the statistical-analysis and query-processing phrases. IMG aims to remedy these two problems by grouping k components into D modalities.

We divide k components into D groups to satisfy two requirements: (1) the correlation between modalities is minimized, and (2) the number of features in each modality is not too large. The first requirement maximizes modality independence. The second requirement avoids the problem of curse-of-dimensionality. To decide on D, we place a soft constraint on the number of components that a modality can have. We set the soft constraint as 30 because several prior works [7, 18, 19] indicate that when the number of dimensions exceeds 20 to 30, the curse starts to kick in. Since only the data can tell us exactly at what dimension the curse starts to take effect, the selection of D must go through a corss-validation process: we pick a small number of candidate D values and rely on experiments to select the best D.

For a given D, we employ a clustering approach to divide k into D groups. Ding et al. [35] provided theoretical analysis to show that minimizing inter-subgraph similarities and maximizing intra-subgraph similarities always lead to more balanced graph partitions. Thus, we apply *minimizing inter-group feature correlation* and *maximizing intra-group feature correlation* as our feature-grouping criteria to determine independent modalities. Suppose we have D modalities $M_1, M_2 \cdots, M_D$, each containing a number of feature components. The inter-group feature correlation between two modalities M_i and M_j is defined as

$$C(M_i, M_j) = \sum_{\forall S_i \in M_i, \forall S_j \in M_j} C(S_i, S_j), \tag{6.3}$$

where S_i and S_j are features belonging to modalities M_i and M_j respectively, and $C(S_i, S_j)$ is the normalized feature correlation between S_i and S_j. $C(S_i, S_j)$ can be calculated using Eqs.(6.1) and (6.2). The intra-group feature correlation within modality M_i is defined as

$$C(M_i) = C(M_i, M_i). \tag{6.4}$$

To minimize inter-group feature correlation while maximizing intra-group feature correlation at the same time, we can formulate the following objective function for grouping all the features into D modalities,

$$\min \sum_{\substack{i=1 \\ j>i}}^{D} \left[\frac{C(M_i, M_j)}{C(M_i)} + \frac{C(M_i, M_j)}{C(M_j)} \right]. \tag{6.5}$$

Solving this objective function yields D modalities, with minimal inter-modality correlation and balanced features in each modality.

6.4 Super-Kernel Fusion

Once D modalities have been identified by our independent modality analysis, we need to fuse multimodal information optimally. Suppose we train for the dth modality classifier f_d. We need to combine these D classifiers to perform class prediction for query instance $\mathbf{x}_q$. The fusion architecture is depicted in Figure 6.4.

After f_d, $d = 1 \cdots D$ have been trained, the information can be fused in several ways. Let f denote the fused classification function. The product-combination rule can be formulated as

$$f = \prod_{d=1}^{D} f_d.$$

And the most widely used weighted-sum rule can be depicted as

$$f = \sum_{d=1}^{D} \mu_d f_d,$$

where μ_d is the weight for individual classifier f_d. As we have discussed in Section 6.2, both these popular models suffer from several shortcomings, including being sensitive to prediction error and being limited by the linear-model complexity. (Please consult Section 6.2 for detailed discussion.) To overcome these shortcomings, we propose using *super-kernel fusion* to aggregate f_d's.

The algorithm of super-kernel fusion is summarized in Figure 6.4, which consists of the following three steps:

1. Train individual classifiers $\{f_d\}$. The inputs to the algorithm are the n training instances $\{\mathbf{x}_1, \mathbf{x}_2, \cdots, \mathbf{x}_n\}$ and their corresponding labels $\{y_1, y_2, \cdots, y_n\}$. After the independent modality analysis (IMA), the m-dimensional features are divided

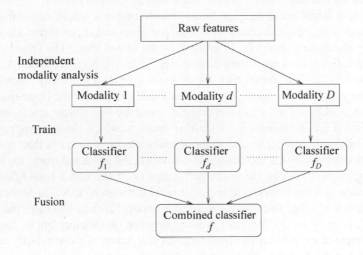

Fig. 6.3: Fusion architecture

into D modalities. Each training instance $\mathbf{x}_i$ is represented by $\{\mathbf{x}_i^1, \mathbf{x}_i^2, \cdots, \mathbf{x}_i^D\}$, where $\mathbf{x}_i^d$ is the feature representation for x_i in the dth modality. All the training instances are divided into D matrices $\{M_1, M_2, \cdots, M_D\}$, where each M_d is an $n \times |M_d|$ matrix, and $|M_d|$ is the number of features in the dth modality ($d = 1, 2, \cdots, D$). To train classifier f_d, we use M_d and the the label information. Though many learning algorithms can be employed to train f_d, we employ an SVM as our base-classifier because of its effectiveness. For training each f_d, the kernel function and kernel parameters are carefully chosen via cross validation (steps 1 to 3 in Figure 6.4).

2. Estimate posterior probability. Once we have trained D classifiers for the D modalties, we create a super-kernel matrix K for modality fusion. This matrix is created by passing each training instance to each of the D classifiers to estimate its posterior probability. We use Platt's formula [36] to convert an SVM score to probability. As a result of this step, we obtain an $n \times D$ matrix consisting of n entries of D class-prediction probability (steps 4 to 6 in Figure 6.4).

3. Fuse the classifiers. The *super-kernel* algorithm treats K a matrix of n training instances, each with a vector of D elements. Next, we again employ SVMs to train the super-classifier. The inputs to SVMs include K, training labels, a selected kernel function, and kernel parameters. At the end of the training process, we yield function f to perform class prediction. The complexity of the fusion model depends on the kernel chosen. For instance, we can select a polynomial, RBF or Laplacian function (steps 7 to 8 in Figure 6.4).

Remark 6.1. A context-based query can be represented by a discriminative function f derived from the above supervised-learning process. Given a candidate data $\mathbf{x}$, the output of $f(\mathbf{x})$ indicates the degree of relevance that $\mathbf{x}$ has to the query. We apply f to the dataset and return top-k most relevant data as the query result.

At first it might seem that non-linear transformations would suffer from high model and computational complexity. But our proposed super-kernel fusion successfully avoids these problems by employing the *kernel trick*. (The kernel trick has been applied to several algorithms in statistics, including Support Vector Machines and kernel PCA.) The kernel trick let us generalize data similarity measurement to operate in a *projected space*, usually nonlinearly related to the *input space*. The *input space* (denoted as $\mathscr{I}$) is the original space in which data are located, and the *projected space* (denoted as $\mathscr{P}$) is that space to which the data are projected, linearly or non-linearly. The advantage of using the *kernel trick* is that, instead of explicitly determining the coordinates of the data in the projected space, the distance computation in $\mathscr{P}$ can be efficiently performed in $\mathscr{I}$ through a *kernel function* [2]. Specifically, given two vectors $\mathbf{x}_i$ and $\mathbf{x}_j$, kernel function $K(\mathbf{x}_i, \mathbf{x}_j)$ is defined as the inner product of $\Phi(\mathbf{x}_i)$ and $\Phi(\mathbf{x}_j)$, where Φ is a basis function that maps the vectors $\mathbf{x}_i$ and $\mathbf{x}_j$ from $\mathscr{I}$ to $\mathscr{P}$. The inner product between two vectors can be thought of as a measure of their similarity. Therefore, $K(\mathbf{x}_i, \mathbf{x}_j)$ returns the similarity of $\mathbf{x}_i$ and $\mathbf{x}_j$ in $\mathscr{P}$. Since a kernel function can be either linear or nonlinear our super-kernel fusion scheme can model non-linear combinations of individual kernels.

One can employ any supervised learning algorithm is the function *Train* in the algorithm (line 2 in the figure). Algorithms that work well with kernel methods are Support Vector Machines [37] and Kernel Discriminative Analysis [38].

Proposition 6.1. *Fused kernel matrix* $\underline{\mathbf{K}}$ *is a mathematically valid kernel, which is symmetric and positive semi-definite.*

Proof. From Figure 6.4, obviously, vectors $\{\mathbf{x}_1, \mathbf{x}_2, \cdots, \mathbf{x}_n\}$ have the same dimensions of D. Therefore, we can use traditional kernel functions such as *Gaussian radial basis kernel function*, *Laplacian kernel function*, and *Polynomial kernel function* to calculate the similarity between these vectors and to build the kernel matrix $\underline{\mathbf{K}}$. Those kernel functions have already been proven to be a mathematically valid kernel satisfying symmetric and positive semi-definite conditions [37]. The resulting kernel matrix $\underline{\mathbf{K}}$ is valid too.   ■

Finally, once the class-prediction function f has been trained, we can use the function to predict the class membership of a query point $\mathbf{x}_q$. Assume $\mathbf{x}_q$ is an m-dimensional feature vector in original feature space, we can convert it to an ICA feature representation $WU^T\mathbf{x}_q$, where W and U are transformation matrices obtained from PCA and ICA process, respectively (Section 6.3). Then, $WU^T\mathbf{x}_q$ is further divided into D modalities (information obtained from the IMG process), named as $\{\mathbf{x}_q^1, \mathbf{x}_q^2, \cdots, \mathbf{x}_q^D\}$. The class-predition function for query point $\mathbf{x}_q$ can be written as

---

[2] Given a kernel function K, we can construct a corresponding kernel matrix $\mathbf{K}$, where $\mathbf{K}(i, j) = K(\mathbf{x}_i, \mathbf{x}_j)$.

Algorithm Super-kernel Fusion
Input:
$X = \{\mathbf{x}_1, \mathbf{x}_2, \cdots, \mathbf{x}_n\};$ /* A set of training data
$Y = \{y_1, y_2, \cdots, y_n\};$ /* Labels of training data

Output:
   $f;$ /* Class-prediction function

Variable:
$\{f_1, f_2, \cdots, f_D\};$ /* A set of discriminative functions
$\{M_1, M_2, \cdots, M_D\};$ /* A set of $n \times |M_d|$ matrices
$K;$ /* Super-kernel matrix with dimension of $n \times D$

Function calls:
$f_d(\mathbf{x}_i^d);$ /* Prediction score of $\mathbf{x}_i^d$ from f_d
$Train(Matrix, Y);$ /* Train a discriminative function
$IMA(X);$ /* Independent modality analysis
$Prob(score);$ /* Convert an SVM score to probability

Begin:
      /* Independent modality analysis to get D modality
1) $\{M_1, M_2, \cdots, M_D\} \leftarrow IMA(X);$
      /* Train classifiers for each modality
2)   **for** each $d = 1, 2, \cdots, D$
3) $f_d \leftarrow Train(M_d, Y);$
      /* Create super-kernel matrix K
4)   **for** each data $\mathbf{x}_i \in X$
5)   **for** each discriminative function f_d
6) $K(i,d) \leftarrow Prob(f_d(\mathbf{x}_i^d));$
      /* Super-kernel Fusion
7) $f \leftarrow Train(K, Y);$
8)   **return** $f;$
End

Fig. 6.4: Super-kernel fusion algorithm

$$\hat{y}_q = f(f_1(\mathbf{x}_q^1), f_2(\mathbf{x}_q^2), \cdots, f_d(\mathbf{x}_q^D)).$$

6.5 Experiments

Our experiments were designed to evaluate the effectiveness of using *independent modality analysis* and *multimodal kernel fusion* to determine the optimal multimodal information fusion for multimedia data retrieval. Specifically, we wanted to answer the following questions:

1. Can independent modality analysis improve the effectiveness of multimedia data analysis?
2. Can super-kernel fusion improve fusion performance?

We conducted our experiments on two real-world datasets: one is a 2K image dataset, and the other is TREC-2003 video track benchmark. We randomly selected a percentage of data from the dataset to be used as training examples. The remaining data were used for testing. For each dataset, the training/testing ratio was empirically chosen via cross-validation so that the sampling ratio worked best in our experiments. To perform independent modality analysis, we applied traditional PCA and ICA algorithms[3] onto the given features (including all the training and testing data) to get the independent components following the steps described in Section 6.3. To perform class prediction, we employed the one-per-class (OPC) ensemble [39], which trains all the classifiers, each of which predicts the class membership for one class. The class prediction on a testing instance is decided by voting among all the classifiers. The results presented here were the average of 10 runs.

- **Dataset #1: 2K image dataset**.
  The image dataset was collected from the Corel Image CDs. Corel images have been widely used by the computer vision, image processing, and multimedia research communities for conducting various experiments. This set contains 2K representative images from fourteen categories: *architecture, bears, clouds, elephants, fabrics, fireworks, flowers, food, landscape, people, textures, tigers, tools*, and *waves*. We tried different kernel functions, kernel parameters and training/testing ratios. Laplacian kernel with $\gamma = 0.001$ and 80% of the dataset as training data gave us the best results on the experiments of using raw features. We used the Laplacian kernel with $\gamma = 0.001$ for all subsequent experiments on this 2K image dataset. We randomly picked 80% of images for training and the remaining 20% were used for testing data. For each image, we extracted 144 features (documented in [40]) including color and texture features. This small dataset is used to provide insights into understanding the effectiveness of our methods, and the tradeoffs between design factors.
- **Dataset #2: TREC-2003 Video Track.**
  TREC-2003 video track used 133 hours digital video (MPEG-1) from ABC and CNN news. The task is to detect the presence of the specified concept in video shots. The ground-truth of the presence of each concept was assumed to be binary (either present or absent in the data). Sixteen concepts are defined in the benchmark, including *airplane, animal, building, female speech, madeleine albright, nature vegetation, news subject face, news subject monologue, NIST non-studio setting, outdoors, people, physical violence, road, sport event, vehicle*, and *weather news*. The video concept detection benchmark is summarized as follows: 60% of the video shots were randomly chosen from the corpus to be used solely for the development of classifiers. The remaining 40% were used for concept

---

[3] InfoMax was chosen as our ICA algorithm because of its robustness, though other ICA algorithms could also be applied.

Table 6.2: Classification accuracy (%) of image dataset

Category	Method 1	Method 2	Method 3	Method 4	Method 5
Architecture	88.00	89.95	90.77	95.38	**96.92**
Bears	74.70	76.72	75.00	75.00	**81.56**
Clouds	84.60	87.61	87.27	90.91	**92.32**
Elephants	83.90	84.67	84.83	87.21	**89.91**
Fabrics	85.10	85.90	87.22	87.82	**87.93**
Fireworks	93.50	95.69	94.91	96.46	**99.50**
Flowers	91.30	**95.53**	92.21	93.49	95.23
Food	92.20	95.58	93.36	95.76	**97.48**
Landscape	78.80	72.79	79.48	79.63	**81.82**
People	82.30	85.50	87.45	86.27	**89.36**
Textures	**96.50**	91.62	91.22	95.00	96.30
Tigers	91.50	92.34	91.13	92.64	**94.80**
Tools	99.50	98.15	96.74	**100.00**	99.20
Waves	86.10	89.49	84.71	87.27	**91.42**
Average	87.71	88.82	88.66	90.20	**92.70**

validation[4]. RBF kernels with $\gamma = 0.0001$ gave us the best results on the experiments, so we used the same parameter settings in all subsequent experiments on this video dataset. For each video shot, we extracted a number of features [4]: *color histogram, edge orientation histogram, wavelet texture, color correlogram, co-occurrence texture, motion vector histogram, visual perception texture, Mel-frequency Cepstral coefficients, speech,* and *closed caption.*

6.5.1 Evaluation of Modality Analysis

The first set of experiments examined the effectiveness of independent modality analysis on the 2K image dataset. Table 6.2 compares five methods based on the classification accuracy results of 14 concepts: original 144 dimensional features before any analysis (Method 1), super-kernel fusion using 108 dimensional color features and 36 dimensional texture features as 2 modalities (Method 2), 58 dimensional features after PCA (Method 3), 58 dimensional features after ICA (Method 4) and super-kernel fusion after IMG (Method 5). As shown in the table, treating color and texture as two modalities improved the accuracy by around 1.0% compared to using raw feature representation. However, the accucary was 4.0% lower than super-kernel fusion after IMG. This observation indicates that improvement can be made by using super-kernel fusion to cover the inter-dependency relationship between features. Moreover, after analyzing the statistical relationships between feature di-

---

[4] IBM research center won most of the best concept models in the final TREC-2003 video concept competition. For the purpose of comparison, we employed the same training and testing data used by IBM.

mensions and getting rid of noise, super-kernel fusion can improve the performance much more. PCA improved accuracy by around 1.0% compared to the original feature format by reducing noise from features. ICA worked better than PCA, improving accuracy by 2.5% compared to the original feature format. However, the improvement is not significant, compared to the performance of super-kernel fusion after IMG. Independent modality analysis plus super-kernel fusion improved classification accuracy around 5.0% compared to the original feature representation. The result shows that the feature sets from independent modality analysis can better interpret the concepts, and super-kernel fusion can further incorporate information from multiple modalities. Next, we evaluated how to select optimal D and compared super-kernel fusion with other fusion methods.

6.5.2 Evaluation of Multimodal Kernel Fusion

The second set of experiments evaluated kernel fusion methods of combining multiple modalities. We grouped the "independent" components after PCA/ICA into independent modalities and trained individual classifiers for each modality. We evaluated the effectiveness of multimodal kernel fusion on the $2k$-image dataset and TREC-2003 video benchmark.

The optimal number of independent modalities D was decided by considering the tradeoff between dimensionality-curse and feature inter-dependency. Once D had been determined, feature components were grouped using the IMG algorithm in Section 6.3.3. When $D = 1$, all the feature components were treated as one vector representation, suffering from the curse of dimensionality. When D became larger, the curse of dimensionality was alleviated, but inter-modality correlation increased[5]. From our 58-dimensional feature data, the optimal modality D is 2 or 3, which enjoys the highest class-prediction accuracy. Table 6.3 shows the optimal D for different concepts (the second column).

Next, we compared different fusion models. Table 6.3 compares the class-prediction accuracy of product combination (PC), linear combination (LC), and super-kernel fusion (SKF). D indicates the number of independent modalities that the 58 independent components have been divided into. We found that super-kernel fusion performed on average 6.5% better than product-combination models and 4.5% better than linear-combination models. Note that the worst results were achieved when using the product rule, 2.0% worse than linear-combination models and 6.5% worse than those of super-kernel fusion. The reason is that if any of the classifiers reports the correct class *a posterior probability* as zero, the output will be zero, and the correct class cannot be identified. Therefore, the final result reported by the combiner in such cases is either a wrong class (worst case) or a reject (when all of the classes are assigned zero *a posterior probability*).

---

[5] The inter-modality correlation for all the D modalities is the summation of inter-modality correlations between every pair of modalities, which is $\sum_{i=1\ j>i}^{D} C(M_i, M_j)$.

Table 6.3: Classification accuracy (%) of image dataset

Category	D	PC	LC	SKF
Architecture	2	96.40	96.53	**96.92**
Bears	2	76.10	75.35	**81.56**
Clouds	3	82.71	89.77	**92.32**
Elephants	2	86.11	80.91	**89.91**
Fabrics	2	85.11	87.46	**87.93**
Fireworks	2	97.63	99.13	**99.50**
Flowers	3	82.29	86.14	**95.23**
Food	2	93.45	89.53	**97.48**
Landscape	2	77.55	74.24	**81.82**
People	2	**90.71**	89.57	89.36
Textures	2	74.51	94.27	**96.30**
Tigers	3	87.31	**95.00**	94.80
Tools	2	91.48	94.20	**99.20**
Waves	2	86.92	82.13	**91.42**
Average	2.3	86.31	88.16	**92.70**

Table 6.4: AP (%) of video concept detection

Concept	IBM	PC	LC	SKF
Airplane	**24.93**	10.60	23.52	24.31
Animal	6.09	6.75	**8.59**	8.2
Building	8.02	7.92	4.68	**8.42**
Female Speech	67.23	49.10	67.23	**67.33**
Madeleine Albright	**47.41**	16.54	33.93	43.27
Nature Vegetation	37.84	31.02	33.65	**39.39**
News Subject Face	**8.12**	1.37	7.89	7.05
News Subject Mono.	**20.41**	3.1	8.87	13.48
NIST Non-Studio	69.1	69.65	66.38	**69.88**
Outdoors	65.16	**69.81**	53.87	66.16
People	11.82	12.95	16.41	**18.91**
Physical Violence	**3.04**	1.06	1.42	1.8
Road	10	7.72	**12.42**	8.38
Sport Event	48.45	24.20	40.49	**52.8**
Vehicle	**20.81**	14.05	15.63	16.54
Weather News	53.64	29.73	53.64	**86.7**
Average	31.38	22.28	28.04	**33.29**

Finally, we conducted fusion experiments on the video dataset. For this TREC video dataset, we got only probability outputs from single-modality classifiers through IBM. Therefore, we evaluated only fusion schemas on this video dataset. Table 6.4 compares the best results from IBM (IBM), product combination (PC), linear combination (LC), and super-kernel fusion (SKF) based on Average Precision of video concept detection. The numbers of modalities for sixteen concepts ranged from 2 to 6. Here we chose the NIST Average Precision (the sum of the precision at each relevant hit in the hitlist divided by the total number of relevant documents in the collection) as the evaluation criteria. Average Precision (AP) was used by NIST to evaluate retrieval systems in TREC-2003 video track competition. For TREC-2003 video track, a maximum of $1,000$ entries[6] were returned and ranked according to the highest probability of detecting the presence of the concept. The ground-truth of the presence of each concept was assumed to be binary (either present or absent in the data). For the 16 concepts in TREC-2003 video benchmark, super-kernel fusion performed around 5.2% better than the linear-combination models on average, 11.3% better than product-combination models. Super-kernel fusion also performed around 2.0% better than the best results provided by IBM.

6.5.3 Observations

After our extensive empirical studies on the two datasets, we can answer the questions proposed at the beginning of this section.

1. To deal with high-dimensional features from multiple media sources, it is necessary to do statistical analysis to reduce noise and find the most representative feature-components. Independent modality analysis can improve the effectiveness of multimedia data analysis by achieving a tradeoff between dimensionality curse and modality independency.
2. Super-kernel fusion is superior in its performance because its high model complexity can explore inter-dependencies between modalities.

6.6 Concluding Remarks

In this chapter, we have presented a framework of optimal multimodal information fusion for multimedia data analysis. First, we constructed *statistically independent modalities* from the given feature set from multiple media sources. Next, we proposed *super-kernel fusion* to learn the optimal combination of multimodal information. We carefully analyzed the tradeoffs between three design factors that affect fusion performance: *modality independence*, *curse of dimensionality*, and *fusion-model complexity*. Empirical studies show that our methods achieved markedly improved performance on a 2K image dataset and TREC-Video 2003 benchmarks.

This chapter shows a discriminative approach for fusing metadata of multiple modalities. In Chapters 7 and 8, we present a couple of generative approaches for conducting multimodal fusion. A discriminative approach tends to work more effectively, but it is difficult to interpret its results. On the contrary, a generative approach

---

[6] This number was chosen in the IBM's work [4] for evaluation.

[41, 42] may have to rely on an assumed statistical model, but one can explain the yielded relationship between features and semantics.

References

1. Wu, Y., Chang, E.Y., Chang, K.C.C., Smith, J.R. Optimal multimodal fusion for multimedia data analysis. In *Proceedings of ACM Multimedia*, pages 572–579, 2004.
2. Rui, Y., Huang, T.S., Chang, S.F. Image retrieval: Past, present, and future. *Journal of Visual Communication and Image Representation*, 10:1–23, 1997.
3. Datta, R., Joshi, D., Li, J., Wang, J.Z. Image retrieval: Ideas, influences, and trends of the new age. *ACM Comput. Surv.*, 40:1–60, May 2008.
4. Adams, B., Amir, A., Dorai, C., Ghosal, S., Iyengar, G., Jaimes, A., Lang, C., Lin, C.Y., Natsev, A., Naphade, M., Neti, C., Nock, H.J., Permutery, H.H., Singhx, R., Smith, J.R., Srinivasany, S., Tseng, B.L., Ashwin, T.V., Zhang, D.Q. IBM Research TREC-2002 video retrieval system. 2002.
5. Stegmaier, F. Interoperable and unified multimedia retrieval in distributed and heterogeneous environments. In *Proceedings of ACM International Conference on Multimedia*, pages 1705–1706, 2010.
6. Anguera, X., Xu, J., Oliver, N. Multimodal photo annotation and retrieval on a mobile phone. In *Proceeding of ACM International Conference on Multimedia Information Retrieval*, pages 188–194, 2008.
7. Beyer, K., Goldstein, J., Ramakrishnan, R., Shaft, U. When is "nearest neighbor" meaningful? In *Proceedings of ICDT*, pages 217–235, 1999.
8. Bellman, R. *Adaptive Control Processes*. Princeton, 1961.
9. Goh, K., Chang, E., Cheng, K.T. Svm binary classifier ensembles for multi-class image classification. In *Proc. of ACM International Conference on Information and Knowledgement Management (CIKM)*, pages 395–402, 2001.
10. Cascia, M.L., Sethi, S., Sclaroff, S. Combining textual and visual cues for content-based image retrieval on the world wide web. In *Proc. of the IEEE Workshop on Content-based Access of Image and Video Libaries*, pages 24–28, 1998.
11. Hansen, L., Larsen, J., Kolenda, T. On independent component analysis for multimedia signals. *Multimedia Image and Video Processing, CRC Press*, 2000.
12. Kolenda, T., Hansen, L.K., Larsen, J., Winther, O. Independent component analysis for understanding multimedia content. In *Proc. of IEEE Workshop on Neural Networks for Signal Processing*, pages 757–766, 2002.
13. Vinokourov, A., Hardoon, D.R., Shawe-Taylor, J. Learning the semantics of multimedia content with application to web image retrieval and classification. In *Proceedings of Fourth International Symposium on Independent Component Analysis and Blind Source Separation*, 2003.
14. Hershey, J., Movellan, J. Using audio-visual synchrony to locate sounds. In *Proceedings of NIPS*, pages 813–819, 2001.
15. J.W. III, F., Darrell, T., Freeman, W., Viola, P. Learning joint statistical models for audio-visual fusion and segregation. In *Proceedings of NIPS*, pages 772–778, 2000.
16. Smaragdis, P., Casey, M. Audio/visual independent components. *International Symposium on Independent Component Analysis and Blind Source Separation*, pages 709–714, 2003.
17. Vinokourov, A., Shawe-Taylor, J., Cristianini, N. Inferring a semantic representation of text via cross-language correlation analysis. In *Proceedings of NIPS*, pages 1473–1480, 2002.
18. Donoho, D.L. High-dimensional data analysis: The curses and blessings of dimensionality. *American Math. Society Lecture—Match Challenges of the 21st Century*, 2000.
19. Fagin, R., Lotem, A., .Naor, M. Optimal aggregation algorithms for middleware. In *Proc. of ACM PODS*, 2001.

20. Flickner, M., Sawhney, H., Ashley, J., Huang, Q., Dom, B., Gorkani, M., Hafner, J., Lee, D., Petkovic, D., Steele, D., Yanker, P. Query by image and video content: the qbic system. *IEEE Computer*, 28(9):23–32, 1995.
21. Smith, J.R., Chang, S.F. Visualseek: A fully automated content-based image query system. In *Proceedings of ACM Multimedia*, pages 87–98, 1996.
22. Rui, Y., Huang, T.S., Mehrotra, S. Content-based image retrieval with relevance feedback in mars. In *Proc. of IEEE Int. Conf. on Image Processing*, pages 815–818, 1997.
23. Velivelli, A., Ngo, C.W., Huang, T.S. Detection of documentarty scene changes by audio-visual fusion. In *Proceedings of CIVR*, pages 227–237, 2003.
24. Hyvarinen, A., Oja, E. A fast fixed-point algorithm for independent component analysis. In *Proceedings of NIPS*, pages 1483–1492, 1997.
25. Amari, S., Cichocki, A., Yang, H.H. A new learning algorithm for blind signal separation. In *Proceedings of NIPS*, pages 757–763, 1996.
26. Bell, A.J., Sejnowski, T.J. An information-maximization approach to blind separation and blind deconvolution. In *Proceedings of NIPS*, pages 1129–1159, 1995.
27. Bach, F.R., Jordan, M.I. Kernel independent component analysis. *Journal of Machine Learning Research*, 3:1–48, 2002.
28. Kittler, J., Hatef, M., Duin, R.P.W. Combining classifiers. In *Proc. Intl. Pattern Recognition*, pages 897–901, 1996.
29. Tax, D.M., van. Breukelen, M., Duin, R.P., Kittler, J. Combing multiple classifiers by averaging or by multiplying. *Pattern Recognition*, 33(9):1475–1485, 2000.
30. Ting, K.M., Witten, I.H. Issues in styacked generalization. *Journal of Artificial Intelligence Research*, 10:271–289, 1999.
31. Yan, R., Hauptmann, A.G. The combination limit in multimedia retrieval. In *Proceedings of ACM Multimedia*, pages 339–342, 2003.
32. Joliffe, I. *Principal Component Analysis*. Springer-Verlag, New York, 1986.
33. Lukic, A.S., Wernick, M.N., Hansen, L.K., Strother, S.C. An ICA algorithm for analyzing multiple data sets. In *Proc. of IEEE Int. Conf. on Image Processing*, pages 821–824, 2002.
34. Bartlett, M.S., Lades, H.M., Sejnowski, T.J. Independent component representation for face recognition. In *Proc. of the SPIE Conf. on Human Vision and Electronic Imaging III*, pages 528–539, 1998.
35. Ding, C.H.Q., He, X., Zha, H., Gu, M., Simon, H.D. A min-max cut algorithm for graph partitioning and data clustering. In *Proceedings of IEEE ICDM*, pages 107–114, 2001.
36. Platt, J. Probabilistic outputs for support vector machines and comparison to regularized likelihood methods. *Advances in Large Margin Classifiers, MIT Press*, pages 61–74, 2000.
37. Burges, C.J.C. A tutorial on support vector machines for pattern recognition. In *Proceedings of ACM KDD*, pages 121–167, 1998.
38. Roth, V., Steinhage, V. Nonlinear discriminant analysis using kernel functions. In *Proceedings of NIPS*, pages 568–574, 1999.
39. Dietterich, T., Bakiri, G. Solving multiclass learning problems via error-correcting output codes. *Journal of Artifical Intelligence Research*, 2:263–286, 1995.
40. Li, B., Chang, E. Discovery of a perceptual distance function for measuring image similarity. *ACM Multimedia Journal Special Issue on Content-Based Image Retrieval*, 8(6):512–522, 2003.
41. Wu, Y., Chang, E.Y., Tseng, B.L. Multimodal metadata fusion using causal strength. In *Proceedings of ACM Multimedia*, pages 872–881, 2005.
42. Chen, W., Zhang, D., Chang, E.Y. Combinational collaborative filtering for personalized community recommendation. In *Proceedings of ACM KDD*, pages 115–123, 2008.

Chapter 7
Fusing Content and Context with Causality

Abstract This chapter[†] presents a generative framework that uses *influence diagrams* to fuse metadata of multiple modalities for photo annotation. We fuse contextual information (location, time, and camera parameters), visual content (holistic and local perceptual features), and semantic ontology in a synergistic way. We use *causal strengths* to encode causalities between variables, and between variables and semantic labels. Through analytical and empirical studies, we demonstrate that our fusion approach can achieve high-quality photo annotation and good interpretability, substantially better than traditional methods.

Keywords: Causality, multimodal fusion, semantic gap.

7.1 Introduction

To help users better organize and search their photos, it would be desirable to provide each photo with useful semantic labels such as time (when), objects (who), location (where), and event (what). In this chapter, we present a photo annotation framework which uses an *influence diagram* to fuse *context*, *content*, and *semantics* in a synergistic way, to generate keywords for organizing and searching photos.

Obtaining the "when and where" information is easy. Already cameras can provide time, and we can easily infer location from GPS or CellID (will be available with all cameraphones). However, determining the "what and who" requires contextual information in addition to time, location, and photo content. The contextual information of cameras can be categorized into *time*, *location*, and *camera parameters*. Photo content consists of perceptual features extracted from photos, which can be categorized into *holistic perceptual features* (color, shape and texture characteristics of a photo) [2, 3] and *local perceptual features* (edges and salient points of regions or objects in a photo) [4, 5]. Besides context and content, another impor-

---

[†] ©ACM, 2005. This chapter is a minor revision of the author's work with Yi Wu and Belle Tseng [1] published in MULTIMEDIA'05. Permission to publish this chapter is granted under copyright license #2587660180893.

tant source of information is the relationship between semantic labels (which we refer to as *semantic ontology*). For instance, let us consider two semantic labels: *outdoor* and *sunset*. We can infer the *outdoor* label from contextual information such as lighting conditions [6], and we can infer *sunset* from time and location [7]. Notice that inferring *outdoor* and *sunset* might not rely on any common contextual modality ($\{lighting\} \cap \{time, location\} = \emptyset$). However, we can say with certainty that a *sunset* photo is outdoors (but not the reverse). Therefore, by considering semantic relationships between labels, we can leverage contextual information in a "transitive" way.

Our fusion framework uses an *influence diagram* [8, 9] to conduct fusion and semantic inferencing. The variables on the diagram can be either *decision variables* (causes), or *chance variables* (effects). For image annotation, decision variables include time, location, camera parameters, and perceptual features. Chance variables are semantic labels. However, some variables may play both roles. For instance, the time of day can affect some camera parameters (such as exposure time and flash on/off), and hence these camera parameters are both decision and chance variables. Furthermore, different photo concepts might have varied influence diagram structures. Finally, the influence diagram connects decision variables with chance variables using arcs weighted by *causal strength*.

To construct an influence diagram, we rely on both domain knowledge and data. In general, learning such a probabilistic graphical model from data is an NP hard problem [10]. For photo annotation, however, we have abundant prior knowledge about the relationships between context, content, and semantic labels, so we can use them to substantially reduce the hypothesis space to search for the right model. For instance: time and location are independent of each other (no arc exists between them in the diagram). Camera parameters such as exposure time and flash on/off depend on time (fixed arcs can be determined). The semantic ontology provides us the relationships between keywords. The only causal relationships that we must learn from data are those between context/content and semantic labels.

Once causal relationships have been learned, causal strengths must be accurately accounted for. Traditional probabilistic graphical models such as Bayesian networks use conditional probability to quantify the correlation between two variables. Unfortunately, conditional probability characterizes *covariation*, not *causation* [10, 11, 12]. A basic tenet of classical statistics is that correlation does not imply causation. Instead, we use recently developed *causal-power* theory [13] to account for causation. We show that fusing context and content using causation achieves superior results over using correlation.

Let us preview a model generated by our algorithm to illustrate the advantages of our fusion model. Figure 7.1 shows the learned diagram for two semantic labels: *outdoor* and *sunset*, which we have just briefly discussed. In addition to the arcs that show the causal relationships of variable-to-variable and label-to-variable, we see that the relationship between *outdoor* and *sunset* can also be encoded. The advantages of our fusion method over traditional fusion methods (presented in Section 7.2) are threefold:

1. Our fusion method provides a general framework to fuse context, content, and semantics for annotating photos. By considering relationships between semantics, we can leverage contextual information in a "transitive" way.
2. Our influence diagram employs causal-power theory to encode causalities between variables, and between variables and semantic labels. The structure-learning algorithm takes advantage of domain knowledge. The causal-strength computation is based on psychological principles, a novel approach that leads to better results.
3. Our fusion method readily handles missing modalities or noisy data. This feature is important because not all photos have metadata of all modalities.

The rest of the chapter is organized into five sections. Section 7.2 surveys related work. Section 7.3 introduces multimodal metadata that we generate for photo annotation. Section 7.4.1 presents an efficient algorithm to learn an influence diagram. In Section 7.4.2 we explain how we accurately quantify causal strengths based on causality principles. In Section 7.4.3, we present a case study for the effectiveness of inferring causal strengths in photo annotation. Section 7.4.4 introduces how to deal with missing attributes. We report our experimental results in Section 7.5. Especially we show the effectiveness of using causal strengths to perform modality fusion in preference to conditional probability. We also show that our model can deal with missing modalities. Finally, we offer our concluding remarks in Section 7.6.

7.2 Related Reading

We discuss related work in two aspects: work in annotating photos, and work in learning a model such as an influence diagram.

7.2.1 Photo Annotation

The prior work in photo annotation can be divided into two main categories: *content-based* and *context-based*. Example works of the content-based approach can be further divided into perception-based [14], object-based [15], and region-based [16]. These methods use different perceptual features and employ either a generative or discriminative model of machine learning to map features to semantics. However, at least a couple of obstacles make content-based annotation difficult. First, images are taken under varying environmental conditions (e.g., lighting, movements, etc.) and with varied resolutions. (Chapter 2 deals with feature-extraction issues.) The same Eiffel tower pictures taken by two different models of cameras, from two different angles, and at different times of day can seem different enough to be mislabeled. Thus, keyword propagation via image content can be unreliable. Second, although some progress has been made, reliable object segmentation is not attainable in the

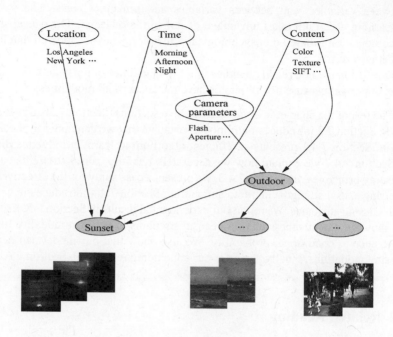

Fig. 7.1: Example influence diagram

foreseeable future [15]. Only when we are able to segment objects with high accuracy, will it be possible to recognize objects with high reliability.

Some recent works [17, 18, 19, 20, 7] annotate digital images using the metadata of spatial and temporal context. Nevertheless, these methods can be improved in several ways. In [17, 7], GPS devices are used to infer the location where a photo is taken. However, the objects in a photo may be far away from the camera, not at the GPS coordinate. For instance, one can take a picture of the Bay Bridge from many different locations in the cities of Berkeley, Oakland, or San Francisco. The GPS information of the camera cannot definitely infer landmarks in a photo. Second, a landmark in a photo might be occluded and hence not deemed important. For instance, a vehicle might occlude a bridge; a person might occlude a building, etc. Also, a person can be inside or outside of a landmark. A robust system must analyze content to identify the landmarks. The work of [7] uses temporal context to find time-of-day, weather, and season information. Annotating vast quantities of other semantics requires additional metadata and effective fusion models.

Neither the *content-based* nor the *context-based* photo-annotation approach alone can be effective [21]. It is clear that a combined approach is the only satisfactory remedy. Our novelty lies in the idea of combining not only context and content, but also semantic ontology (as illustrated in Figure 7.1).

7.2.2 Probabilistic Graphical Models

It is well known that learning a general graph from data is NP hard. Reducing the graph to be directional and acyclic (such as Bayesian networks or BNs) does not reduce its computational complexity. The work of [6] suggests using BNs to combine content with camera parameters to infer outdoor/indoor. However, the scope of that work is limited, considering just one kind of contextual information. Furthermore, the structure in that paper was manually generated, not learned from the data. Our approach can support the fusion of a large number of variables and states, by carefully considering prior knowledge.

More importantly, traditional Bayesian networks use conditional probability to quantify an arc between two nodes in the graph. For inferring semantics, however, conditional probability can be misleading. In this chapter, we provide indepth treatment on the subject of *causality* vs. *correlation*, and conduct extensive experiments to show that causality is a much better measure for fusing metadata of multiple modalities. Notice that the discriminative approach presented in Section 7 may be good to model *content*. To fuse *content* and *context*, which may have features of discrete values, a generative approach is both natural and interpretable. We discuss the details in Section 7.4.2.

7.3 Multimodal Metadata

The multimodal metadata that our system generates or utilizes include contextual information, perceptual content, and semantic ontology, which are described as follows.

7.3.1 Contextual Information

The contextual metadata of cameras can be categorized into *location*, *time*, and *camera parameters*.
Location. The location recorded for each photo is the location of the camera at the moment of exposure, which can be captured by GPS or CellID (for cell phones), and translated into a location tuple, e.g., (country, state, county, city). In the photo annotation scenario, we are more interested in the subjects (landmarks) in the photo,

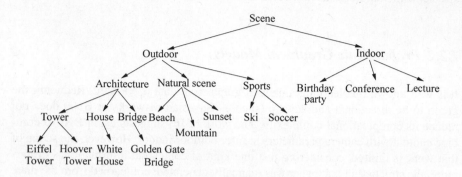

Fig. 7.2: Semantic ontology

which may or may not be the location of the camera. However, the location information can reduce the number of subjects possibly appearing in the photo. For example,
the Eiffel Tower cannot possibly appear in the city of Palo Alto, California, U.S.A.
Time. The data/time stamp is recorded when a photo is taken by the cell phone or
digital camera. Many cell phones have the functionality of updating time when a
person moves to a different time zone. For digital camera, every new camera has its
initial clock setting. In most cases, users set their camera clock just once according
to the time zone of their home. The difficulty appears when the user travels to other
time zones. This problem can be solved given the location information where each
photo was taken, and the original time zone where the clock time was set. The local
time for each photo can then be computed [20]. However, it is also possible the clock
of a camera was never set. To timestamp all the photos taken with the camera, we
can calculate the exact time $t_c = (t_{cc} - t_{ci}) + t_i$, where t_{cc} is the current time shown
on the camera, t_{ci} is the initial clock on the camera, and t_i is the initial real time.
Camera parameters. Digital cameras can record information related to the photo
capture conditions. The Exif standard specifies hundreds of camera parameters [22].
Some useful parameters for photo annotation are *presence or absence of flash, subject distance, exposure time, aperture value,* and *focal length.* Some of these parameters can help distinguish various scenes. For example, flash tends to be used more
frequently with indoor photos than with outdoor photos. Also these parameters are
sometimes mutually related. For example, if flash is turned off for an indoor photo,
the brightness value is low.

7.3.2 Perceptual Content

Photo content consists of perceptual features, which can be categorized into holistic perceptual features and local perceptual features:

1. *Holistic perceptual feature.* For each photo, we can extract global features, which include *color, texture, shape*, etc. The color feature is one of the most widely used visual features in image/photo retrieval. It is relatively robust to background complication and independent of image size and orientation. Commonly used color feature representations include color histogram [23], color moments [23] and color sets [24]. Texture contains important information about the visual patterns, structural arrangement of surfaces and their relationship to the surrounding environment. Shape representations try to capture objects in the image invariant to translation, rotation, and scaling. Commonly used shape representations include boundary-based representation using only the outer boundary of the shape, and region-based representation using the entire shape region [25].
2. *Local perceptual features.* Scale Invariant Feature Transform (SIFT) has become a very popular local feature representation in recent years [4, 5]. SIFT extracts local, distinctive, invariant features. SIFT consists of four major stages: (1) scale-space extreme detection, (2) keypoint localization, (3) orientation assignment, and (4) keypoint descriptor. Features generated by SIFT have been shown to be robust in matching across an affine distortion, change in 3D viewpoint, addition of noise, and change of illumination. Ke and Sukthankar recently proposed a modified version of SIFT, called the PCA-SIFT [26]. PCA-SIFT uses the Principal Components Analysis (PCA) to normalize the gradient patch based on the output of SIFT. The result is shown to be robust to image deformations.

7.3.3 Semantic Ontology

Historically, domain-dependant ontologies have been employed to achieve better performance in the image/photo retrieval system. A semantic ontology defines a set of representative concepts and the inter-relationships among these concepts. Khan and McLeod [27] proposed an ontology for sports news. Smith and Chang [28] proposed an ontology structure for image and video subjects. Naaman et al. [20] conducted a survey to identify the concepts on what people care about when searching for photos. Recently, ImageNet is being developed to provide an ontology to relate objects in and semantics of images [29].

Based on those previously proposed ontology structures and the photo-concept surveys, Figure 7.2 illustrates an example of semantic ontology that we are developing for photo topics. The ontology is described by a directed acyclic graph (DAG). Here, each node in the DAG represents a concept. An interaction relationship between concept C_i and C_j goes from a more generic concept C_i to a more specific concept C_j, which is represented by a directional arc. For example, "outdoor" is more generic, covering the concept of "sunset;" and "indoor" is more generic, covering "lecture."

7.4 Influence Diagrams

We use an *influence diagram* to model causes and effects. The variables on the diagram can be divided into two classes: *decision variables D* and *chance variables U*. Each decision variable has a set of choices or states. An *act* corresponds to a choice of a state for each decision variable (e.g., flash light on or off). Each chance variable can also be in one of a set of states (e.g., outdoor or indoor). In addition, ξ is the state of the world. A decision problem, according to [9], is described by the variables U, D, and ξ. Now, for each chance variable X_i in U, we identify a set of parents $Pa(X_i) \in U \cup D - \{X_i\}$ that renders X_i and $U \cup D - \{X_i\}$ conditionally independent. This can be written as

$$p(X_i|U \cup D - \{X_i\}, \xi) = p(X_i|Pa(X_i), \xi). \tag{7.1}$$

Using the chain rule of probability, we have

$$p(U|D, \xi) = p(X_1, \cdots, X_n|D, \xi) = \prod_{i=1}^{n} p(X_i|U \cup D - \{X_i\}, \xi). \tag{7.2}$$

Combining Eqs. (7.1) and (7.2), we obtain

$$p(U|D, \xi) = \prod_{i=1}^{n} p(X_i|Pa(X_i), \xi). \tag{7.3}$$

Our system needs to learn a directed, acyclic graph[1] wherein $G = \{V, E\}$, $V = \{X_i\}$, and E is a subset of edges $\{(X_i, X_j)|0 \le i, j < n, i \ne j\}$ from data. We denote the dataset L as a collection of n-tuples $(d_0^k, d_1^k, \cdots, d_{n-1}^k)$, $0 \le k < |L|$, where d_i can take on one of the legal values (or states) for X_i, i.e., $x_{ij}, 0 \le j < s_i$. The diagram-learning problem then comprises two stages: (1) structure learning to determine the subset of edges to be included in the diagram, and (2) content learning to specify the causal strengths on all the edges, given the inferred diagram. These steps are described in detail below.

---

[1] We use "network" and "graph" interchangeably to refer to "influence diagram." The major difference between a network, a graph, and an influence diagram (which will become evident in Section 7.4.2) lies in how the weights of the edges are measured. Otherwise, an influence diagram or a probabilistic causal model under the assumption of the *causal Markov condition* is a Bayesian network [30].

7.4.1 Structure Learning

The semantic ontology provides us the causal structure between different semantic concepts. The only causal structure that we must learn from data are those between context/content and semantic labels.

Boutell et al. [6] employed a fixed structure of context and content for various photo concepts. However, various concepts have different relationships among variables. For example, the "sunset" concept relies on the contextual feature *time*, and this feature influences camera features such as *brightness* and *flash*. For the "indoor" concept, no relationship has been established between *time* and *brightness*.

We apply the idea of Bayesian Multi-net for structure learning. Bayesian Multi-nets were first introduced in [31] and then studied in [32] as a type of classifiers. A Bayesian multi-net is composed of the prior probability distribution of the class node and a set of local networks, each corresponding to a value that the class node can take. Bayesian multi-nets allow the relations among the features to be different for different values the class node takes on. The features can form different local networks with different structures.

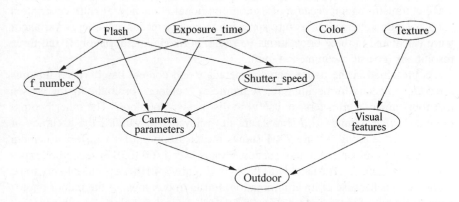

Fig. 7.3: Outdoor influence diagram

We formulate the diagram-learning problem as follows: The space of all possible diagram (or graph) structures given n vertices and $n(n-1)$ directed edges is represented as a vector of random variables of length $n(n-1)$, or $Y = (Y_0, Y_1, \cdots, Y_{n(n-1)-1})$, where Y_k is 1 if the directed edge (X_i, X_j) is selected, and 0 otherwise, where $i = k/n$, $j = k\%n$, and % is the modulo operator, or $k\%n = k - (k/n) \times n$. To solve the diagram-learning problem, we need to specify a goodness (or likelihood) measure

$L(G(Y)|L)$, and develop a search procedure to examine the space of Y efficiently to locate a best diagram. The diagram-learning problem is similar to Bayesian-network learning, except that at the end we will label edges with causalities rather than conditional probabilities.

Figure 7.3 presents an example of structure learning for the "outdoor" semantic. For this example and in our experiments, we utilized *time* and *camera parameters* as *context information*, in which camera parameters include *exposure time, aperture, f-number, shutter speed, flash and focal length*. These camera features were shown to be useful for scene classification in [6]. The first four features reflect the brightness of a photo. Natural lighting is stronger than artificial lighting. This causes outdoor scenes to be brighter than indoor scenes, even under overcast skies, and they, therefore, have a shorter exposure time and a larger brightness value. Because of the lighting differences described above, (automatic and manual) camera flash is used on a much higher percentage of images of indoor scenes than that of outdoor scenes. Focal length is related to subject distance in less direct and intuitive ways through camera zoom.

For this example and in our experiments, we utilized *holistic perceptual feature* as *content information*. we extracted holistic perceptual features including color, texture, and shape from each photo. We first divided color into twelve color bins, of which eleven were for the eleven culture colors [33] and the remaining bin for outlier colors. For each culture color, we extracted color features of multiple resolutions, which included color histograms, color means and variances in HSV channels, color spreadness, and color-blob elongation. For texture extraction, we used discrete wavelet transformation because of its computational efficiency. Texture descriptors such as texture elongation, texture spreadness, energy means and energy variances were calculated in three orientations (vertical, horizontal, and diagonal) and three resolutions (coarse, medium, and fine).

As presented in the figure, camera metadata and content-based visual features are complementary for deciding the semantic of "outdoor." From the figure, we see that four camera parameters are useful to distinguish "outdoor." The other two parameters, *aperture* and *focal length*, are not relevant. To validate the accuracy of the structure learning, Figure 7.4.1 shows the distribution of six different camera metadata for outdoor and indoor scenes. Photos over $1/60$ (0.017) second of exposure time (Figure 7.4.1(a)) and with flash on (Figure 7.4.1(b)) are more likely to be indoor scenes because of low illumination. Furthermore, most of the indoor photos have 6 seconds of shutter speed (Figure 7.4.1(c)) and a low value of f-number (Figure 7.4.1(d)). However, the distribution of aperture (Figure 7.4.1(e)) for both indoor and outdoor scenes is quite similar; and the same can be said about the distribution of focal length (Figure 7.4.1(f)). Thus, neither *aperture* nor *focal length* is useful for inferring whether a photo is taken outdoors.

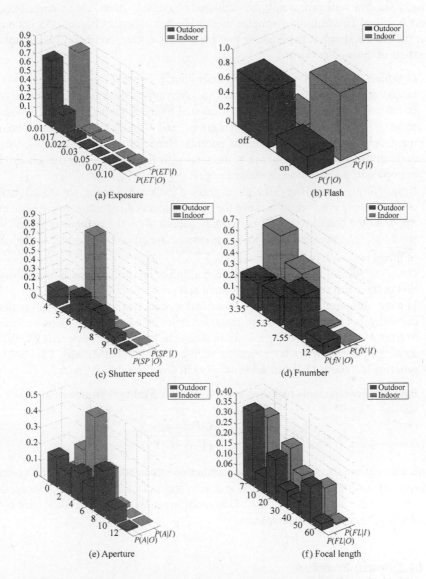

Fig. 7.4: Camera metadata distribution for outdoor and indoor scenes (See color insert)

7.4.1.1 Likelihood Measure

Bayesian networks received the name mainly because of the use of the Bayesian inference scheme. Bayesian inference uses an elegant likelihood definition which states that a good model is one that is likely and can explain data well. Whether a model fits data well can usually be validated objectively based on an agreed-upon cost function. In addition, a simpler model is often a better choice. (Occam's razor prefers the simplest explanation that fits the data.) Our likelihood function thus consists of three components:

1. *Goodness of fitting.* A good edge, by definition, is one that captures the causal relationship between a cause (a decision variable) and an effect (a chance variable). In Information Theory, causality is measured by the reduction in the uncertainty (entropy) of a random variable given others, and $H(X_i|Pa(X_i))$ measures the remaining uncertainty of X_i given its parents. Hence a reasonable definition of a good fit for a Bayesian network is the total entropy reduction (or the amount of entropy left) given the set of chosen edges, $-\sum_{i=0}^{n-1} H(X_i|Pa(X_i))$.
2. *Cost of fitting.* Using the principle expressed in Occam's razor, we would prefer using a simple network to achieve the same amount of reduction in network entropy. As the complexity of a network is measured by the number of edges, we would prefer a network that has fewer connections if at all possible, or the cost of fitting is inversely proportional to $-\sum_{k=0}^{n(n-1)-1} Y_k$.
3. *Other constraints and prior.* We capture all other constraints and prior information in this third category. As constraints and prior information can be expressed in many different ways, there is no unified mechanism to express them mathematically. For our purpose, we order all variables on a list with decision variables in front of chance variables, and some dual-role variables in between. We allow an arc to point only to a later element on the list, not the other way. This *vertice ordering* technique [34] can substantially cut down the search space.

Putting all components together, we arrive at our likelihood function as

$$L(G(Y)|L) = \frac{-\sum_{i=0}^{n-1} H(X_i|Pa(X_i)) - \sum_{k=0}^{n(n-1)-1} Y_k}{\prod_C \delta_C(G(Y))} \qquad (7.4)$$

Note that entropy is measured by the number of bits needed to encode the system entropy whereas the number of edges in a network is just that, a number. Hence, the two terms in the numerator of Eq.(7.4) have different units. We will show later how to reconcile these two measurements.

7.4.1.2 Search Procedure

Armed with the likelihood function in Eq.(7.4), we are ready to evaluate the space of Y to look for diagrams that maximize this likelihood measure, given training data. The number of possible Y's is $2^{n(n-1)}$, a very large space to be examined exhaustively. Furthermore, it is well known that finding general Bayesian network structures is an NP-hard problem. Hence, we employ some efficient search algorithms to examine the space of Y (in addition to taking advantage of the domain knowledge to prune some unlikely edges [35]).

Many sampling techniques exist, and they can roughly be put into two categories: independent sampling, and dependent (chain) sampling. Independent sampling techniques, such as importance sampling and rejection sampling, generate independent samples of the search space. However, independent sampling techniques are inadequate for high-dimensional search spaces. Chain sampling (such as Metropolis sampling, Metropolis-Hasting sampling, and Gibbs sampling that underlie the Markov Chain Monte Carlo (MCMC) methods) can generate dependent samples. In order to use the dependent samples to approximate independent distributions $P(Y|L)$, the chain sampling has to satisfy a certain Markov invariance property and has to run for a sufficiently long time. While it is hard to diagnose or detect convergence in chain sampling techniques, it is possible to speed up convergence using advanced techniques such as Hamiltonian Monte Carlo, over-relaxation, and simulated annealing.

We use Gibbs sampling in our work. This is because in order to explore a large search space, efficiency in sample generation is of paramount importance, and Gibbs sampling provides such efficiency. In Gibbs sampling, a new sample is generated using a sequence of proposal distributions that are defined in terms of conditional distributions of the joint distribution $P(Y|L)$. This is advantageous because sampling directly $P(Y|L)$ is expensive, yet it costs much less to sample the conditional distributions $P(\{Z\}|Y - \{Z\})$.

More specifically, if independent samples are generated from the space of Y, the likelihood of each sample (i.e., a new diagram) needs to be evaluated using Eq.(7.4). This will require the evaluation of many $H(X_i|Pa(X_i))$ terms and the verification of the new diagram structure against the constraints and prior knowledge. On the other hand, if new samples are conditionally generated, such as in Gibbs sampling, only a few edges will differ from one sample to the next. It is comparatively easy to examine the effect of the insertion (or deletion) of a few edges.

In Gibbs sampling, only one edge, Y_k, is different from one sample to the next, and $P(Y_k|Y - \{Y_k\})$ is a simple Bernoulli distribution. To choose between the two possibilities (include or not include Y_k), we need only to compare the likelihood of the two diagrams, which can be efficiently performed as follows: Denote the particular edge represented by Y_k as $(X_i, X_j) = (X_{k/n}, X_{k\%n})$. Assume that this edge was not included in the current Bayesian network $G(Y)$. Then its inclusion will decrease the entropy of the whole network by a factor

$$H(X_j|Pa(X_j)) - H(X_j|Pa(X_j) \cup \{X_i\}), \tag{7.5}$$

while increasing the network complexity by one. So the entropy decrease per edge is given in Eq.(7.5). How good is this measure?

We can compare this entropy decrease to the average decrease that can be expected by including one more edge for such a network. Theoretically, the maximum entropy in a network with n vertices is $\sum_{i=0}^{n-1} H(X_i|L)$ where no edges (or causality relationships) are exploited. The minimum entropy is where all $n(n-1)$ edges are included, so the minimum entropy is $\sum_{i=0}^{n-1} H(X_i|X - \{X_i\}, L)$. Because the reduction of entropy is achieved by including all edges, per-edge drop in network entropy is thus

Table 7.1: Pseudo-code MCMC structure learning algorithm

Initiation:
    $Y \leftarrow$ tree structures from an MST algorithm
Loop until maximum iteration count is reached:
    for each $Y_t \in Y$
        $Y_t^{new} \leftarrow$ choose Y_t using Eq. 7.5 and Eq. 7.6
        $Y \leftarrow Y - \{Y_t\} \cup \{Y_t^{new}\}$
        $\#Y \rightarrow \#Y + 1$
    end for

$$\frac{\sum_{i=0}^{n-1}[H(X_i|L) - H(X_i|X - \{X_i\},L)]}{n(n-1)}. \tag{7.6}$$

Hence, the merit of including the particular edge (X_i, X_j) (Eq.(7.5)) can be compared to the expected average expressed in Eq.(7.6). If the drop in system entropy is larger than the average, we accept this addition; otherwise, we do not. A similar case can be made for the decision that an edge is to be deleted from the network.

The effort in computing Eqs.(7.5) and (7.6) theoretically can be high (though Eq.(7.6) need only be computed once, and that can be done off-line). This is because $X - \{X_i\}$ contains $s_0 \cdots s_{i-1} \ s_{i+1} \cdots s_{n-1}$ bins, and that many conditional distributions of X_i need to be computed and averaged. However, in reality we can simplify the computation by having the pair-wise conditional entropy, $H(X_j|X_i)$, serve as the upper bound for how much an attribute X_i can tell us about X_j. (Because similar information about X_j might be derived from other attributes, the contribution from X_i might be less.) Hence, we compute and tabulate all pair-wise conditional entropy $H(X_j|X_i), 0 \le i, j < n, i \ne j$ (an $O(n^2)$ operation), and eliminate spurious or weak connections with low mutual information. Say, if we accept only the top 10% of $H(X_i|X_j)$, then the expected number of parents per vertice drops from n to $n/3$. We can further impose a hard threshold on the number of parents that a vertice is allowed to have, to guarantee a minimal level of performance for sample generation even in the worst case.

Finally, we mention in passing that because Gibbs sampling can be viewed as a special kind of Metropolis method, which satisfies the *detailed balance* property, the probability distribution of the sequence of states generated will approach $P(Y|L)$. To ensure faster convergence, we propose to use initial configurations that are known to be good. For example, both Prim'a and Kruskal'a minimum-cost spanning tree (MST) algorithms can be used to generate initial diagram configurations in the form of trees. Optimal MST algorithms have been developed for which each vertice can have at most one parent; they run in $O(n^2 \log n)$ time. The pseudo-code MCMC algorithm for structure learning is presented in Table 7.1.

7.4.2 Causal Strength

Once the topology of the influence diagram has been determined, we assign causal strengths to the edges of the diagram. Usually, the edges of the diagram have been assigned as conditional probabilities. Conditional probability reflects the correlation between a cause variable and a decision variable, not the causation between them [11]. For instance, statistics collected in several school districts recently indicated a strong correlation between *asian students* and *good in math*. Many parents concluded that "Asians are good in math." The study performed by the John Hopkins talent-youth program concludes differently. The study shows that first-generation immigrants from both Asia and Europe are both competitive in math. Since recently numerous immigrants have arrived from Asia, strong correlation exists between *Asian* and *good in math*. However, according to the study, the second generation Asians do not do as well in math as the first generation. Another example is that *eating ice cream* and *swimming* may exhibit covariation. But until we have also considered *temperature* as a cause, the picture of causation will not be complete. (High temperature causes people to swim and consume more ice cream.) These examples show that correlation reflects only the covariation between variables in the data collected; they cannot be soundly interpreted as causation.

To precisely account for causation, researchers must collect data that contain all identifiable causes, and then compute causal strengths between the causes and the effect in an accurate way. However, even with the best effort, one may not have the knowledge to identify all decision variables completely. In other words, there might be a confounding variable or some hidden variable that mediates between two variables. (In the example of *eating ice cream* and *swimming*, *temperature* is a hidden cause, and some other variables, e.g., *humidity*, *income*, etc., could be unknown confounding causes of *swimming* or *eating ice cream*.) Nevertheless, causal analysis can be conducted effectively as long as the information left over is not significant. For photo annotation, our domain knowledge suffices to consider most variables to ensure the process to be effective.

How to we account for causal strength? Essentially, for a decision variable $d \in D$ and a chance variable $u \in U$, the condition probability $P(u|d, \xi)$, where d is independent of ξ, can be computed from the data[2]. Once we have computed $P(u|d, \xi)$, the causal strength is defined[3] by [13] as

$$CS_{u|d} = \frac{P(u|d, \xi) - P(u|\xi)}{1 - P(u|\xi)}. \tag{7.7}$$

The key of causal strength is that it considers not only the generative strength of d but also the generative strength when d is absent $(P(u|\xi))$. Intuitively, if both the presence and the absence of d give u an equal chance to occur, we cannot say

---

[2] In general, when two variables u and d are dependent, we cannot tell which causes which. For photo annotation, we can determine the direction of the arcs based on domain knowledge.

[3] We changed the term $P(u|\bar{d}, \xi)$ in [13] to $P(u|\xi)$ in the formula, because $\bar{d}$ could be interpreted as the negation (instead of absence) of d.

that d causes u. (Factors in ξ other than d have caused u.) In addition, in order to isolate the effect of d, the causal strength is normalized by $1 - P(u|\xi)$, which is the maximum amount which d can contribute to cause u.

When $P(u|d,\xi) - P(u|\xi) > 0$, we can say d causes u with generative strength of $CS_{u|d}$. When $P(u|d,\xi) - P(u|\xi) < 0$, the presence of d actually prevents the occurrence of u. The preventive strength of d on u is

$$CS_{u|d} = \frac{P(u|\xi) - P(u|d,\xi)}{P(u|\xi)}, \tag{7.8}$$

and $CS_{u|d} < 0$.

When multiple causes are involved, we assume that they are independent of each other, and hence we can compute causal strength without considering interactions between causes.

7.4.3 Case Study

We present a case study on landmark recognition using the influence diagram. Recognizing objects (landmarks or people) in a photo remains a very challenging computer-vision research problem. However, with available contextual information, such as time, location, and a person's social network, recognizing objects among an accordingly limited set of candidates is not such a daunting task. For example, given the photo-taking location as the Stanford campus, the interesting landmarks in the photo are limited. With candidate objects being limited, matching becomes easier, and matching time becomes shorter, so we can afford to employ more expensive matching techniques for improving accuracy.

For landmark recognition, we use two modalities: *location* and *local features*. Based on the domain knowledge, we can construct an inference diagram with *location* and *local features* as the two causes, and *landmarks* as the effect. (For semantics that are more complicated, we must learn their influence diagrams from data. Section 7.5 presents several influence diagrams learned from training data.) However, we still need to compute causal strengths between *location* and *landmarks*, and between *local features* and *landmarks* based on training data, which we will discuss shortly.

For local features, we use SIFT presented in Section 7.3.2. Our testbed was obtained from the Internet and a Stanford collection [7]. The dataset was constructed by collecting photographs taken by visitors to the Stanford Visitor Center. All these Stanford photos were annotated with GPS information. From this dataset, we used a subset containing $1,000$ images, and added 60 more taken on Stanford campus, and $13,500$ images of landmarks from all over the world, downloaded from the Internet. All images were rescaled to 320×240, before SIFT-feature extraction was performed.

We made three queries on the dataset: "Hoover Tower," "Memory Church," and "Rodin Garden." Each query used 10 images containing the landmark taken at different times, and from different angles. For each query, we used one of the ten images as the query, and the remaining nine were mingled in the 14,530-image dataset. We performed this leave-one-out query for each of the ten images, and for three landmarks.

When considering just *location*, the landmark-recognition accuracy is 30%. This is because the location information in this case can narrow the number of candidate landmarks to three only. Also, a photo might not contain a landmark or a landmark might be occluded by a tree or other objects. Let lm_i denote the ith landmark ($i = 1, 2, 3,$), and α denote *location*. We obtained $P(lm_i|\alpha) = 0.3$. This low probability seems to imply that the location information is not very helpful for landmark recognition. We will show that *causal strength* is a much better measure for accessing the contribution of a cause.

Next, let us examine the effectiveness of SIFT features. We processed each image in the dataset individually, and extracted SIFT features. Given a query image, we extracted its SIFT features and matched the features with those of images that contained candidate landmarks. We used a distance threshold to separate likely matches from non-matches. If the distance (computed using the SIFT features) between the query image and a landmark sample came within the threshold, we considered the landmark to be a possible match. If no possible match was found (after comparing with all landmark samples), we concluded that the image contained no landmark. Otherwise, the best match was used to annotate the image. Let β denote the SIFT features. We obtain conditional probability $P(lm_i|\beta) = 0.75$. In other words, we can recognize the three query landmarks with 75% accuracy without *location*.

In the final experiment, when we combined *location* (α) and SIFT (β) for recognizing landmarks, the accuracy improved to 95%. To account for causal strengths, we attribute the strength of *location* as

$$CS_{(lm_i|\alpha)} = \frac{P(lm_i|\alpha, \beta) - P(lm_i|\beta)}{1 - P(lm_i|\beta)} = 80\%,$$

and the strength of SIFT as

$$CS_{(lm_i|\beta)} = \frac{P(lm_i|\beta, \alpha) - P(lm_i|\alpha)}{1 - P(lm_i|\alpha)} = 93\%.$$

Discussion

Let us compare the effectiveness of using conditional probability and using causal strength for measuring causality. Using conditional probability, we find that the number of candidate landmarks (denoted as $|lm|$) affects $P(lm_i|\alpha)$ significantly. When $|lm| = 1$, the conditional probability approaches one; when $|lm| \geq 10$, the conditional probability dips below 10%. The conditional probability of $P(lm_i|\alpha)$

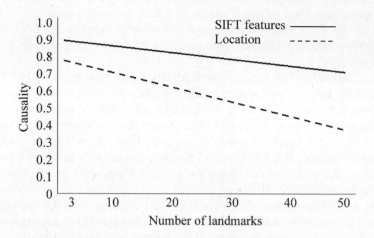

Fig. 7.5: Causal strengths of location & SIFT

is independent of the number of candidate landmarks. The conditional probability $P(lm_i|\beta) = 75\%$ is independent of $|lm|$. Conditional probability implies that when $|lm| > 10$, the location information may not be a reliable indicator for landmark recognition. It also implies that SIFT always maintains the same predictive power independent of the number of candidate landmarks. These interpretations based on conditional probability do not seem reasonable.

When we use causal strength to explain the inference diagram, we get much better insight as to how *location* or SIFT affects landmark recognition. Figure 7.5 plots causal strength with respect to $|lm|$ for *location* and SIFT, respectively. When $|lm| = 3$, $P(lm_i|\beta, \xi = \alpha) = 95\%$; when $|lm| = 50$, $P(lm_i|\beta, \xi = \alpha) = 85\%$. The causal strength of *location* decreases from 0.8 to about 0.6 when the number of landmarks increases from 3 to 25. This result means that even if we just know the city (instead of a more specific location) where a photo is taken, the location information is still very useful. (Most cities do not have more than 25 landmarks.) Next, we observe that the causal strength of SIFT reduces as $|lm|$ increases. This is also intuitively understandable, since the larger the number of landmarks, the more difficult for the SIFT features to tell different landmarks apart. Causality strength is much more intuitive for explaining the relationships between causes and effects. As a result, we can accurately quantify the *confidence* of a landmark annotation, which leads to more effective fusion and retrieval performance. Our empirical study in Section 7.5 will show more evidence and explanation of improved results.

7.4.4 Dealing with Missing Attributes

As we have discussed in Section 7.1, some metadata might not be available in a photo, e.g., camera parameters were not collected when a photo was taken. Bayesian network generates a probability distribution over all possible existent values. The probability distribution represents the implicit uncertainty in the estimation of missing values. Either the entire distribution or the most common value that has the highest probability may then be selected as the replacement for the missing value [36]. Instead of substituting the missing values, we can estimate the posterior probability of the missing values by prior probability. Given that f_i is unknown, the posterior probability $P(C|f_i)$ of the data belonging to class C can be replaced with the prior probability $P(C)$. These methods cannot provide exact prediction for a missing attribute, but they provide the most accurate guesses statistically, and hence can improve annotation accuracy.

7.5 Experiments

Our experiments were designed to evaluate the effectiveness of using *influence diagram* for fusing context, content and semantics in the photo annotation task. Specifically, we wanted to answer the following questions:

1. Can varied influence structures of different semantic concepts be successfully learned?
2. Can causal strength improve the effectiveness of photo annotation compared with conditional probability?
3. Can semantic ontology help improve annotation accuracy through transitive inference?
4. How well can an inference diagram cope with missing metadata?

We conducted our experiments on two real-world photo datasets: one is a small dataset with 3k photos and the other is a larger one with 24k photos. Our datasets were obtained from the Internet and personal collections.

Dataset #1: 3k photo dataset. The first dataset contains 3k photos with completed content and contextual information, collected from photos taken by friends and family. We made seven queries on the dataset: "outdoor", "indoor", "beach", "sunset", "ski", "soccer", and "lecture." Figure 7.6 lists the number of photos in each category in terms of the semantic ontology structure.

Dataset #2: 24k photo dataset. The second dataset contains the 3k photos in the first dataset and 21k extra photos with content information only. The context information for these 21k is missing.

In our experiments, we utilized *time* and *camera parameters* as contextual information, in which camera parameters included *exposure time, aperture, f-number, shutter speed, flash and focal length*. These camera features were shown to be useful for scene classification in [6]. The first four features reflect the brightness of a photo.

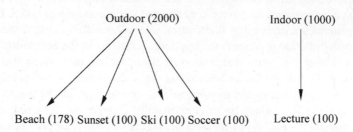

Fig. 7.6: Number of photos in each category

Natural lighting is stronger than artificial lighting. This causes outdoor scenes to be brighter than indoor scenes, even under overcast skies, and they, therefore, have a shorter exposure time and a larger brightness value. Because of the lighting differences described above, (automatic and manual) camera flash is used on a much higher percentage of indoor scenes than of outdoor scenes. Focal length is related to subject distance in less direct and intuitive ways through camera zoom.

We utilized *holistic perceptual feature* as content information. We extracted holistic perceptual features including color, texture, and shape from each photo. We first divided color into twelve color bins, of which eleven were for the eleven culture colors [33] and the remaining bin for outlier colors. For each culture color, we extracted color features of multiple resolutions, which included color histograms, color means and variances in HSV channels, color spreadness, and color-blob elongation. For texture extraction, we used discrete wavelet transformation because of its computational efficiency. Texture descriptors such as texture elongation, texture spreadness, energy means and energy variances were calculated in three orientations (vertical, horizontal, and diagonal) and three resolutions (coarse, medium, and fine). The semantic ontology in Figure 7.2 was employed as the domain knowledge for deriving the semantic relationships between our seven query categories.

We randomly selected a percentage of data from each dataset to be used as training examples. The remaining data were used for testing. For each dataset, the training/testing ratio was empirically chosen via cross-validation so that the sampling ratio worked best in our experiments. To perform class prediction, we employed the one-per-class (OPC) ensemble [37], which trains all the classifiers, each of which predicts the class membership for one class. We employed NIST *Average Precision* [38] to evaluate the accuracy of each concept model (classifier). The results presented here were the average of 10 runs.

7.5.1 Experiment on Learning Structure

This set of experiments reported the influence diagrams of different query concepts by using Bayesian multi-net. Figure 7.7 lists the influence diagrams for seven concepts[4]

The input of *visual features* is pseudo probabilistic by applying a sigmoid function to the output of a Support Vector Machine [39]. The input of *camera parameters* is either binary (e.g., flash on or off) or discrete (e.g., focal length is quantized into discrete intervals, and time is segmented into morning, noon, afternoon, evening).

Figure 7.7 depicts different semantics in different influence diagram structures. Use Figure 7.7(a) as an example to illustrate. Camera metadata and content-based features are complementary for deciding the semantic of "outdoor." Although this resulting structure did not surprise us, the figure shows that only four camera parameters are useful to distinguish "outdoor." This runs counter to the assumption made in [6] that all camera parameters should be used for annotating photos, independent of their semantics. Figure 7.7 shows that different semantics are best inferred by different combinations of camera parameters, together with the perceptual features. The learning process helps identify the true *causes*. More importantly, being able to learn an inference diagram allows us to use *causal Markov condition* to simplify causal inference. Let us revisit Figure 7.7(a). We can draw an inference on whether an image is "outdoor" using nodes *camera parameters* and *visual features* on the diagram. According to *causal Markov condition*, the parents of these two nodes are conditionally independent of "outdoor."

7.5.2 Experiment on Causal Strength Inference

This set of experiments examined the effectiveness of multimodal fusion on the 3k photo dataset by applying causal strength. We randomly picked 60% of images for training; the remaining 40% were used for testing data. (The 60%/40% ratio was determined through cross validation.) For each concept, we utilized the influence diagram structure learned in the previous section and fused content information and contextual information. We compared the fusion results when employing conditional probability and causal strength respectively. This experiment was used to provide insights into understanding the effectiveness of causal strength in comparison to conditional probability.

Table 7.2 lists the average precision in each of seven categories when using multimodal fusion for recognizing photos. For each category, employing causal strength for fusion (CP_{Fusion}) consistently improved the accuracy compared to using conditional probability for fusion (CS_{Fusion}). The improvement is 4.7% on average. This

---

[4] To conserve space, we draw the influence diagrams only using context and content features. Relationships between semantic labels can be found in Figure 7.2.

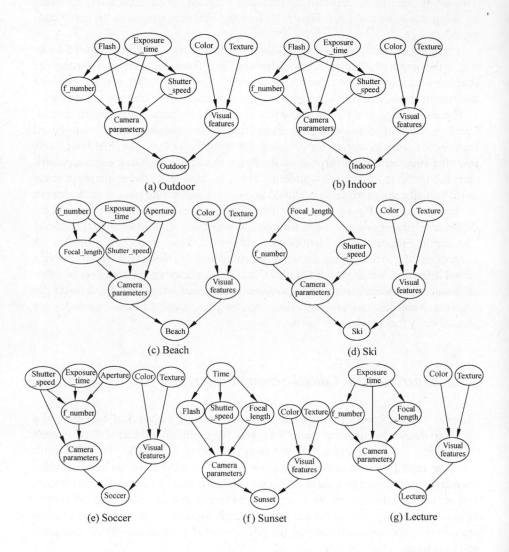

Fig. 7.7: Inference diagrams

observation again indicates that improvement can be made by using causal strength to better quantify the confidence of photo annotation.

To further illustrate the effectiveness of causal strength for photo recognition, Figures 7.8 and 7.9 present the precision/recall curves for "indoor" and "soccer"

Table 7.2: Average precisions of photo recognition on 3k dataset using Bayesian multimodal fusion

Category	CP_{Fusion}	CS_{Fusion}
Outdoor	0.9284	0.9691
Indoor	0.8691	0.9103
Beach	0.8877	0.9118
Sunset	0.7850	0.8791
Ski	0.9823	1.0000
Soccer	0.8968	0.9423
Lecture	0.9241	0.9534
Average	0.8912	0.9380

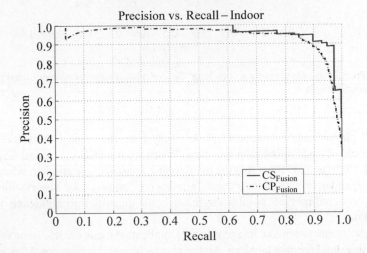

Fig. 7.8: Precision/recall curves for recognizing "Indoor" photos when using conditional probability and causal strength

photo recognition, respectively. (Because of space limitations, we presented only two categories.) In both figures, the precision/recall curves perform better when employing causal strength compared to those when employing conditional probability.

To provide more detailed information, Figure 7.10 depicts the confidence distributions for "soccer" photo recognition using $CP_{Context}$, $CS_{Context}$, $CP_{Content}$ and $CS_{Content}$. $CP_{Context}$ represents employing conditional probability on contextual information; $CP_{Content}$ represents employing conditional probability on content information; $CS_{Context}$ represents employing causal strength on contextual information; $CS_{Content}$ represents employing causal strength on content information. Figures 7.10 (a) and (c) show that when using $CP_{Context}$ and $CP_{Content}$, significant overlaps occur between the confidence scores of "soccer" and "non-soccer". In other words, many

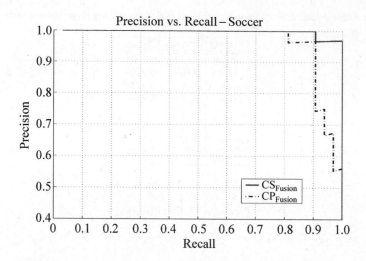

Fig. 7.9: Precision/recall curves for recognizing "Soccer" photos when using conditional probability and causal strength

"soccer" photos and "non-soccer" photos may have the same confidence scores, which could degrade annotation accuracy. When we used $CS_{Context}$ and $CS_{Content}$, Figure 7.10 (b) and (d) show that the overlapping areas become smaller, which leads to a better separation of "soccer" and "non-soccer" photos. Our observations from these figures confirm that causal strength can better quantify the confidence of photo annotation.

For photo annotation and image retrieval applications, one critical issue is how to select a threshold for data labeling. All the photos (images) whose confidence scores larger than the threshold will be annotated as belonging to a specific concept. Figure 7.11 illustrates the threshold variances (y-axis) for different semantic concepts. The x-axis in the figure represents four situations: $CP_{Context}$, $CP_{Content}$, $CS_{Context}$ and $CS_{Content}$. The four points connected by the line in the figure represent mean thresholds for seven categories under four different situations. We also plotted the standard deviations of thresholds for seven categories. We see that the mean threshold attained by using conditional probability is around 0.31 for contextual information and 0.39 for content information, which doesn't seem reasonable. (Intuitively, we will annotate a photo as belonging to a concept when the confidence score is sufficiently high, say higher than 0.5.) In contrast, the mean threshold selected by using causal strength makes more sense, which is around 0.52 for contextual information and 0.56 for content information. Furthermore, the standard deviations of thresholds for different concepts when using conditional probability are very high. This observation indicates the difficulty of selecting a good threshold for photo annotation due to the instability. By employing causal strength, the thresholds for different concepts are much more stable, making them more suitable for real applications.

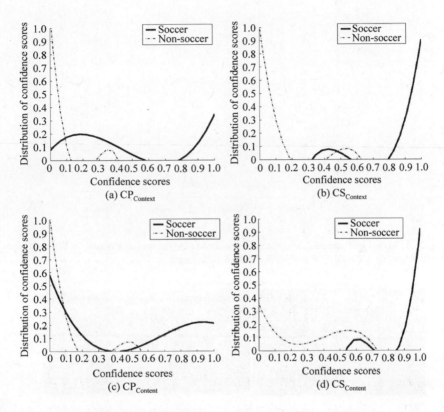

Fig. 7.10: Distribution of confidence scores for "Soccer"

7.5.3 *Experiment on Semantic Fusion*

The third set of experiments examined the effectiveness of using semantic ontology for photo annotation. Each single concept detector is learned independently by fusing context and content information as shown in the previous section. Then, the confidence scores from individual concept detectors will be adjusted by connecting the influence relations between concepts based on the predefined semantic ontology. For example, Figure 7.12(a) shows a photo being misclassified as "sunset." The confidence score for this photo from "sunset" photo detector is 0.55 (the threshold is 0.51 for "sunset" photo detector). The confidence score for this photo from "outdoor" classifier is very low, 0.13. Since a "sunset" photo has to be an "outdoor" photo, if we take the influence from "outdoor" detector, this photo should not be classified as a "sunset" photo.

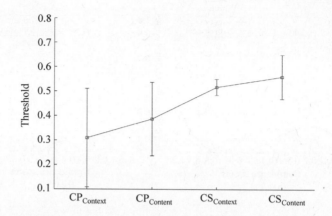

Fig. 7.11: Means and standard deviations of thresholds for recognizing photos in different categories when using conditional probability and causal strengths

(a) Misclassified "sunset"   (b) Misclassified "lecture"

Fig. 7.12: Misclassified photos

Figure 7.12(b) shows a photo being misclassified as "lecture." The confidence score for this photo from "lecture" photo detector is 0.61 (the threshold is 0.60 for "lecture" photo detector). The confidence score for this photo from "indoor" classifier is 0.20. Since a "lecture" photo has to be an "indoor" photo, if we take the influence from "indoor" detector, this photo should not be classified as a "lecture" photo.

We employed ontology-based multi-classification learning [40] to estimate the new confidence score of data s belonging to a concept C_i ($P'(s|C_i)$) by a linear interpolation of all hierarchy nodes from the root to the concept C_i (see Eq.(7.9)).

$$P'(s|C_i) = P(s|C_i) + \sum_{j=1}^{n} \lambda_j P(s|C_j). \qquad (7.9)$$

Table 7.3: Average precisions of photo recognition on 3k dataset using semantic fusion

Category	$Before_{SF}$	$After_{SF}$
Beach	0.9118	0.9321
Sunset	0.8791	0.8912
Ski	1.0000	1.0000
Soccer	0.9423	0.9590
Lecture	0.9534	0.9625
Average	0.9373	0.9471

$P(s|C_i)$ is the original confidence score of s belonging to C_i and $\{C_j, j = 1, 2, \cdots, n\}$ is a set of concepts that are the ancestors of C_i in the semantic ontology. The weighting parameters $\{\lambda_j, j = 1, 2, \cdots, n\}$ are calculated based on the correlation of confidence scores between C_i and C_j. Table 7.3 lists the average precision in each of five categories (these five categories are located at leaf nodes in the semantic ontology) after we employed the semantic fusion. The *mean average precision* of photo recognition on the five categories was improved by utilizing semantic fusion.

7.5.4 Experiment on Missing Features

We evaluated the effectiveness of fusion for missing features on the 3k dataset and the 24k dataset. Section 7.4.4 lists a couple of ways to fill in the missing attributes in a Bayesian network. In our experiment, we replaced missing values with the most common values.

For the 3k photo dataset, we used the same training data and testing data as in Section 7.5.2. We randomly deleted *contextual information* from $m\%$ of the testing data, and formulated new testing data M_1. Similarly, we deleted *content information* from the randomly selected photos, and formulated testing dataset M_2.

Figure 7.13 compares the mean average precision (the mean of average precisions for 7 categories) of photo recognition using causal strength for Bayesian fusion when the *contextual information* was partially missing (M_1) and when the *content information* (M_2) was partially missing. The values of m varied from 10 to 50. From the figure, we can see that missing *content information* ($M_2(\text{CS}_{\text{Fusion}})$) or missing *contextual information* ($M_1(\text{CS}_{\text{Fusion}})$) result in lower performance compared to having all the information present (the third column in Table 7.2). Missing *content information* causes worse performance than missing the same percentage of *contextual information*. This observation indicates that when the causal strength is weak for a certain attribute, missing that attribute might not affect accuracy as much as missing another attribute with a much higher causal strength.

For testing the 24k large photo dataset (which does not have any contextual information), we randomly selected 80% from the 3k photos with both contextual

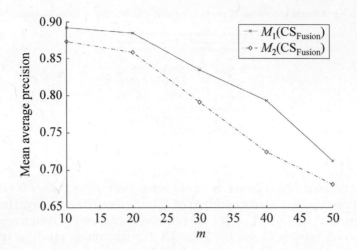

Fig. 7.13: Mean average precision of photo recognition on 3k dataset with $m\%$ missing attributes.

Table 7.4: Average precisions of photo recognition on 24k dataset with missing attributes

Category	$CP_{Content}$	CP_{Fusion}	CS_{Fusion}
Outdoor	0.6615	0.7216	0.7429
Indoor	0.5703	0.6033	0.6387
Beach	0.7509	0.7793	0.8654
Sunset	0.6758	0.7010	0.6943
Ski	0.7016	0.7932	0.8247
Soccer	0.8689	0.8564	0.9030
Lecture	0.8537	0.9043	0.8745
Average	0.7261	0.7656	0.7919

and content information as training data. Table 7.4 lists the average precision in recognizing different photo categories using multimodal fusion. Compared to using content only with conditional probability ($CP_{Content}$), employing conditional probability for fusion (CP_{Fusion}) can improve the accuracy by 4.0% on average. Employing causal strength for fusion (CS_{Fusion}) improves the accuracy by around 6.4% on average.

7.6 Concluding Remarks

We have presented a powerful fusion framework using multimodal information to annotate photos. We showed how an influence diagram can be learned for each semantic label, and how causalities can be accurately accounted for. Through our empirical study, we demonstrated not only the high annotation accuracy achieved by our fusion technique, but also its better capability to assess annotation quality. Our fusion framework can be installed in the cameras, at the cell-phone providers, or users' desktops to provide useful metadata for photos, and we believe that this fusion technique will play a critical role in enabling effective personal photo organization and search applications.

This work can be extended in several directions. First, more effective visual features such as the ones depicted in Chapter 2 can be used to describe photos. Second, causal models can be extended to account for high-order interactions between causes, although treating causes conditionally independent of each other seems to suffice for photo annotation.

References

1. Wu, Y., Chang, E.Y., Tseng, B.L. Multimodal metadata fusion using causal strength. In *Proceedings of ACM Multimedia*, pages 872–881, 2005.
2. Manjunath, B.S., Ma, W.Y. Texture features for browsing and retrieval of image data. *IEEE Transactions on Pattern Analysis and Machine Intelligence*, 18:837–842, 1996.
3. Rui, Y., Huang, T.S., Chang, S.F. Image retrieval: Current techniques, promising directions and open issues. *Journal of Visual Communication and Image Representation*, 1999.
4. Lowe, D.G. Object recognition from local scale-invariant features. In *Proceedings of IEEE ICCV*, pages 1150–1157, 1999.
5. Lowe, D.G. Distinctive image features from scale-invariant keypoints. *International Journal of Computer Vision*, 60(2):91–100, 2004.
6. Boutell, M., Luo, J. Bayesian fusion of camera metadata cues in semantic scene classification. In *Proceedings of IEEE CVPR*, pages 623–630, 2004.
7. Naaman, M., Paepcke, A., Garcia-Molina, H. From where to what: Metadata sharing for digital photographs with geographic coordinates. In *Proceedings of the International Conference on Cooperative Information Systems (CoopIS)*, pages 196–217, 2003.
8. Chang, E.Y. Extent: Fusing context, content, and semantic ontology for photo annotation. In *Proceedings of ACM Workshop on Computer Vision Meets Databases (CVDB) in conjunction with ACM SIGMOD*, pages 5–11, 2005.
9. Heckerman, D., Shachter, R. Decision-theoretic foundations for causal reasoning. *Microsoft Technical Report MSR-TR-94-11*, 1994.
10. Heckerman, D. A bayesian approach to learning causal networks. In *Proceedings of the Conference on Uncertainty in Artificial Intelligence*, pages 107–118, 1995.
11. Pearl, J. *Causality: Models, Reasoning and Inference*. Cambridge University Press, 2000.
12. Pearl, J. Causal inference in the health sciences: A conceptual introduction. *Special issue on causal inference, Kluwer Academic Publishers, Health Services and Outcomes Research Methodology*, 2:189–220, 2001.
13. Novick, L.R., Cheng, P.W. Assessing interactive causal influence. *Psychological Review*, 111(2):455–485, 2004.
14. Tong, S., Chang, E. Support vector machine active learning for image retrieval. In *Proceedings of ACM International Conference on Multimedia*, pages 107–118, October 2001.
15. Barnard, K., Forsyth, D. Learning the semantics of words and pictures. pages 408–415, 2000.
16. Wang, J.Z., Li, J., Wiederhold, G. Simplicity: Semantics-sensitive integrated matching for picture libraries. In *Proceedings of ACM Multimedia*, pages 483–484, 2000.

17. Davis, M., King, S., Good, N., Sarvas, R. From context to content: Leveraging context to infer media metadata. In *Proceedings of the ACM International Conference on Multimedia*, pages 188–195, 2004.
18. Dey, A.K. Understanding and using context. *Personal and Ubiquitous Computing Journal*, 5(1):4–7, 2001.
19. Diomidis, D.S. Position-annotated photographs: a geotemporal web. *IEEE Pervasive Computing*, 2(2), 2003.
20. Naaman, M., Harada, S., Wang, Q., Garcia-Molina, H., Paepcke, A. Context data in geo-referenced digital photo collections. In *Proceedings of ACM International Conference on Multimedia*, pages 196–203, 2004.
21. Jain, R., Sinha, P. Content without context is meaningless. In *Proceedings of ACM Multimedia*, pages 1259–1268, 2010.
22. http://www.exif.org.
23. Stricker, M., Orengo, M. Similarity of color images. In *Proc. SPIE Storage and Retrieval for Image and Video Databases*, 1995.
24. Smith, J.R., Chang, S.F. Tools and techniques for color image retrieval. In *SPIE Proceedings Storage and Retrieval for Image and Video Databases IV*, 1995.
25. Rui, Y., She, A.C., Huang, T.S. Modified fourier descriptors for shape representations- a practical approach. In *Proc. of First International Workshop on Image Databases and Multi Media Search*, 1996.
26. Ke, Y., Sukthankar, R. Pca-sift: A more distinctive representation for local image descriptors. In *Proceedings of IEEE CVPR*, 2004.
27. Khan, L., McLeod, D. Effective retrieval of audio information from annotated text using ontologies. In *Proceedings of Workshop of Multimedia Data Mining with ACM SIGKDD*, pages 37–45, 2000.
28. Smith, J.R., Chang, S.F. Visually searching the web for content. *IEEE Multimedia*, 4(3):12–20, 1997.
29. Deng, J., Dong, W., Socher, R., Li, L., Li, K., Li, F.F. Imagenet: A large-scale hierarchical image database. In *Proceedings of IEEE CVPR*, pages 156–161, 2009.
30. Williamson, J. *Causality, in Dov Gabbay & F. Guenthner (eds.): Handbook of Philosophical Logic*. Kluwer, 2005.
31. Geiger, D., Heckerman, D. Knowledge representation and inference in similarity networks and bayesian multinets. *Artificial Intelligence*, 82:45–74, 1996.
32. Friedman, N., Geiger, D., Goldszmidt, M. Bayesian network classifiers. *Machine Learning*, 29:131–161, 1997.
33. Goldstein, E.B. *Senstation and Perception (5th edition)*. Wadsworth, 1999.
34. Friedman, N., Koller, D. Learning bayesian networks from data (tutorial). In *Proceedings of NIPS*, 2000.
35. Tenenbaum, J.B., Griffiths, T.L. Generalization, similarity, and bayesian inference. *Behavioral and Brain Sciences*, 24:629–641, 2001.
36. Doshi, P.J., Greenwald, L.G., Clarke, J.R. Using bayesian networks for cleansing trauma data. In *Proc. of FLAIRS Conference*, pages 72–76, 2003.
37. Dietterich, T., Bakiri, G. Solving multiclass learning problems via error-correcting output codes. *Artifical Intelligence Research*, 2:263–286, 1995.
38. NIST. Common evaluation measures. *Appendix in Special Publication 500-250 (TREC 2001)*, 2001.
39. Platt, J. Probabilistic outputs for svms and comparisons to regularized likelihood methods. In *Advances in Large Margin Classifiers*. MIT press, 1999.
40. Wu, Y., Tseng, B.L., Smith, J.R. Ontology-based multi-classification learning for video concept detection. In *Proceedings of the IEEE International Conference on Multimedia and Expo*, pages 1003–1006, 2004.

Chapter 8
Combinational Collaborative Filtering, Considering Personalization

Abstract For the purpose of multimodal fusion, collaborative filtering can be regarded as a process of finding relevant information or patterns using techniques involving collaboration among multiple views or data sources. In this chapter[†], we present a collaborative filtering method, *Combinational Collaborative Filtering* (CCF), to perform recommendations by considering multiple types of co-occurrences from different information sources. CCF differs from the approaches presented in Chapters 6 and 7 by constructing a latent layer in between the recommended objects and multimodal descriptions of these objects. We use community recommendation throughout this chapter as an example to illustrate critical design points. We first depict a community by two modalities: a collection of documents and a collection of users, respectively. CCF fuses these two modalities through a latent layer. We show how the latent layer is constructed, how multiple modalities are fused, and how the learning algorithm can be both effective and efficient in handling massive amount of data. CCF can be used to perform virtually any multimedia-data recommendation tasks such as recommending labels to images (annotation), recommending images to images (clustering), and images to users (personalized search).

Keywords: Collaborative filtering, multimodal fusion, personalization, recommendation systems, semantic gap, social media.

8.1 Introduction

Collaborative filtering is a method of filtering for information or patterns using techniques involving collaboration among multiple agents, viewpoints, data sources, etc. (defined in Wikipedia). Collaborative filtering can be regarded as a push model of search. A search engine provides recommendations to the users proactively based on information collected from the user and her social networks. In this chapter,

---

we tackle the problem of *community recommendation* for social networking sites. (One can substitute community with any data object such as image, video, or music, and the techniques remain the same.) What differentiates our work from prior work is that we propose a fusion method, which combines information from multiple sources. We name our method CCF for *Combinational Collaborative Filtering*. CCF views a community from two simultaneous perspectives: *a bag of users* and *a bag of words*. A community is viewed as a bag of participating users; and at the same time, it is viewed as a bag of words describing that community. Traditionally, these two views are independently processed. Fusing these two views provides two benefits. First, by combining *bags of words* with *bags of users*, CCF can perform *personalized* community recommendations, which the *bags of words* alone model cannot. Second, augmenting *bags of users* with *bags of words*, CCF improves data density and hence can achieve better personalized recommendations than the *bags of users* alone model.

A practical recommendation system must be able to handle large-scale data sets and hence demands scalability. We devise two strategies to speed up training of CCF. First, we employ a hybrid training strategy, which combines Gibbs sampling with the Expectation-Maximization (EM) algorithm. Our empirical study shows that Gibbs sampling provides better initialization for EM, and thus can help EM to converge to a better solution at a faster pace. Our second speedup strategy is to parallelize CCF to take advantage of the distributed computing infrastructure of modern data centers.

Though in this chapter we use community recommendation as our target application, one can simply replace communities with multimedia data e.g., images. An image can be depicted as a bag of pixels, a bag of contextual information, or a bag of users who have assessed the image. CCF can then fuse these modalities to perform recommendation tasks such as recommending labels to images (annotation), recommending images to images (clustering), and images to users (personalized search). The techniques presented in Chapters 6 are suitable for fusing metadata at the lowest, syntactic layer such as color, shape, and texture descriptions. Chapter 7 presents a model to fuse content with context. This chapter considers fusion with semantics, which is a layer above the syntactic ones.

The remainder of this chapter is organized as follows. In Section 8.2, we discuss the related work on probabilistic latent aspect models. In Section 8.3, we present CCF, including its model structure and semantics, hybrid training strategy, and parallelization scheme. In Section 8.4, we present our experimental results on both synthetic and Orkut data sets. We provide concluding remarks and discuss future work in Section 8.5.

8.2 Related Reading

Several algorithms have been proposed to deal with either *bags of words* or *bags of users*. Specifically, Probabilistic Latent Semantic Analysis (PLSA) [2] and Latent Dirichlet Allocation (LDA) [3] model document-word co-occurrence, which is similar to the *bags of words* community view. Probabilistic Hypertext Induced Topic Selection (PHITS) [4], a variant of PLSA, models document-citation co-occurrence, which is similar to the *bags of users* community view. However, a system that considers just bags of users cannot take advantage of content similarity between communities. A system that considers just bags of words cannot provide personalized recommendations: all users who joined the same community would receive the same set of recommendations. We propose CCF to model multiple types of data co-occurrence simultaneously. CCF's main novelty is in fusing information from multiple sources to alleviate the information sparsity problem of a single source.

Several other algorithms have been proposed to model publication and email data[1]. For instance, the *author-topic* (AT) model [5] employs two factors in characterizing a document: the document's authors and topics. Modeling both factors as variables within a Bayesian network allows the AT model to group the words used in a document corpus into semantic topics, and to determine an author's topic associations. For emails, the *author-recipient-topic* (ART) model [6] considers email recipient as an additional factor. This model can discover relevant topics from the sender-recipient structure in emails, and enjoys an improved ability to measure role-similarity between users. Although these models fit publication and email data well, they cannot be used to formulate personalized community recommendations, whereas CCF can.

8.3 CCF: Combinational Collaborative Filtering

We start by introducing the baseline models. We then show how our CCF model combines baseline models. Suppose we are given a collection of co-occurrence data consisting of communities $C = \{c_1, c_2, ..., c_N\}$, community descriptions from vocabulary $D = \{d_1, d_2, ..., d_V\}$, and users $U = \{u_1, u_2, ..., u_M\}$. If community c is joined by user u, we set $n(c, u) = 1$; otherwise, $n(c, u) = 0$. Similarly, we set $n(c, d) = R$ if community c contains word d for R times; otherwise, $n(c, d) = 0$. The following models are latent aspect models, which associate a latent class variable $z \in Z = \{z_1, z_2, ..., z_K\}$.

Before modeling CCF, we first model community-user co-occurrences (C-U), shown in Figure 8.1(a); and community-description co-occurrences (C-D), shown in Figure 8.1(b). Our CCF model, shown in Figure 8.1(c), builds on C-U and C-D models. The shaded and unshaded variables in Figure 8.1 indicate latent and observed variables, respectively. An arrow indicates a conditional dependency between variables.

---

[1] We discuss only related model-based work since the model-based approach has been proven to be superior to the memory-based approach.

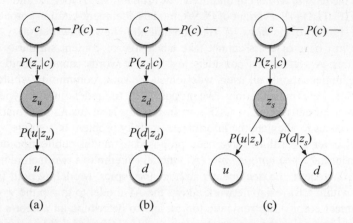

Fig. 8.1: (a) Graphical representation of the Community-User (C-U) model. (b) Graphical representation of the Community-Description (C-D) model. (c) Graphical representation of Combinational
Collaborative Filtering (CCF) that combines both bag of users and bag of words information.

8.3.1 C-U and C-D Baseline Models

The C-U model can be derived from PLSA for community-user co-occurrence analysis. The co-occurrence data consists of a set of community-user pairs (c,u), which
are assumed to be generated independently. The key idea is to introduce a latent
class variable z to every community-user pair, so that community c and user u are
rendered conditionally independent. The resulting model is a mixture model that
can be written as follows:

$$P(c,u) = \sum_z P(c,u,z) = P(c) \sum_z P(u|z)P(z|c), \qquad (8.1)$$

where z represents the topic for a community. For each community, a set of users
is observed. To generate each user, a community c is chosen uniformly from the
community set, then a topic z is selected from a distribution $P(z|c)$ that is specific to
the community, and finally a user u is generated by sampling from a topic-specific
distribution $P(u|z)$.

The second model is for community-description co-occurrence analysis. It has a
similar structure to the C-U model with the joint probability written as:

$$P(c,d) = \sum_z P(c,d,z) = P(c) \sum_z P(d|z)P(z|c), \qquad (8.2)$$

where z represents the topic for a community. Each community's interests are modeled with a mixture of topics. To generate each description word, a community c is

chosen uniformly from the community set, then a topic z is selected from a distribution $P(z|c)$ that is specific to the community, and finally a word d is generated by sampling from a topic-specific distribution $P(d|z)$. (One can model C-U and C-D using LDA. Please see Chapter 12 for indepth discussion on LDA.)

8.3.2 CCF Model

In the C-U model, we consider only *links*, i.e., the observed data can be thought of as a very sparse binary $M \times N$ matrix W, where $W_{i,j} = 1$ indicates that user i joins (or linked to) community j, and the entry is unknown elsewhere. Thus, the C-U model captures the linkage information between communities and users, but not the community content. The C-D model learns the topic distribution for a given community, as well as topic-specific word distributions. This model can be used to estimate how similar two communities are in terms of topic distributions. Next, we introduce our CCF model, which combines both the C-U and C-D.

For the CCF model (Figure 8.1(c)), the joint probability distribution over community, user, and description can be written as:

$$P(c,u,d) = \sum_z P(c,u,d,z)$$
$$= P(c) \sum_z P(u|z)P(d|z)P(z|c). \tag{8.3}$$

The CCF model represents a series of probabilistic generative processes. Each community has a multinomial distribution over topics, and each topic has a multinomial distribution over users and descriptions, respectively.

8.3.3 Gibbs & EM Hybrid Training

Given the model structure, the next step is to learn model parameters. There are some standard learning algorithms, such as Gibbs sampling [7], Expectation-Maximization (EM) [8], and Gradient descent. For CCF, we propose a hybrid training strategy: We first run Gibbs sampling for a few iterations, then switch to EM. The model trained by Gibbs sampling provides the initialization values for EM. This hybrid strategy serves two purposes. First, EM suffers from a drawback in that it is very sensitive to initialization. A better initialization tends to allow EM to find a "better" optimum. Second, Gibbs sampling is too slow to be effective for large-scale data sets in high-dimensional problems [9]. A hybrid method can enjoy the advantages of Gibbs and EM.

Gibbs sampling

Gibbs sampling is a simple and widely applicable Markov chain Monte Carlo algorithm, which provides a simple method for obtaining parameter estimates and allows for combination of estimates from several local maxima of the posterior distribution. Instead of estimating the model parameters directly, we evaluate the posterior distribution on z and then use the results to infer $P(u|z)$, $P(d|z)$ and $P(z|c)$.

For each user-word pair, the topic assignment is sampled from:

$$P(z_{i,j} = k|u_i = m, d_j = n, \mathbf{z}_{-i,-j}, U_{-i}, D_{-j}) \propto$$
$$\frac{C_{mk}^{UZ} + 1}{\sum_{m'} C_{m'k}^{UZ} + M} \frac{C_{nk}^{DZ} + 1}{\sum_{n'} C_{n'k}^{DZ} + V} \frac{C_{ck}^{CZ} + 1}{\sum_{k'} C_{ck'}^{CZ} + K}, \tag{8.4}$$

where $z_{i,j} = k$ represents the assignment of the ith user and j^{th} description word in a community to topic k. $u_i = m$ represents the observation that the ith user is the mth user in the user corpus, and $d_j = n$ represents the observation that the jth word is the nth word in the word corpus. $\mathbf{z}_{-i,-j}$ represents all topic assignments not including the ith user and the jth word. Furthermore, C_{mk}^{UZ} is the number of times user m is assigned to topic k, not including the current instance; C_{nk}^{DZ} is the number of times word n is assigned to topic k, not including the current instance; C_{ck}^{CZ} is the number of times topic k has occurred in community c, not including the current instance.

We analyze the computational complexity of Gibbs sampling in CCF. In Gibbs sampling, one needs to compute the posterior probability

$$P(z_{i,j} = k|u_i = m, d_j = n, \mathbf{z}_{-i,-j}, U_{-i}, D_{-j})$$

for user-word pairs $(M \times L)$ within N communities, where L is the number of words in community description (Note $L \geq V$). Each $P(z_{i,j} = k|u_i = m, d_j = n, \mathbf{z}_{-i,-j}, U_{-i}, D_{-j})$ consists of K topics, and requires a constant number of arithmetic operations, resulting in $O(K \cdot N \cdot M \cdot L)$ for a single Gibbs sampling. During parameter estimation, the algorithm needs to keep track of a topic-user $(K \times M)$ count matrix, a topic-word $(K \times V)$ count matrix, and a community-topic $(N \times K)$ count matrix. From these count matrices, we can estimate the topic-user distributions $P(u_m|z_k)$, topic-word distributions $P(d_n|z_k)$ and community-topic distributions $P(z_k|c_c)$ by:

$$P(u_m|z_k) = \frac{C_{mk}^{UZ} + 1}{\sum_{m'} C_{m'k}^{UZ} + M},$$

$$P(d_n|z_k) = \frac{C_{nk}^{DZ} + 1}{\sum_{n'} C_{n'k}^{DZ} + V},$$

$$P(z_k|c_c) = \frac{C_{ck}^{CZ} + 1}{\sum_{k'} C_{ck'}^{CZ} + K}, \tag{8.5}$$

where $P(u_m|z_k)$ is the probability of containing user m in topic k, $P(d_n|z_k)$ is the probability of using word n in topic k, and $P(z_k|c_c)$ is the probability of topic k

occurring in community c. The estimation of parameters by Gibbs sampling replaces the random seeding in EM's initialization step.

Expectation-Maxmization algorithm

The CCF model is parameterized by $P(z|c)$, $P(u|z)$, and $P(d|z)$, which are estimated using the EM algorithm to fit the training corpus with community, user, and description by maximizing the log-likelihood function:

$$L = \sum_{c,u,d} n(c,u,d) \log P(c,u,d), \tag{8.6}$$

$$n(c,u,d) = n(c,u)n(c,d) = \begin{cases} R, & \text{if community } c \text{ has user } u \\ & \text{and contains word } d \text{ for } R \text{ times;} \\ 0, & \text{otherwise.} \end{cases} \tag{8.7}$$

Starting with the initial parameter values from Gibbs sampling, the EM procedure iterates between Expectation (E) step and Maximization (M) step:

- **E-step**: where the probability that a community c has user u and contains word d explained by the latent variable z is estimated as:

$$P(z|c,u,d) = \frac{P(u|z)P(d|z)P(z|c)}{\sum_{z'} P(u|z')P(d|z')P(z'|c)}. \tag{8.8}$$

- **M-step**: where the parameters $P(u|z)$, $P(d|z)$, and $P(z|c)$ are re-estimated to maximize L in Eq.(refequ:ucfLikelihood):

$$P(u|z) = \frac{\sum_{c,d} n(c,u,d)P(z|c,u,d)}{\sum_{c,u',d} n(c,u',d)P(z|c,u',d)}, \tag{8.9}$$

$$P(d|z) = \frac{\sum_{c,u} n(c,u,d)P(z|c,u,d)}{\sum_{c,u,d'} n(c,u,d')P(z|c,u,d')}, \tag{8.10}$$

$$P(z|c) = \frac{\sum_{u,d} n(c,u,d)P(z|c,u,d)}{\sum_{u,d,z'} n(c,u,d)P(z'|c,u,d)}. \tag{8.11}$$

We analyze the computational complexity of the E-step and the M-step. In the E-step, one needs to compute the posterior probability $P(z|c,u,d)$ for M users, N communities, and V words. Each $P(z|c,u,d)$ consists of K values, and requires a constant number of arithmetic operations to be computed, resulting in $O(K \cdot N \cdot M \cdot V)$ operations for a single E-step. In the M-step, the posterior probabilities are accumulated to form the new estimates for $P(u|z)$, $P(d|z)$ and $P(z|c)$. Thus, the M-step also requires $O(K \cdot N \cdot M \cdot V)$ operations. Typical values of K in our experiments range from 28 to 256. The community-user (c,u) and community-description (c,d) co-occurrences are highly sparse, where $n(c,u,d) = n(c,u) \times n(c,d) = 0$ for a large percentage of the triples (c,u,d). Because the $P(z|c,u,d)$ term is never separated from the $n(c, u, d)$ term in the M-step, we do not need to compute $P(z|c, u, d)$ for $n(c, u, d) = 0$ in the E-step. We compute only $P(z|c, u, d)$ for $n(c, u, d) \neq 0$. This greatly reduces computational complexity.

8.3.4 Parallelization

The parameter estimation using Gibbs sampling and the EM algorithm described in the previous sections can be divided into parallel subtasks. We consider Message Passing Interface (MPI) for implementation as it is more suitable for parallelizing iterative algorithms than MapReduce. Since standard MPI implementations (MPICH2) cannot be directly ported to our system, we implemented our own system by modifying MPICH2 [10].

Parallel Gibbs sampling

We distribute the computation among machines based on community IDs. Thus, each machine i only deals with a specified subset of communities c_i, and is aware of all users u and all descriptions d. We then perform Gibbs sampling simultaneously on each machine independently and update local counts. Afterward, each machine *reduces* the local difference $(C_{m_i k}^{UZ} - C_{mk}^{UZ}, C_{n_i k}^{DZ} - C_{nk}^{DZ})$ to a specified root, then the root *broadcasts* the global difference (sum of all local differences) to other machines to update global counts (C_{mk}^{UZ} and C_{nk}^{DZ}) [11]. This is an *MPI_AllReduce* operation in MPI. We summarize the process in Figure 8.3.4.

Parallel EM algorithm

The parallel EM algorithm can be applied in a similar fashion. We describe the procedure below and summarize the process in Figure 8.3.

- **E-step:** Each machine i computes the $P(z|c_i, u, d)$ values, the posterior probability of the latent variables z given communities c_i, users u and descriptions d, using the current values of the parameters $P(z|c_i)$, $P(u|z)$ and $P(d|z)$. As this posterior computation can be performed locally, we avoid the need for communications between machines in the E-step.
- **M-step:** Each machine i computes the local parameters $P(z|c_i)$, $P(u_i|z)$ and $P(d_i|z)$ using the previously calculated values $P(z|c_i, u, d)$. After that, each machine *reduces* the local parameters $(P(u_i|z), P(d_i|z))$ to a specified root, and the root *broadcasts* the global parameters to other machines. This is done through a *MPI_AllReduce* operation in MPI.

We analyze the computational and communication complexities for both algorithms using distributed machines. Assuming that there are P machines, the com-

Input: $N \times M$ community-user matrix; $N \times V$ community-description matrix; I: number of iterations; P: number of machines
Output: $P(u|z)$, $P(d|z)$, $P(z|c)$
Variables:
x_{ic}: the ith row of community-user matrix with community id c
y_{ic}: the ith row of community-word matrix with community id c
1: **for** $i = 0$ to $N - 1$ **do**
2:     Load x_{ic} into machine $c\%P$.
3:     Load y_{ic} into machine $c\%P$.
4: **end for**
5: Gibbs sampling initialization.
6: **for** $iter = 0$ to $I - 1$ **do**
7:     **for** each <user, word> pair **do**
8:         Each machine i performs Gibbs sampling as in Eq. (8.4) and updates local counts $C_{m_ik}^{UZ}$, $C_{n_ik}^{DZ}$ and $C_{c_ik}^{CZ}$.
9:     **end for**
10:     Each machine *reduces* the local difference to a specified root, and root *broadcasts* the global difference to others to update global counts:
11:         $C_{mk}^{UZ} = C_{mk}^{UZ} + \sum_i (C_{m_ik}^{UZ} - C_{mk}^{UZ})$
12:         $C_{nk}^{DZ} = C_{nk}^{DZ} + \sum_i (C_{n_ik}^{DZ} - C_{nk}^{DZ})$
13: **end for**

Fig. 8.2: Parallel Gibbs Sampling of CCF

putational complexity of each training algorithm reduces to $O((K \cdot N \cdot M \cdot L)/P)$ (for Gibbs) and $O((K \cdot N \cdot M \cdot V)/P)$ (for EM) since P machines share the computations simultaneously. For communication complexity, two variables are reduced and broadcasted among P machines for next iteration training: C_{mk}^{UZ}, C_{nk}^{DZ} in Gibbs sampling, and $P(u|z)$, $P(d|z)$ in EM. The communication cost is $O(\alpha \cdot \log P + \beta \cdot \frac{P-1}{P} K(M+V) + \gamma \cdot \frac{P-1}{P} K(M+V))$, where α is the startup time of a transfer, β is the transfer time per byte, and γ is the computation time per byte for performing the reduction operation locally on any machine.

8.3.5 Inference

Once we have learned the model parameters, we can infer three relationships using Bayesian rules, namely user-community relationship, community similarity, and user similarity. We derive these three relationships as follows:

- **User-community relationship:** Communities can be ranked for a given user according to $P(c_j|u_i)$, *i.e.* which communities should be recommended for a given user? Communities with top ranks and communities that the user has not yet joined are good candidates for recommendations. $P(c_j|u_i)$ can be calculated using Eq.(8.12):

Input: $N \times M$ community-user matrix; $N \times V$ community-description matrix; I: number of iterations; P: number of machines; $P(u|z)$, $P(d|z)$, $P(z|c)$ of Gibbs sampling
Output: $P(u|z)$, $P(d|z)$, $P(z|c)$
Variables:
x_{ic}: the ith row of community-user matrix with community id c
y_{ic}: the ith row of community-word matrix with community id c
1: Load $P(u|z)$, $P(d|z)$, $P(z|c)$ of Gibbs sampling.
2: **for** $i = 0$ to $N-1$ **do**
3:      Load x_{ic} into machine $c\%P$.
4:      Load y_{ic} into machine $c\%P$.
5: **end for**
6: **for** $iter = 0$ to $I-1$ **do**
7:      **for** $k = 0$ to $K-1$ **do**
8:          **E-step**:
9:          Each machine i computes $P(z_k|u,c_i,d)$
10:         **M-step**:
11:         Each machine i computes parameters $P(z_k|c_i)$, $P(u_i|z_k)$, $P(d_i|z_k)$, and *reduces* local parameters to a specific root, then root *broadcasts* the global parameters to others:
12:              $P(u|z_k) = \sum_i P(u_i|z_k)$
13:              $P(d|z_k) = \sum_i P(d_i|z_k)$
14:     **end for**
15: **end for**

Fig. 8.3: Parallel EM algorithm of CCF

$$
\begin{aligned}
P(c_j|u_i) &= \frac{\sum_z P(c_j, u_i, z)}{P(u_i)} \\
&= \frac{P(c_j) \sum_z P(u_i|z) P(z|c_j)}{P(u_i)} \\
&\propto \sum_z P(u_i|z) P(z|c_j),
\end{aligned}
\qquad (8.12)
$$

where we assume that $P(c_j)$ is a uniform prior for simplicity.

- **Community similarity**: Communities can also be ranked for a given community according to $P(c_j|c_i)$. We calculate $P(c_j|c_i)$ using Eq.(8.13):

$$
\begin{aligned}
P(c_j|c_i) &= \frac{\sum_z P(c_j, c_i, z)}{P(c_i)} \\
&= \frac{\sum_z P(c_j|z) P(c_i|z) P(z)}{P(c_i)} \\
&= P(c_j) \sum_z \frac{P(z|c_j) P(z|c_i)}{P(z)} \\
&\propto \sum_z \frac{P(z|c_j) P(z|c_i)}{P(z)},
\end{aligned}
\qquad (8.13)
$$

where we assume that $P(c_j)$ is a uniform prior for simplicity.

- **User similarity:** Users can be ranked for a given user according to $P(u_j|u_i)$, *i.e.* which users should be recommended for a given user? Similarly, we can calculate $P(u_j|u_i)$ using Eq.(8.14):

$$
\begin{aligned}
P(u_j|u_i) &= \frac{\sum_z P(u_j, u_i, z)}{P(u_i)} \\
&= \frac{\sum_z P(u_j|z)P(u_i|z)P(z)}{P(u_i)} \\
&= P(u_j) \sum_z \frac{P(z|u_j)P(z|u_i)}{P(z)} \\
&\propto \sum_z \frac{P(z|u_j)P(z|u_i)}{P(z)},
\end{aligned}
\tag{8.14}
$$

where we assume that $P(u_j)$ is a uniform prior for simplicity.

8.4 Experiments

We divided our experiments into two parts. The first part was conducted on a relatively small synthetic dataset with ground truth to evaluate the Gibbs & EM hybrid training strategy. The second part was conducted on a large, real-world dataset to test out CCF's performance and scalability. Our experiments were run on up to 200 machines at our distributed data centers. While not all machines are identically configured, each machine is configured with a CPU faster than 2GHz and memory larger than 4GB (a typical Google configuration in 2007).

8.4.1 Gibbs + EM vs. EM

To precisely account for the benefit of Gibbs & EM over the EM-only training strategy, we used a synthetic dataset where we know the ground truth. The synthetic dataset consists of $5,000$ documents with 10 topics, a vocabulary size $10,000$, and a total of $50,000,000$ word tokens. The true topic distribution over each document was pre-defined manually as the ground truth. We conducted the comparisons using the following two training strategies: (1) EM-only strategy (without Gibbs sampling as initialization) where the number of EM iterations is 10 through 100 respectively, (2) Gibbs & EM stragety where the number of Gibbs sampling iterations is 5, 10, 15 and 20, and the number of EM iterations is 10 through 70, respectively. We used Kullback-Leibler divergence (K-L divergence) to evaluate model performance since the K-L divergence is a good measure for the difference between the true topic distribution (P) and the estimated topic distribution (Q) defined as follows:

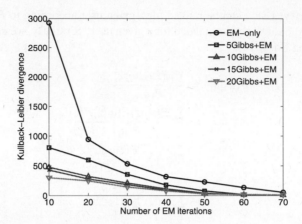

Fig. 8.4: The Kullback-Leibler divergence as a function of the number of iterations

$$D_{KL}(P||Q) = \sum_i P(i) \log \frac{P(i)}{Q(i)}. \tag{8.15}$$

The smaller the K-L divergence is, the better the estimated topic distribution approximates the true topic distribution.

Figure 8.4 compares the average K-L divergences over 10 runs. It shows that more rounds of Gibbs sampling can help EM reach a solution that enjoys a smaller K-L divergence. Since each iteration of Gibbs sampling takes longer than EM, we must also consider *time*. Figure 8.5 shows the values of K-L divergence as a function of the training time, where EM-only strategy began with 20 EM iterations. We can make two observations. First, given a large amount of time, both EM and the hybrid scheme can reach very low K-L divergence. On this dataset, when the training time exceeded 350 seconds, the value of K-L divergence approached zero for all strategies. Nevertheless, on a large dataset, we cannot afford a long training time, and the Gibbs & EM hybrid strategy provides a earlier point to stop training, and hence reduces the overall training time.

The second observation is on the number of Gibbs iterations. As shown in both figures, running more iterations of Gibbs before handing over to EM takes longer to yield a better initial point for EM. In other words, spending more time in the Gibbs stage can save time in the EM stage. Figure 8.5 shows that the best performance was produced by 10 iterations of Gibbs sampling before switching to EM. Finding the "optimal" switching point is virtually impossible in theory. However, the figure shows that different Gibbs iterations can all outperform the EM-only strategy to obtain a better solution early, and a reasonable number of Gibbs iterations can be obtained through an empirical process like our experiment. Moreover, the figure shows that a range of number of iterations can achieve similar K-L divergence (e.g., at time 250). This indicates that though an empirical process may not be able to pin

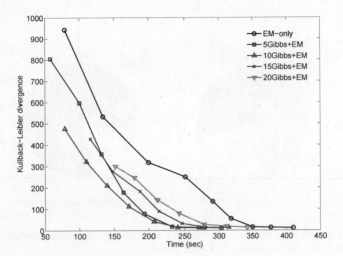

Fig. 8.5: The Kullback-Leibler divergence as a function of the training time

down the "optimal" number of iterations (because of e.g., new training data arrival), the hybrid scheme can work well on a range of Gibbs-sampling iterations.

8.4.2 The Orkut Dataset

Orkut is an extremely active community site with more than two billion page views a day world-wide. The dataset we used was collected on July 26, 2007, which contains two types of data for each community: community membership information and community description information. We restrict our analysis to English communities only. We collected 312,385 users and 109,987 communities[2]. The number of entries in the community-user matrix, or the number of community-user pairs, is 35,932,001. As the density is around 0.001045, this matrix is extremely sparse. Figure 8.6(a) shows a distribution of the number of users per community. About 52% of all communities have less than 100 users, whereas 42% of all communities have more than 100 but less than 1,000 users.

For the community description data, after applying downcasing, stopword filtering, and word stemming, we obtained a vocabulary of 191,034 unique English words. The distribution of the number of description words per community is displayed in Figure 8.6(b). On average, there are 27.64 words in each community description after processing. In order to establish statistical significance of the findings, we repeated all experiments 10 times with different random seeds and parameters, such as the number of latent aspects (ranging from 28 to 256), the number of Gibbs

---

[2] All user data were anonymized, and user privacy is safeguarded, as performed in [12].

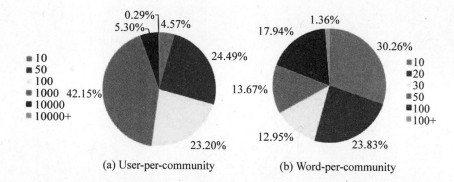

Fig. 8.6: (a) Distribution of the number of users per community, and (b) distribution of the number of description words per community (See color insert)

sampling iterations (ranging from 10 to 30) and the number of EM iterations (ranging from 100 to 500). The reported results are the average performance over all runs.

Results

Community recommendation: $P(c_j|u_i)$. We use two standard measures from information retrieval to measure the recommendation effectiveness: *precision* and *recall*, defined as follows:

$$Precision = \frac{|\{recommendation\ list\} \cap \{joined\ list\}|}{|\{recommendation\ list\}|},$$

$$Recall = \frac{|\{recommendation\ list\} \cap \{joined\ list\}|}{|\{joined\ list\}|}. \tag{8.16}$$

Precision takes all recommended communities into account. It can also be evaluated at a given cut-off rank, considering only the topmost results recommended by the system. As it is possible to achieve higher recall by recommending more communities (note that a recall of 100% is trivially achieved by recommending all communities, albeit at the expense of having low precision), we limit the size of our community recommendation list to at most 200.

To evaluate the results, we randomly deleted one joined community for each user in the community-user matrix from the training data. We evaluated whether the deleted community could be recommended. This evaluation is similar to *leave-one-out*. Figure 8.7 shows the precision and recall as functions of the length (up

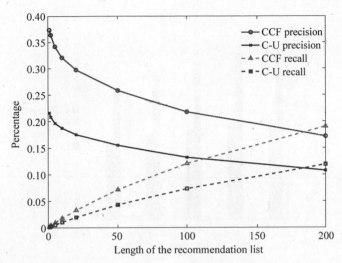

Fig. 8.7: The precision and recall as functions of the length of the recommendation list (See color insert)

to 200) of the recommendation list for both C-U and CCF. We can see that CCF always outperforms C-U for all lengths. Figure 8.8 presents precision and recall for the top 20 recommended communities. As both precision and recall of CCF are nearly twice higher than those of C-U, we can conclude that CCF enjoys better prediction accuracy than C-U. This is because C-U only considers community-user co-occurrence, whereas CCF considers users, communities, and descriptions. By taking other views into consideration, the information is denser for CCF to achieve higher prediction accuracy.

Figure 8.9 depicts the relationship between the precision of the recommendation for a user and the number of communities that the user has joined. The more communities a user has joined, the better both C-U and CCF can predict the user's preferences. For users who joined around 100 communities, the precision is about 15% for C-U and 27% for CCF. However, for users who joined just 20 communities, the precision is about 7% for C-U, and 10% for CCF. This is not surprising since it is very difficult for latent-class statistical models to generalize from sparse data. For large-scale recommendation systems, we are unlikely to ever have enough direct data with sufficient coverage to avoid sparsity. However, at the very least, we can try to incorporate indirect data to boost our performance, just as CCF does by using bags of words information to augment bags of users information.

Remark: Because of the nature of leave-one-out, our experimental result can only show whether a joined community could be recovered. The low precision/recall reflects this necessary, restrictive experimental setting. (This setting is necessary for objectivity purpose as we cannot obtain ground-truth of all users' future pref-

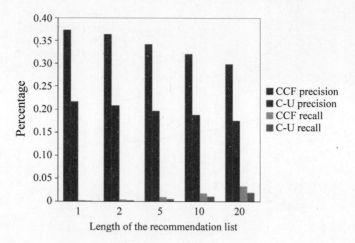

Fig. 8.8: The precision and recall as functions of the length (up to 20) of the recommendation list (See color insert)

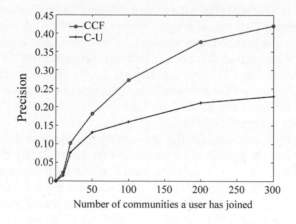

Fig. 8.9: The precision as a function of the number of communities a user has joined. Here, the length of the recommendation list is fixed at 20

erences.) The key observation from this study is not the absolute precision/recall values, but is the relative performance between CCF and C-U.

Community similarity: $P(c_j|c_i)$. We next report the results of community similarities calculated by the three models. We used *community category* (available at Orkut websites) as the ground-truth for clustering communities. We also assigned

Table 8.1: The comparison results of the three models using Normalized Mutual Information (NMI)

Model	C-U	C-D	CCF
NMI	0.4508	0.3127	0.4526

Table 8.2: The top recommended users using the C-U and CCF models for the query user 79. The number of communities that user 79 joined is 339. (Note that the "Communities" field contains three numbers: the first number n is the total number of communities a user joined; the second number k is the number of overlapping communities between the recommended user and the query user, and the last number is percentage of $\frac{k}{n}$.)

Model	Rank 1st		Rank 2nd		Rank 3rd	
	User ID	Communities	User ID	Communities	User ID	Communities
C-U	2390	551 (102, 18.5%)	8207	456 (100, 21.9%)	6734	494 (95, 19.2%)
CCF	7931	518 (106, 20.5%)	10968	680 (102, 15.0%)	6776	680 (91, 13.4%)

each community an estimated label for the latent aspect with the highest probability value. We treated communities with the same estimated label as members of the same *community cluster*. We then compared the difference between community clusters and categories using the Normalized Mutual Information (NMI).

NMI between two random variables CAT (category label) and CLS (cluster label) is defined as $NMI(CAT;CLS) = \frac{I(CAT;CLS)}{\sqrt{H(CAT)H(CLS)}}$, where $I(CAT;CLS)$ is the mutual information between CAT and CLS. The entropies $H(CAT)$ and $H(CLS)$ are used for normalizing the mutual information to be in the range $[0,1]$. In practice, we made use of the following formulation to estimate the NMI score [13]:

$$NMI = \frac{\sum_{s=1}^{K} \sum_{t=1}^{K} n_{s,t} \log \left(\frac{n \cdot n_{s,t}}{n_s \cdot n_t} \right)}{\sqrt{\left(\sum_s n_s \log \frac{n_s}{n} \right) \left(\sum_t n_t \log \frac{n_t}{n} \right)}}, \tag{8.17}$$

where n is the number of communities, n_s and n_t denote the numbers of community in category s and cluster t, $n_{s,t}$ denotes the number of community in category s as well as in cluster t. The NMI score is 1 if the clustering results perfectly match the category labels and 0 for a random partition. Thus, the larger this score, the better the clustering results.

Table 8.1 shows that CCF slightly outperforms both C-U and C-D models, which indicates the benefit of incorporating two types of information.

User similarity: $P(u_j|u_i)$. An interesting application is friend suggestion: finding users similar to a given user. Using Eq.(8.14), we can compute user similarity for all pairs of users. From these values, we derive a ranking of the most similar users for a given query user. Due to privacy concerns, we were not able to obtain the friend

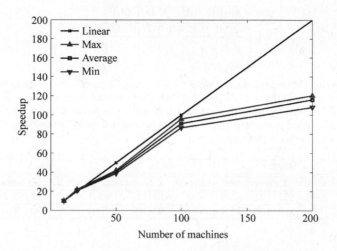

Fig. 8.10: Speedup analysis for different number of machines

graph of each user to evaluate accuracy. Table 8.2 shows an example of this ranking for a given user.

"Similar" users typically share a significant percentage of commonly-joined communities. For instance, the query user also joined 18.5% of the communities joined by the top user ranked by C-U, compared to 20.5% for CCF. It is encouraging to see that CCF's top ranked user has more overlap with the query user than C-U's top ranked user does. We believe that, again, incorporating the additional word co-occurrences has improved information density and hence yields higher prediction accuracy.

8.4.3 Runtime Speedup

In analyzing runtime speedup for parallel training, we trained CCF with 20 latent aspects, 10 Gibbs sampling, and 20 EM iterations. As the size of a dataset is large, a single machine cannot store all the data — ($P(u|z)$, $P(d|z)$, $P(z|c)$, and $P(z|c,u,d)$) — in its local memory, we cannot obtain the running time of CCF on one machine. Therefore, we use the runtime of 10 machines as the baseline and assume that 10 machines can achieve 10 times speedup. This assumption is reasonable as we will see shortly that our parallelization scheme can achieve linear speedup on up to 100 machines. Table 8.3 and Figure 8.10 report the runtime speedup of CCF using up to 200 machines. The Orkut dataset enjoys a linear speedup when the number of ma-

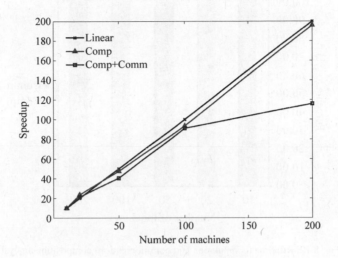

Fig. 8.11: Speedup and overhead analysis

Table 8.3: Runtime comparisons for different number of machines

Machines	Time (sec.)	Speedup
10	9,233	10
20	4,326	21.3
50	2,280	40.5
100	1,014	91.1
200	796	116

chines is up to 100. After that, adding more machines receives diminishing returns. This result led to our examination of overheads for CCF, presented next.

No parallel algorithm can infinitely achieve linear speedup because of the Amdahl's law. When the number of machines continues to increase, the communication cost starts to dominate the total running time. The running time consists of two main parts: computation time (Comp) and communication time (Comm). Figure 8.11 shows how Comm overhead influences the speedup curves. We draw on the top the computation only line (Comp), which approaches the linear speedup line. The speedup deteriorates when communication time is accounted for (Comp + Comm). Figure 8.12 shows the percentage of Comp and Comm in the total running time. As the number of machines increases, the communication cost also increases. When the number of machines exceeds 200, the communication time becomes even larger than the computation time.

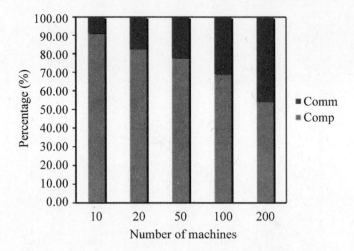

Fig. 8.12: Runtime (computation and communication) composition analysis

Though the Amdahl's law eventually kicks in to forbid a parallel algorithm to achieve infinite speedup, our empirical study draws two positive observations.

1. When the dataset size increases, the "saturation" point of the Amdahl's law is deferred, and hence we can add more machines to deal with larger sets of data.
2. The speedup that can be achieved by parallel CCF is very significant to enable near-real-time recommendations. As shown in the table, the parallel scheme reduces the training time from one day to less than 14 minutes. The parallel CCF can be run every 14 minutes to produce a new model to adapt to new access patterns and new users.

8.5 Concluding Remarks

This chapter has presented a generative graphical model, Combinational Collaborative Filtering (CCF), for collaborative filtering based on both bags of words and bags of users information. CCF uses a hybrid training strategy that combines Gibbs sampling with the EM algorithm. The model trained by Gibbs sampling provides better initialization values for EM than random seeding. We also presented the parallel computing required to handle large-scale data sets. Experiments on a large Orkut data set demonstrate the approaches to successfully produce better quality recommendations, and accurately cluster relevant communities/users with similar semantics.

There are several directions for future research. First, one can consider expanding CCF to incorporate more types of co-occurrence data. More types of co-occurrence data would help to overcome sparsity problem and make better recommendation. Second, in our analysis, the community-user pair value equals one, $i.e.$ $n(u_i, c_j) = 1$ (if user u_i joins community c_j). An interesting extension would be to give this count a different value, $i.e.$ $n(u_i, c_j) = f$, where f is the frequency of the user u_i visiting the community c_j. Third, as we have mentioned, one can replace PLSA with LDA (Chapter 12) or the causality strength model (Chapter 7) to conduct inference. Finally, CCF, as a general framework of combining multiple types of co-occurrence data, has many applications in information retrieval, social network mining, and other related areas.

References

1. Chen, W., Zhang, D., Chang, E.Y. Combinational collaborative filtering for personalized community recommendation. In *Proceedings of ACM SIGKDD*, pages 115–123, 2008.
2. Hofmann, T. Probabilistic latent semantic analysis. In *Proc. of the 15th UAI Conference*, pages 289–296, 1999.
3. Blei, D.M., Ng, A.Y., Jordan, M.I. Latent Dirichlet allocation. *Journal of Machine Learning Research*, 3:993–1022, 2003.
4. Cohn, D., Chang, H. Learning to probabilistically identify authoritative documents. In *Proc. of the 17th ICML Conference*, pages 167–174, 2000.
5. Steyvers, M., Smyth, P., Rosen-Zvi, M., Griffiths, T. Probabilistic author-topic models for information discovery. In *Proc. of the 10th ACM SIGKDD Conference*, pages 306–315, 2004.
6. McCallum, A., Corrada-Emmanuel, A., Wang, X. The author-recipient-topic model for topic and role discovery in social networks: Experiments with enron and academic email. Technical report, Computer Science, University of Massachusetts Amherst, 2004.
7. Geman, S., Geman, D. Stochastic relaxation, gibbs distributions, and the bayesian restoration of images. *IEEE Transactions on Pattern Analysis and Machine Intelligence*, 6:721–741, 1984.
8. Dempster, A.P., Laird, N.M., Rubin, D.B. Maximum likelihood from incomplete data via the EM algorithm. *Journal of the Royal Statistical Society. Series B (Methodological)*, 39(1):1–38, 1977.
9. Blei, D.M., Jordan, M.I. Variational methods for the Dirichlet process. In *Proc. of the 21st ICML Conference*, pages 373–380, 2004.
10. Gropp, W., Lusk, E., Skjellum, A. *Using MPI-2: Advanced Features of the Message-Passing Interface*. MIT Press,, 1999.
11. Newman, D., Asuncion, A., Smyth, P., Welling, M. Distributed inference for latent Dirichlet allocation. In *Proceedings of NIPS*, 2007.
12. Spertus, E., Sahami, M., Buyukkokten, O. Evaluating similarity measures: a large-scale study in the orkut social network. In *Proc. of the 11th ACM SIGKDD Conference*, pages 678–684, 2005.
13. Strehl, A., Ghosh, J. Cluster ensembles – a knowledge reuse framework for combining multiple partitions. *Journal on Machine Learning Research*, 3:583–617, 2002.

Chapter 9
Imbalanced Data Learning

Abstract An imbalanced training dataset can pose serious problems for many real-world data-mining tasks that conduct supervised learning. In this chapter[†], we present a kernel-boundary-alignment algorithm, which considers training-data imbalance as prior information to augment SVMs to improve class-prediction accuracy. Using a simple example, we first show that SVMs can suffer from high incidences of false negatives when the training instances of the target class are heavily outnumbered by the training instances of a non-target class. The remedy we propose is to adjust the class boundary by modifying the kernel matrix, according to the imbalanced data distribution. Through theoretical analysis backed by empirical study, we show that the kernel-boundary-alignment algorithm works effectively on several datasets.
Keywords: Imbalanced data, kernel alignment, SVMs.

9.1 Introduction

In many data-mining tasks, finding rare objects or events is of primary interest [2]. Some examples include identifying fraudulent credit-card transactions [3], diagnosing medical diseases, and recognizing suspicious activities in surveillance videos [4]. The task of finding rare objects or events is usually formulated as a supervised learning problem. Training instances are collected for both target and non-target events, and then a classifier is trained on the collected data to predict future instances. Researchers in the data-mining community have been using Support Vector Machines (SVMs) as the learning algorithm, since SVMs have strong theoretical foundations and excellent empirical successes in many pattern-recognition applications such as handwriting recognition [5], image retrieval [6], and text classification [7]. However, for rare-object detection and event mining, when the training

---
[†] ©IEEE, 2005. This chapter is written based on the author's work with Gang Wu [1] published in IEEE TKDE 17(6). Permission to publish this chapter is granted under copyright license #2587680962412.

instances of the target class are significantly outnumbered by the other training instances, the class-boundary learned by SVMs can be severely skewed toward the target class. As a result, the false-negative rate can be excessively high in identifying important target objects (e.g., a surveillance event or a disease-causing agent), and can result in catastrophic consequences.

Skewed class boundary is a subtle but serious problem that arises from using an SVM classifier — in fact from using *any* classifier — for real-world problems with imbalanced training data. To understand the nature of the problem, let us consider it in a binary classification setting (positive vs. negative). We know that the Bayesian framework estimates the posterior probability using the class conditional and the prior [8]. When the training data are highly imbalanced, the results naturally tend to favor the majority class. Hence, when ambiguity arises in classifying a particular sample because of similar class-conditional densities for the two classes, the Bayesian framework will rely on the large class prior favoring the majority class to break the tie. Consequently, the decision boundary will skew toward the minority class.

To illustrate this skew problem graphically, Figure 9.1 shows a 2D checkerboard example. The checkerboard divides a 200×200 square into four quadrants. The top-left and bottom-right quadrants are occupied by negative (majority) instances, but the top-right and bottom-left quadrants contain only positive (minority) instances. The lines between the classes represent the "ideal" boundary that separates the two classes. In the rest of this chapter, we will use *positive* when referring to minority instances, and *negative* when referring to majority instances.

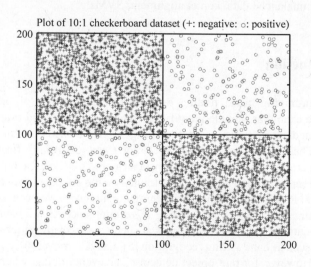

Fig. 9.1: Checkerboard experiment

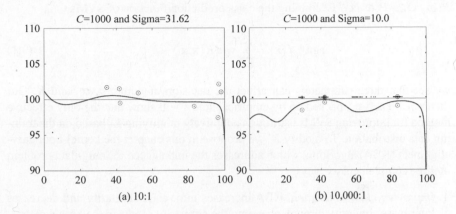

Fig. 9.2: Boundaries of different ratios. We use a Gaussian RBF kernel for training. For better illustration, we zoom into the area around the ideal boundary ($y = 100$) between left two quadrants. Only support vectors are shown in the figures.

Figure 9.2 exhibits the boundary distortion between the two left quadrants in the checkerboard under two different negative/positive training-data ratios, where a black dot with a circle represents a support vector, and its radius represents the weight value α_i of the support vector. The bigger the circle, the larger the α_i. Figure 9.2(a) shows the SVM class boundary when the ratio of the number of negative instances (in the quadrant above) to the number of positive instances (in the quadrant below) is 10:1. Figure 9.2(b) shows the boundary when the ratio increases to 10,000:1. The boundary in Figure 9.2(b) is much more skewed towards the positive quadrant than the boundary in Figure 9.2(a), thus causing a higher incidence of false negatives.

Although in a theoretical sense the Bayesian framework gives the optimal results (in terms of the smallest average error rate), we must be careful in applying it to real-world applications. In a real-world application such as security surveillance or disease diagnosis, the risk (or consequence) of mispredicting a positive event (a false negative) far outweighs that of mispredicting a negative event (a false positive). It is well known that in a binary classification problem, Bayesian risks are defined as:

$$R(\alpha_p|\mathbf{x}) = \lambda_{pp}P(\omega_p|\mathbf{x}) + \lambda_{pn}P(\omega_n|\mathbf{x})$$
$$R(\alpha_n|\mathbf{x}) = \lambda_{np}P(\omega_p|\mathbf{x}) + \lambda_{nn}P(\omega_n|\mathbf{x}),$$

where p refers to the positive events and n to the negative, λ_{np} refers to the risk (or cost) of a false negative, and λ_{pn} the risk of a false positive. The decisions about which action (α_p or α_n) to take — or which action has a smaller risk — are affected not just by the event likelihood (which directly influences the misclassification error), but also by the risk of mispredictions (λ_{np} and λ_{pn}).

How can we factor risk into SVMs to compensate for the effect caused by $P(\omega_n|\mathbf{x}) \gg P(\omega_p|\mathbf{x})$? Examining the class prediction function of SVMs,

$$\text{sgn}\left(f(\mathbf{x}) = \sum_{i=1}^{n} y_i \alpha_i K(\mathbf{x}, \mathbf{x}_i) + b \right), \qquad (9.1)$$

we see that three parameters can affect the decision outcome: b, α_i, and K. Our theoretical analysis, backed up by empirical study, will show that the only effective method for improving SVMs is through adaptively modifying K, based on the training data distribution. To modify K, we propose in this chapter the kernel-boundary-alignment (KBA) algorithm, which addresses the imbalanced training-data problem in three complementary ways:

1. *Improving class separation.* KBA increases intra-class similarity and decreases inter-class similarity through changing the similarity scores in the kernel matrix. Therefore, instances in the same class are better clustered in the feature space $\mathscr{F}$ away from those in the other classes.
2. *Safeguarding overfitting.* To avoid overfitting, KBA uses the existing support vectors to guide its boundary-alignment procedure.
3. *Improving imbalanced ratio.* By properly adjusting the similarity scores between majority instances, KBA can reduce the number of support vectors on the majority side and hence improve the imbalanced support-vector ratio.

Our experimental results on both UCI and real-world image/video datasets show the kernel-boundary-alignment algorithm to be effective in correcting a skewed boundary caused by imbalanced training data.

9.2 Related Reading

Approaches for addressing the imbalanced training-data problem can be divided into two main categories: the data processing approach and the algorithmic approach. The data processing approach [9] can be further sub-divided into two methods: under-sample the majority class, or over-sample the minority class. The one-sided selection proposed by Kubat [10] is a representative under-sampling approach which removes noisy, borderline, and redundant majority training instances. However, these steps typically can remove only a small fraction of the majority instances, so they might not be very helpful in a scenario with a majority-to-minority ratio of more than 100:1 (which is becoming common in many emerging pattern-recognition applications). Multi-classifier training [11] and Bagging [12] are two other under-sampling methods. These methods do not deal with noisy and borderline data directly, but use a large ensemble of sub-classifiers to reduce prediction variance.

Over-sampling [13, 14] is the opposite of the under-sampling approach. It dupli-
cates or interpolates minority instances in the hope of reducing the imbalance. The
over-sampling approach can be considered as a "phantom-transduction" method.
It assumes the neighborhood of a positive instance to be still positive, and the instances between two positive instances positive. The validity of assumptions like
these, however, can be data-dependent.

The algorithmic approach, which is traditionally[1] orthogonal to the data-processing
approach, is the focus of this chapter. Nugroho [15] suggests combining a competitive learning network and a multilayer perceptron as a solution for the class
imbalance problem. Kubat et al. [16, 17, 10, 18] modify the decision-tree gener-
ator to improve its learning performance on imbalanced datasets. For SVMs, few
attempts [19, 20, 21, 22, 23, 24] have dealt with the imbalanced training-data problem. Basically, all those work aims to incorporate into the SVMs the prior knowledge of the risk factors of false negatives and false positives. Karakoulas et al. [21]
proposed an approach to modify the bias (or parameter b) in the class prediction
function (Eq. 9.1). Veropoulos et al. [22, 24, 25] use different pre-defined penalty
constants (based on some prior knowledge) for different classes of data. The effectiveness of this method is limited since the Karush Kuhn Tucker (KKT) conditions
[26] use the penalty constants as the upper bounds, rather than the lower bounds, of
misclassification costs. Moreover, the KKT condition $\sum_{i=1}^{n} \alpha_i y_i = 0$ imposes an equal
total influence from the positive and negative support vectors. The increases in some
α_i's at the positive side will inadvertently increase some α_i's at the negative side to
satisfy the constraint. These constraints can make the increase of C^+ on minority
instances ineffective. (Validation is presented in Section 9.4.)

Another algorithmic approach to improve the SVMs for imbalanced training is
to modify the employed kernel function K or kernel matrix[2] $\mathbf{K}$. In kernel-based
methods, such as SVMs, the kernel K represents a pairwise similarity measurement
among the data. Because of the central role of the kernel, a poor K will lead to a poor
performance of the employed classifier [27, 28]. Our prior work ACT [29] falls into
this category by modifying the K using (quasi-) conformal transformation so as to
change the spatial resolution around the class boundary. However, ACT works only
when data have a fixed-dimensional vector-space representation, since the algorithm
relies on information in the input space. The kernel-boundary alignment algorithm
(KBA) that we propose in this chapter is a more general approach, which does not
require the data to have a vector-space representation. This relaxation is important
so that we can deal with a large class of sequence data (motion trajectories, DNA sequences, sensor-network data, etc.), which may have different length. Furthermore,
KBA provides greater flexibility in adjusting the class boundary.

Recently, several kernel alignment algorithms [20, 27, 30, 23] have been proposed in the Machine Learning community to learn a kernel function or a kernel

---

[1] Although our algorithmic approach focuses on aligning class boundary, it can effectively remove
redundant majority instances as a by-product.

[2] Given a kernel function K and a set of instances $\mathscr{X}_{train} = \{\mathbf{x}_i, y_i\}_{i=1}^{n}$, the kernel matrix (Gram
matrix) is the matrix of all possible inner-products of pairs from X_{train}, $\mathbf{K} = (k_{ij}) = K(\mathbf{x}_i, \mathbf{x}_j)$.

matrix from the training data. The motivation behind these methods is that a good kernel should be data dependent, and a systematic method for learning a good kernel from the data is useful. All these methods are based on the notion of the kernel target alignment proposed by Cristianini et al. [27]. The alignment score is used for measuring the quality of a given kernel matrix. To address the imbalanced training-data problem, Kandola et al. [30] propose an extension to kernel-target alignment by giving the alignment targets of $\frac{1}{n^+}$ to the positive instances and $-\frac{1}{n^-}$ to the negative instances. (We use n^+ and n^- to denote the number of minority and majority instances, respectively.) Unfortunately, when $\frac{n^+}{n^-}$ is small (when n^+ does not remain $O(n^+ + n^-)$), the concentration property upon which that kernel-target alignment relies may no longer hold. In other words, the proposed method can deal only with uneven data that are not very uneven. Our proposed KBA algorithm is based on maximizing the separation margin of the SVMs, and is more effective in its solution.

9.3 Kernel Boundary Alignment

Let us consider a two-class classification problem with training dataset $\mathscr{X}_{train} = \{\mathbf{x}_i, y_i\}_{i=1}^n$, where $\mathbf{x}_i \in \mathfrak{R}^m$ and $y \in \{-1, +1\}$. The basic idea of kernel methods is to map $\mathscr{X}$ from its input space $\mathscr{I}$ to a feature space $\mathscr{F}$, where the data can be separated by applying a linear procedure [5]. The attractiveness of kernel methods is that the mapping from $\mathscr{I}$ to $\mathscr{F}$ can be performed efficiently through the inner product defined in $\mathscr{F}$, or $K(\mathbf{x}_i, \mathbf{x}_j) = \Phi(\mathbf{x}_i)^T \Phi(\mathbf{x}_j)$. Common choices for kernels are polynomial functions $K(\mathbf{x}_i, \mathbf{x}_j) = (\mathbf{x}_i \cdot \mathbf{x}_j + 1)^p$ and Gaussian radial basis functions (RBF) $K(\mathbf{x}_i, \mathbf{x}_j) = \exp\left(-\frac{\|\mathbf{x}_i - \mathbf{x}_j\|^2}{2\sigma^2}\right)$. More generally, $K(\mathbf{x}_i, \mathbf{x}_j)$ can be considered as a similarity measure between instances $\mathbf{x}_i$ and $\mathbf{x}_j$. (Theoretical justifications are presented in [28].) For instance, when a Gaussian RBF function is employed, the value of $K(\mathbf{x}_i, \mathbf{x}_j)$ ranges from 0 to 1, where $K(\mathbf{x}_i, \mathbf{x}_j) = 0$ when $\mathbf{x}_i$ and $\mathbf{x}_j$ are infinitely far away (dissimilar) in input space, and $K(\mathbf{x}_i, \mathbf{x}_j) = 1$ when $\mathbf{x}_i$ and $\mathbf{x}_j$ are infinitely close (almost identical). Thus, the choice of a good kernel is equivalent to the choice of a good distance function for measuring similarity.

To tackle the imbalanced training-dataset problem, we propose to modify the kernel by considering the imbalanced data distribution as the prior information. There are two approaches to modify the kernel. The first approach is to modify the kernel function K directly in input space $\mathscr{I}$. The second approach is to modify the kernel matrix $\mathbf{K}$ generated by a kernel function (for vector data) or a similarity measurement (for non-vector data) on the training set $\mathscr{X}$ in feature space $\mathscr{F}$. The first approach relies on the data information in $\mathscr{I}$, and hence the fixed-dimensional input space must exist. However, the second approach to modify the kernel matrix in $\mathscr{F}$ can bypass this limitation by only relying on the mapped data information in the feature space. Indeed, as long as the resulting kernel matrix $\mathbf{K}$ maintains the positive (semi-) definite property, the modification is mathematically valid.

Table 9.1: Notations used in ACT and KBA

Symbol	Meaning
$D(\mathbf{x})$	Conformal transformation function
$\tau_k^2,\ \tau_b^2$	Parameters of $D(\mathbf{x})$
M	Nearest neighborhood range
$|\mathbf{SI}|$	Number of support instances
$|\mathbf{SI}^+|$	Number of minority support instances
$|\mathbf{SI}^-|$	Number of majority support instances
$\mathbf{x}^+,\ \Phi(\mathbf{x}^+)$	A minority support instance
$\mathbf{x}^-,\ \Phi(\mathbf{x}^-)$	A majority support instance
$\mathbf{x}_b,\ \Phi(\mathbf{x}_b)$	An interpolated boundary instance
α	Weight parameter of interpolation
$\mathscr{X}_{train}$	Set of the training instances
$\mathscr{X}_b^*$	Sets of the interpolated boundary instances
$\mathscr{X}_{mis}^+$	Set of the misclassified minority test instances
$\mathscr{X}_{mis}^-$	Set of the misclassified majority test instances

In the remainder of this chapter, we first summarize ACT [29], our prior function-modification approach, to set up the context for discussing KBA. (KBA must obey the theoretical justification on which ACT is explicitly founded.) We then propose the kernel-boundary-alignment (KBA) algorithm. This algorithm generalizes the work of ACT, by modifying the kernel matrix in $\mathscr{F}$, to deal with data that have a fixed-dimensional vector-space representation and also data that do not (e.g., sequence data). At the end of Section 9.3 we will discuss the differences between KBA and ACT, and in particular, the additional flexibility that KBA enjoys in adjusting similarity measures. Table 9.1 lists key notations used in this section.

9.3.1 Conformally Transforming Kernel K

Kernel-based methods, such as SVMs, introduce a mapping function Φ which embeds the $\mathscr{I}$ into a high-dimensional $\mathscr{F}$ as a curved Riemannian manifold $\mathscr{S}$ where the mapped data reside [31]. A Riemannian metric $g_{ij}(\mathbf{x})$ is then defined for $\mathscr{S}$, which is associated with the kernel function $K(\mathbf{x}, \mathbf{x}')$ by

$$g_{ij}(\mathbf{x}) = \left(\frac{\partial^2 K(\mathbf{x}, \mathbf{x}')}{\partial x_i \partial x_j'} \right)_{\mathbf{x}' = \mathbf{x}}. \tag{9.2}$$

The metric g_{ij} shows how a local area around $\mathbf{x}$ in $\mathscr{I}$ is magnified in $\mathscr{F}$ under the mapping of Φ. The idea of conformal transformation in SVMs is to enlarge the margin by increasing the magnification factor $g_{ij}(\mathbf{x})$ along the boundary (represented by support vectors) and to decrease it around the other points. This could be implemented by a conformal transformation[3] of the related kernel $K(\mathbf{x}, \mathbf{x}')$ according to Eq.(9.2), so that the spatial relationship between the data would not be affected too much [19]. Such a (quasi-) conformal transformation can be depicted as

$$\tilde{K}(\mathbf{x}, \mathbf{x}') = D(\mathbf{x})D(\mathbf{x}')K(\mathbf{x}, \mathbf{x}'). \tag{9.3}$$

In Eq.(9.3), $D(\mathbf{x})$ is a properly defined positive (quasi-) conformal function. $D(\mathbf{x})$ should be chosen in such a way that the new Riemannian metric $\tilde{g}_{ij}(\mathbf{x})$, associated with the new kernel function $\tilde{K}(\mathbf{x}, \mathbf{x}')$, has larger values near the decision boundary. Furthermore, to deal with the skew of the class boundary caused by imbalanced classes, we magnify $\tilde{g}_{ij}(\mathbf{x})$ more in the boundary area close to the minority class. In [29], we demonstrate that an RBF distance function such as

$$D(\mathbf{x}) = \sum_{k \in \mathbf{SV}} \exp\left(-\frac{|\mathbf{x} - \mathbf{x}_k|}{\tau_k^2}\right) \tag{9.4}$$

is a good choice for $D(\mathbf{x})$.

In Eq.(9.4), we can see that if τ_k^2's are fixed (equal) for all support vectors $\mathbf{x}_k$'s, $D(\mathbf{x})$ would be very dependent on the density of support vectors in the neighborhood of $\mathbf{x}$. To alleviate this problem, we adaptively tune τ_k^2 according to the spatial distribution of support vectors in $\mathscr{F}$. This goal can be achieved by the following equation:

$$\tau_k^2 = \mathrm{AVG}_{i \in \{\|\Phi(\mathbf{x}_i) - \Phi(\mathbf{x}_k)\|^2 < M, \, y_i \neq y_k\}} \left(\|\Phi(\mathbf{x}_i) - \Phi(\mathbf{x}_k)\|^2\right). \tag{9.5}$$

In the above equation, the average on the right-hand side comprises all support vectors in $\Phi(\mathbf{x}_k)$'s neighborhood within a radius of M but having a different class label. If we choose a large M, such as the maximum distance $\|\Phi(\mathbf{x}_i) - \Phi(\mathbf{x}_k)\|^2$, we might not be able to achieve the local spatial distribution of the support vectors in the neighborhood of $\Phi(\mathbf{x})$. On the contrary, if we choose a small M, we might not be able to find enough support vectors in $\Phi(\mathbf{x}_k)$'s neighborhood for density calculation. To alleviate this problem, ACT automatically calculates M as the average distance of support vectors that are nearest and farthest from $\Phi(\mathbf{x}_k)$. Setting τ_k^2 in this way takes into consideration the spatial distribution of the support vectors in $\mathscr{F}$. Moreover, since ACT aims to further increase the margin of SVMs, in Eq.(9.5), we only take into account the support vectors which have different class labels with $\Phi(\mathbf{x}_i)$ while computing τ_k^2. With this method, we could expect to achieve higher magnification around the margin area, compared to the method of counting the support vectors without the constraint $y_i \neq y_k$.

Although the mapping Φ is unknown, we can use the kernel trick to calculate the distance in $\mathscr{F}$:

$$\|\Phi(\mathbf{x}_i) - \Phi(\mathbf{x}_k)\|^2 = K(\mathbf{x}_i, \mathbf{x}_i) + K(\mathbf{x}_k, \mathbf{x}_k) - 2 * K(\mathbf{x}_i, \mathbf{x}_k). \tag{9.6}$$

---

[3] Usually, it is difficult to find a totally-conformal mapping function to transform the kernel. As suggested in [19], we can choose a quasi-conformal mapping function for kernel transformation.

Substituting Eq.(9.6) into Eq.(9.5), we can then calculate the τ_k^2 for each support vector, which can adaptively reflect the spatial distribution of the support vector in $\mathscr{F}$, not in $\mathscr{I}$.

When the training dataset is very imbalanced, the class boundary tends to be skewed towards the minority class in the input space $\mathscr{I}$. We hope that the new metric $\tilde{g}_{ij}(\mathbf{x})$ would further magnify the area far away from a minority support vector $\mathbf{x}_i$ so that the boundary imbalance could be alleviated. Our algorithm thus assigns a multiplier for the τ_k^2 in Eq.(9.5) to reflect the boundary skew in $D(\mathbf{x})$. We tune $\tilde{\tau}_k^2$ as $\eta_p \tau_k^2$ if $\mathbf{x}_k$ is a minority support vector; otherwise, we tune it as $\eta_n \tau_k^2$. Examining Eq.(9.4), we can see that $D(\mathbf{x})$ is a monotonically increasing function of τ_k^2. To increase the metric $\tilde{g}_{ij}(\mathbf{x})$ in an area which is not very close to the support vector $\mathbf{x}_k$, it would be better to choose a larger η_p for the τ_k^2 of a minority support vector. For a majority support vector, we can choose a smaller η_n, so as to minimize influence on the class-boundary. We empirically demonstrate that η_p and η_n are proportional to the skew of support vectors, or η_p as $O(\frac{|\mathbf{SV}^-|}{|\mathbf{SV}^+|})$, and η_n as $O(\frac{|\mathbf{SV}^+|}{|\mathbf{SV}^-|})$, where $|\mathbf{SV}^+|$ and $|\mathbf{SV}^-|$ denote the number of minority and majority support vectors, respectively. (Please refer to [29] for more details on the theoretical justification of ACT.)

9.3.2 Modifying Kernel Matrix $\mathbf{K}$

For data that do not have a fixed-dimensional vector-space representation (e.g., sequence data), it may not be feasible to transform kernel function K conformally directly in a vector space. In this situation, KBA modifies kernel matrix $\mathbf{K}$ based on the training-data distribution in $\mathscr{F}$. Kernel matrix $\mathbf{K}$ encodes all pairwise-similarity information between the instances in the training dataset. Hence, modifying the kernel matrix transforms the kernel function indirectly. (Notice that KBA is certainly applicable to data that *do* have a vector-space representation, since $\mathbf{K} = (k_{\mathbf{xx}'} = K(\mathbf{x}, \mathbf{x}'))$.) Now, because a training instance $\mathbf{x}$ might not be a vector, we introduce a more general term, *support instance*[4], to denote $\mathbf{x}$ if its embedded point via $\mathbf{K}$ is a support vector in $\mathscr{F}$.

In the following subsections, we will first propose a data-dependent way to estimate the "ideal" class boundary in $\mathscr{F}$ (Section 9.3.2.1). We then choose a feasible conformal function $D(\mathbf{x})$, which can assign a larger spatial resolution along the estimated "ideal" boundary in $\mathscr{F}$ (Section 9.3.2.2). Finally, we present KBA's iterative training procedure (Section 9.3.2.3).

---

[4] In the KBA algorithm, if $\mathbf{x}$ is a support instance, we call both $\mathbf{x}$ and its embedded support vector via $\mathbf{K}$ in $\mathscr{F}$ *support instance*.

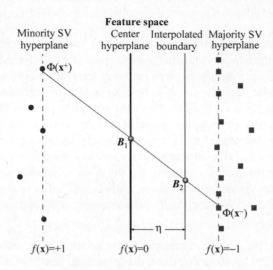

Fig. 9.3: Estimate boundary instances in $\mathscr{F}$

9.3.2.1 Estimation of Boundary

Performing transformation on K or $\mathbf{K}$ aims to magnify the spatial resolution along the decision boundary, thereby improving the class separation. According to the work of [19, 29], maximal magnification should be performed along the class boundary. Unfortunately, locating the class boundary in input space $\mathscr{I}$ is difficult [19]. (When the data do not have a fixed-dimensional vector-space representation, locating the class boundary in $\mathscr{I}$ is impossible.) Instead, KBA locates the class boundary in feature space $\mathscr{F}$ through interpolation. In $\mathscr{F}$, the class boundary learned from the training data is the center hyperplane in the margin. When the training dataset is balanced, the center hyperplane approximates the "ideal" boundary well. However, when the training dataset is imbalanced, the decision boundary is skewed toward the minority class. To compensate for this skew, KBA gives the maximal magnification to an interpolated boundary between the center hyperplane and the hyperplane formed by the majority support instances in $\mathscr{F}$.

Figure 9.3 illustrates how the interpolation procedure works. Let $\Phi(\mathbf{x}^+)$ and $\Phi(\mathbf{x}^-)$ denote a minority support instance and a majority support instance, respectively. A boundary instance $\Phi(\mathbf{x}_b)$ on the "ideal" boundary should reside between the center hyperplane (the thick line in the middle of Figure 9.3) and the majority support-instance hyperplane (the dash line on the right-hand side of the figure). We can thus estimate the location of $\Phi(\mathbf{x}_b)$ by interpolating the positions of $\Phi(\mathbf{x}^+)$ and $\Phi(\mathbf{x}^-)$ as follows:

$$\Phi(\mathbf{x}_b) = (1-\beta)\Phi(\mathbf{x}^+) + \beta\Phi(\mathbf{x}^-), \ \frac{1}{2} \le \beta \le 1. \qquad (9.7)$$

When the training dataset is balanced, β is $\frac{1}{2}$, and $\Phi(\mathbf{x}_b)$ lies on the center hyperplane (e.g., point B_1 in the figure). In this balanced case, the estimated "ideal" boundary coincides with the learned boundary. When the training dataset is imbalanced, however, we need to adjust β to estimate the "ideal" boundary. The key research question to answer is: "How to determine β in a data-dependent way?"

We propose a cost function to measure the loss caused by false negatives and false positives when different values of β are introduced. We then choose the β which can achieve the minimal cost. Let $\mathscr{X}_{mis}^{+}$ denote the set of the misclassified minority test-instances and $\mathscr{X}_{mis}^{-}$ the set of the misclassified majority test-instances. We define the cost functional $C(\cdot)$ for any scalar decreasing loss functions $c_p(\cdot)$ and $c_n(\cdot)$ as follows:

$$C(\eta) = \sum_{i=1}^{|\mathscr{X}_{mis}^{+}|} c_p\left(y_i f'(\mathbf{x}_i)\right) + \sum_{i=1}^{|\mathscr{X}_{mis}^{-}|} c_n\left(y_i f'(\mathbf{x}_i)\right), \qquad (9.8)$$

where $f'(\mathbf{x}_i) = f(\mathbf{x}_i) + \eta$, and $0 \le \eta \le 1$. In the equation above, $f(\mathbf{x}_i)$ is the SVM predication score for test instance $\mathbf{x}_i$, η is the offset of the interpolated boundary from the center hyperplane, as shown in Figure 9.3, and $y_i f'(\mathbf{x}_i)$ is the associated margin in $\mathscr{F}$ for instance $\mathbf{x}_i$ with respect to the interpolated class boundary. The loss functions $c_p(\cdot)$ and $c_n(\cdot)$ are used to penalize the misclassified[5] minorities (false negative) and majorities (false positive), respectively. Each loss function, $c_p(\cdot)$ or $c_n(\cdot)$, can be chosen as any scalar decreasing function of the margin $y_i f'(\mathbf{x}_i)$ according to the prior knowledge. When no prior knowledge is available, usually, we can choose the exponential loss function as $c_p(\cdot)$ and the log-likelihood loss function as $c_n(\cdot)$, i.e,

$$c_p(y_i f'(\mathbf{x}_i)) = \exp(-y_i f'(\mathbf{x}_i)),$$
$$c_n(y_i f'(\mathbf{x}_i)) = \ln(1 + \exp(-y_i f'(\mathbf{x}_i))).$$

The justification of choosing them as the loss functions comes from boosting [32], where the exponential loss criterion concentrates much more influence (exponentially) on observations with large negative margins ($y_i f'(\mathbf{x}_i) < 0$), and the log-likelihood loss concentrates relatively less influence (linearly) on such observations. Since KBA aims to concentrate on false negatives, we use the exponential loss as $c_p(\cdot)$ and the log-likelihood loss as $c_n(\cdot)$. Notice that since both exponential and log-likelihood loss functions are convex [32], our cost formulation in Eq.(9.8) is also convex with respect to η.

The optimal η^* is then chosen by minimizing the total loss induced by all test instances falling into the margin of SVMs,

$$\eta^* = \arg\min_{\eta} C(\eta), \quad 0 \le \eta \le 1.$$

---

[5] In KBA, we only consider the misclassified test instances among the margin so as to reduce the influence from the outliers. Their SVM scores $f(\mathbf{x})$ range from -1 to $+1$.

The optimal η^* can be calculated from $\frac{\partial C(\eta)}{\partial \eta} = 0$ and truncated between 0 and 1. The above optimization procedure involves with only one unknown variable η. It can thus be efficiently solved using many numerical analysis methods such as the conjugate gradient algorithm.

After the optimal position η^* of the interpolated boundary is calculated, we can obtain β in Eq.(9.7) as follows:

$$\beta = \frac{1+\eta^*}{2}.$$

9.3.2.2 Selection of $D(\mathbf{x})$

After interpolating a boundary in the margin, we then magnify the spatial resolution along the boundary by modifying the Riemannian metric $g_{ij}(\mathbf{x})$ according to Eqs.(9.2) and (9.3). When given a prior kernel, $g_{ij}(\mathbf{x})$ is determined by the conformal function $D(\mathbf{x})$. As what we discussed in Section 9.3.1, a good $D(\mathbf{x})$ function should be larger when $\mathbf{x}$ is closer to the boundary in $\mathscr{F}$ so as to achieve a larger spatial resolution around the boundary. According to this criteria, we choose $D(\mathbf{x})$ as a set of Gaussian functions:

$$D(\mathbf{x}) = \frac{1}{|\mathscr{X}_b^*|} \sum_{\mathbf{x}_b \in \mathscr{X}_b^*} \exp\left(-\frac{\|\Phi(\mathbf{x}) - \Phi(\mathbf{x}_b)\|^2}{\tau_b^2}\right), \tag{9.9}$$

where τ_b^2 is a parameter controlling the magnitude of each exponential function in $D(\mathbf{x})$. For a given instance $\mathbf{x}$, $D(\mathbf{x})$ is calculated as the average of all exponential functions, each of which is related with one interpolated boundary instance $\Phi(\mathbf{x}_b)$ in $\mathscr{X}_b^*$ set. In addition, $\|\Phi(\mathbf{x}) - \Phi(\mathbf{x}_b)\|^2$ is calculated via the kernel trick as follows

$$\|\Phi(\mathbf{x}) - \Phi(\mathbf{x}_b)\|^2 \tag{9.10}$$
$$= \|\Phi(\mathbf{x}) - (1-\beta)\Phi(\mathbf{x}^+) - \beta\Phi(\mathbf{x}^-)\|^2$$
$$= k_{\mathbf{xx}} + (1-\beta)^2 k_{\mathbf{x}^+\mathbf{x}^+} + \beta^2 k_{\mathbf{x}^-\mathbf{x}^-} - 2(1-\beta)k_{\mathbf{xx}^+} - 2\beta k_{\mathbf{xx}^-} + 2\beta(1-\beta)k_{\mathbf{x}^+\mathbf{x}^-},$$

where $k_{\mathbf{xx}'}$ is from the kernel matrix $\mathbf{K}$. When the instance $\mathbf{x}$ is an unseen test instance, $k_{\mathbf{xx}'}$ is computed using the pre-defined similarity measurement which generates the kernel matrix $\mathbf{K}$.

According to [19, 28], we have the following corollary to guarantee the kernel transformation induced by $D(\mathbf{x})$, as defined in Eq.(9.9), performs a mathematically valid conformal transformation.

Corollary 9.1. *The function $D(\mathbf{x})$ defined in Eq.(9.9) gives a valid conformal transformation on feature space $\mathscr{F}$ induced by the pre-defined kernel matrix $\mathbf{K}$.*

Proof. Suppose the mapped vector of an input instance $\mathbf{x}$ is $\Phi(\mathbf{x})$ before transformation and $\Psi(\mathbf{x})$ after transformation. Eq.(9.12) defines the kernel transformation in $\tilde{k}_{\mathbf{xx}'} = D(\mathbf{x})D(\mathbf{x}')k_{\mathbf{xx}'}$. Thus, the cosine value of the angle between two mapped vectors $\Psi(\mathbf{x})$ and $\Psi(\mathbf{x}')$ can be written as follows [28]:

$$\cos\left(\angle\left(\Psi(\mathbf{x}), \Psi(\mathbf{x}')\right)\right)$$

$$= \frac{< \Psi(\mathbf{x}), \Psi(\mathbf{x}') >}{\|\Psi(\mathbf{x})\| \cdot \|\Psi(\mathbf{x}')\|}$$

$$= \frac{D(\mathbf{x})D(\mathbf{x}')k_{\mathbf{xx}'}}{\sqrt{D(\mathbf{x})D(\mathbf{x})k_{\mathbf{xx}}}\sqrt{D(\mathbf{x}')D(\mathbf{x}')k_{\mathbf{x}'\mathbf{x}'}}}$$

$$= \frac{k_{\mathbf{xx}'}}{\sqrt{k_{\mathbf{xx}}}\sqrt{k_{\mathbf{x}'\mathbf{x}'}}}$$

$$= \cos\left(\angle(\Phi(\mathbf{x}), \Phi(\mathbf{x}'))\right),$$

where we use the fact that $D(\mathbf{x})$ defined in Eq.(9.9) is a positive function. We can see that the kernel transformation by $D(\mathbf{x})$ defined in Eq.(9.9) does not affect pairwise-angles between the mapped data in feature space and hence is a valid conformal transformation. ∎

In KBA, we adaptively choose τ_b^2 in a data-dependent way as

$$\tau_b^2 = \text{AVG}_{i \in \{Dist^2(\mathbf{x}_i, \mathbf{x}_b)) < M} \left(Dist^2(\mathbf{x}_i, \mathbf{x}_b)\right), \tag{9.11}$$

where the neighborhood range M is a constant. We choose the threshold M as the margin value of SVMs. In ACT, we use the locations of support vectors to approximate the decision boundary. Empirically, we found that selecting different M's for different support vectors works better than using a fixed M, though it incurs higher computational cost. In KBA, we approximate the "ideal" boundary by a set of interpolated boundary instances $\mathbf{x}_b$'s. Since $\mathbf{x}_b$'s are already located on the decision boundary, our empirical study showed that KBA is not very sensitive to M. We thus fix M as the margin value of SVMs in KBA. The distance $Dist^2(\mathbf{x}_i, \mathbf{x}_b)$ between two interpolated boundary instances $\mathbf{x}_i$ and $\mathbf{x}_b$ is $\|\Phi(\mathbf{x}_i) - \Phi(\mathbf{x}_b)\|^2$ and can be computed using Eq.(9.7) and Eq.(9.10). Notice that we do not need to scale τ_b^2 as in Section 9.3.1 for dealing with the imbalanced training-data problem, since we have considered this factor when interpolating the class boundary and selecting $D(\mathbf{x})$. Compared to Eq.(9.5) in ACT, Eq.(9.11) does not include the constraint $y_i \neq y_b$ since the interpolated boundary instance $\Phi(\mathbf{x}_b)$ does not have a label attribute.

We believe that our adjusted interpolation procedure and selection of $D(\mathbf{x})$ enjoy two benefits.

1. *Improved class-prediction accuracy.* In the imbalanced situation, most of mis-classified minority instances fall into the margin area between the center hyperplane and the majority support-vector hyperplane. By maximizing the spatial resolution in this area, we expect to move those ambiguous instances as far away from the decision boundary as possible, so as to improve class-prediction accuracy.

2. *Improved imbalance ratio.* Since the majority support instances are located nearer the interpolated boundary than the minority support instances ($\frac{1}{2} \leq \beta \leq 1$ in Eq.(9.7)), by choosing a proper form of $D(\mathbf{x})$ as in Eq.(9.9), we can increase the degree of similarity between majority support instances and make them close each other in feature space after kernel transformation. This increase can lead to

a reduction of the number of majority support instances, and hence improve the imbalanced support-instance ratio.

9.3.2.3 Retraining

After choosing $D(\mathbf{x})$, KBA modifies the given kernel matrix $\mathbf{K} = (k_{ij})$ in the following way.

$$\tilde{k}_{ij} = D(\mathbf{x}_i) \times D(\mathbf{x}_j) \times k_{ij}. \tag{9.12}$$

The new kernel matrix $\tilde{\mathbf{K}}$ after modification is then put back into the regular SVMs algorithm for retraining. We have the following corollary, supported by the work of [28], to guarantee that the new kernel matrix after transformation in Eq.(9.12) is a valid kernel matrix.

Corollary 9.2. *When given a positive (semi-) definite kernel matrix* $\mathbf{K}$*, the kernel transformation defined in Eq.(9.12) results in a new kernel matrix* $\tilde{\mathbf{K}}$ *which is also positive (semi-) definite.*

Proof. Since $D(\mathbf{x})$ is a scalar function,

$$D(\mathbf{x})D(\mathbf{x}') = < D(\mathbf{x}), D(\mathbf{x}') >$$

is a positive (semi-) definite (psd) kernel function, which is the so-called one-rank kernel in [28]. Denoting $\mathbf{d} = (d_i)$ as an n-dimensional vector with $d_i = D(\mathbf{x}_i)$, where n is the number of training instances, we have a matrix $\mathbf{dd}^T$ which is associated with the psd function $D(\mathbf{x})D(\mathbf{x}')$ for the training set $\mathbf{X}_{train}$. Hence, $\mathbf{dd}^T$ is a psd matrix. On the other hand, Eq.(9.12) can be rewritten as

$$\tilde{\mathbf{K}} = \mathbf{dd}^T \otimes \mathbf{K}.$$

Since the prior kernel $\mathbf{K}$ is also psd, using the closure property of kernels under tensor product $\otimes$ [28], the new kernel matrix $\tilde{\mathbf{K}}$ is a psd matrix, and hence a valid kernel matrix.   ∎

Figure 9.4 summarizes the KBA algorithm. We apply KBA on the training dataset X_{train} for several iterations or until the imbalanced support-instance ratio cannot be further decreased. In each iteration, KBA adaptively calculates τ_b^2 for each interpolated boundary instance (steps 8 to 10), based on the distribution in $\mathcal{F}$. Then KBA updates the training dataset X_{train} using the support-instance set (step 11). Why do we use the support-instance set as the training data in the next iteration? We do so because the decision boundary of SVMs will not change if we just use the support-instance set for retraining [33, 5]. One benefit of doing so is that we can reduce the computational cost of training and we can also reduce the ratio of the majority-over-minority support-instances. Finally, KBA updates the kernel matrix and performs retraining on X_{train} (steps 16 to 18).

Input:
X_{train}, **K**;
θ; /* stopping threshold */
T; /* maximum running iterations */

Output:
$\mathscr{C}$; /* output classifier */

Variables:
SI; /* support-instance set */
$\mathscr{X}_b$; /* interpolated boundary set */
$\mathscr{X}_b^*$; /* subset of $\mathscr{X}_b$ */
M; /* neighborhood range */
x; /* a training instance */
s; /* an interpolated boundary instance */
s.τ; /* parameter of **s** */
$\varepsilon_{old}, \varepsilon_{new}$; /* support-vector ratios */

Function Calls:
SVMTrain(X_{train}, **K**); /* train $\mathscr{C}$ */
ExtractSI($\mathscr{C}$);   /* obtain **SI** from $\mathscr{C}$ */
InterpolateBI(**SI**); /* interpolate $\mathscr{X}_b$ */
ComputeM(**s**, $\mathscr{X}_b$);   /* compute M */
ExtractBI(**x**, $\mathscr{X}_b$, M); /* sampling */

Begin
1) $\mathscr{C} \leftarrow$ SVMTrain(X_{train}, **K**);
2) $\varepsilon_{old} \leftarrow \infty$;
3) $\varepsilon_{new} \leftarrow \frac{|\mathbf{SI}^-|}{|\mathbf{SI}^+|}$;
4) $t \leftarrow 0$;
5) **while** $((\varepsilon_{old} - \varepsilon_{new} > \theta)\&\&(t < T))$ {
6)   **SI** $\leftarrow$ ExtractSI($\mathscr{C}$);
7) $\mathscr{X}_b \leftarrow$ InterpolateBI(**SI**);
8)   **for each s** $\in \mathscr{X}_b$ {
9)   M $\leftarrow$ ComputeM(**s**, $\mathscr{X}_b$);
10)   **s**.$\tau \leftarrow \sqrt{\mathrm{AVG}_{i \in \{Dist^2(s_i,s) < M\}} (Dist^2(s_i, s))}$;}
11) $X_{train} \leftarrow$ **SI**;
12)   **for each x** $\in X_{train}$ {
13)   M $\leftarrow$ ComputeM(**x**, $\mathscr{X}_b$);
14) $\mathscr{X}_b^* \leftarrow$ ExtractBI(**x**, M, $\mathscr{X}_b$);
15) $D(\mathbf{x}) = \frac{1}{|\mathscr{X}_b^*|} \sum_{\mathbf{s} \in \mathscr{X}_b^*} \left(\exp\left(-\frac{\|\Phi(\mathbf{x}) - \Phi(\mathbf{s})\|^2}{\mathbf{s}.\tau^2} \right) \right)$;}
16)   **for each** k_{ij} **in K** {
17) $k_{ij} \leftarrow D(\mathbf{x}_i) \times D(\mathbf{x}_j) \times k_{ij}$;}
18) $\mathscr{C} \leftarrow$ SVMTrain(X_{train}, **K**);
19) $\varepsilon_{old} \leftarrow \varepsilon_{new}$;
20) $\varepsilon_{new} \leftarrow \frac{|\mathbf{SI}^-|}{|\mathbf{SI}^+|}$;
21) $t \leftarrow t + 1$; }
22) **return** $\mathscr{C}$;
End

Fig. 9.4: The KBA algorithm

9.4 Experimental Results

Our empirical study examined the effectiveness of the kernel-boundary-alignment algorithm in two aspects.

1. *Vector-space evaluation.* We compared KBA with other algorithms for imbalanced-data learning. We used six UCI datasets and an image dataset to conduct this evaluation. (We present the datasets shortly.)
2. *Non-vector-space evaluation.* We evaluated the effectiveness of KBA on a set of video surveillance data, which are represented as spatio-temporal sequences that do not have a vector-space representation.

In our experiments, we used C-SVMs as our yardstick to measure how other methods perform. We employed Laplacian kernels of the form $\exp(-\gamma|\mathbf{x} - \mathbf{x}'|)$ as $K(\mathbf{x}, \mathbf{x}')$ of C-SVMs. Then we used the following procedure. The dataset was randomly split into training and test subsets generated in a certain ratio which was empirically chosen to be optimal on each dataset for the regular C-SVMs. Hyperparameters (C and γ) of $K(\mathbf{x}, \mathbf{x}')$ were obtained for each run using 7-fold cross-validation. All training, validation, and test subsets were sampled in a stratified

Table 9.2: Mean and standard deviation of g-Means accuracy on UCI datasets

Dataset	# Attrib	# Pos	# Neg	$\frac{\text{SI}^-}{\text{SI}^+}$	SVMs	SMOTE	ACT	KBA
Segmentation	19	30	180	1.8:1	**98.1±5.1**	**98.1±5.1**	**98.1±5.1**	**98.1±5.1**
Glass	10	29	185	2.0:1	89.9±6.3	91.8±6.5	93.7±6.7	**93.7±6.6**
Euthyroid	24	238	1762	1.5:1	92.8±3.6	92.4±4.3	94.5±3.0	**94.6±2.9**
Car	6	69	1659	1.8:1	99.0±2.2	99.0±2.3	**99.9±0.2**	**99.9±0.2**
Yeast	8	51	1433	3.0:1	59.0±12.1	69.9±10.0	78.5±4.5	**82.2±7.1**
Abalone	8	32	4145	9.0:1	0.0±0.0	0.0±0.0	51.9±7.6	**57.8±5.4**

manner ensuring each of them had the same negative/positive ratio [10]. We repeated this procedure seven times, computed average class-prediction accuracy, and compared the results. For ACT and KBA, we chose the maximum running iterations T as 5. The detailed choices of parameters are presented in Sections 9.4.1.1 and 9.4.1.2.

9.4.1 Vector-Space Evaluation

For this evaluation, we used six UCI datasets and a 116-category image dataset. The six UCI datasets we experimented with are *abalone* (19), *car* (3), *segmentation* (1), *yeast* (5), *glass* (7), and *euthyroid* (1). The class-label in the parentheses indicates the target class we chose. Table 9.2 shows the characteristics of these six datasets organized according to their negative-to-positive training-instance ratios. The top three datasets (segmentation, glass, and euthyroid) are not-too-imbalanced. The middle two (car and yeast) are mildly imbalanced. The bottom dataset (abalone) is the most imbalanced (the ratio is about 130:1).

The image dataset contains 20K images in 116 categories collected from the Corel Image CDs[6]. Each image is represented by a vector of 144 dimensions including color, texture, and shape features [6]. To perform class prediction, we employed the one-per-class (OPC) ensemble [34], which trains 116 classifiers, each of which predicts the class membership for one class. The class prediction on a testing instance is decided by voting among the 116 classifiers.

9.4.1.1 Results on UCI Benchmark Datasets

Tables 9.2 and 9.3 report the experimental results with the six UCI datasets. In addition to conducting experiments with SVMs, ACT, and KBA, we also implemented

---

[6] We exclude from our testbed those categories that cannot be classified automatically, such as "industry", "Rome", and "Boston". (E.g., the Boston category contains various subjects, e.g., architectures, landscapes, and people, of Boston.)

Table 9.3: Mean and standard deviation of AUCs (in %) on UCI datasets

Dataset	SVMs	SMOTE	ACT	KBA
Segmentation	100.0±0.0	100.0±0.0	100.0±0.0	100.0±0.0
Glass	96.9±3.0	97.1±3.1	98.5±2.5	**98.9±2.6**
Euthyroid	96.6±2.2	96.0±2.8	98.2±1.8	**98.8±1.5**
Car	99.8±0.2	99.8±0.2	**99.9±0.1**	**99.9±0.1**
Yeast	89.2±5.4	91.1±5.0	93.8±2.2	**95.2±2.5**
Abalone	62.5±12.1	62.5±12.1	80.2±7.1	**87.4±6.8**

Table 9.4: Training time (in second) on UCI datasets

Dataset	SVMs	SMOTE	ACT	KBA
Segmentation	7	12	33	30
Glass	8	11	33	30
Euthyroid	12	33	155	120
Car	9	10	55	45
Yeast	8	10	52	40
Abalone	10	25	109	83

and tested one popular minority-oversampling strategy SMOTE [?]. We used the L_2-norm RBF function for $D(\mathbf{x})$ in ACT. In each run, the training and test subsets were generated in the ratio 6:1. For SMOTE[7], the minority class was over-sampled at 200%, 400% and 1000% for each of three groups of UCI datasets in Table 9.2, respectively.

We report in Table 9.2 using the Kubat's g-means metric defined as $\sqrt{a^+ \cdot a^-}$, where a^+ and a^- are positive (the target class) and negative testing accuracy, respectively [10]. Means and standard deviations of the experimental results are both reported in the table. In all the six datasets, KBA achieves the highest or ties for the highest accuracy. (The best results are marked in bold.) When the data is very imbalanced (the last row abalone of Table 9.2), both SVMs and SMOTE cannot make accurate predictions. KBA achieves 57.8% mean class-prediction accuracy (in g-means), and shows 5.9 percentile points improvement over ACT.

We also report in Table 9.3 using AUC [35] defined as the area under an ROC curve to compare the four strategies on the six UCI datasets. Means and standard deviations of the AUC scores are reported in the table. For readability, we report AUCs as percentages between 0% and 100%, instead of between 0 and 1. Again, KBA achieves the highest mean AUCs in all six UCI datasets. Compared to ACT, KBA generated better results especially for the last datasets (yeast and abalone),

---

[7] For the datasets in Table 9.2 from top to bottom, for SMOTE, the optimal γ was 0.002, 0.003, 0.085, 0.3, 0.5, and 0.084, respectively. For SVMs, ACT, and KBA, the optimal γ was 0.004, 0.003, 0.08, 0.3, 0.5, and 0.086, respectively. All optimal C's were 1,000.

with 1.4 and 7.2 percentile points improvement, respectively. Such gains bear out the flexibility and superiority of KBA working in feature space $\mathscr{F}$. Statistically, the higher AUCs from KBA means that our KBA algorithm will favor in classifying a positive (target) instance with a higher probability than other algorithms and hence could well tackle the imbalanced training-dataset problem.

Finally, we report in Table 9.4 the total training time of each method. The time is reported in seconds by averaging seven runs of training on different subsets of the training data. Compared to SVMs and SMOTE, both ACT and KBA took longer time to train. This was because some computational cost were spent on modifying the kernel of SVMs in a data-dependent way to deal with the imbalanced-training problem. However, we can see that for ACT, the training time increases only linearly compared to SVMs. For *euthyroid* and *abalone*, which have the largest number of training instances among the six UCI datasets, ACT's training time is 14.5 and 9.9 times longer than that of SVMs . For all six datasets, the average increase in training time is about seven times. In addition, compared to ACT, KBA takes shorter time to train. For the six UCI datasets, KBA's average training time is 16.1% shorter than ACT's. This is expected, since KBA only used the support-instance set from the last iteration as the new training set in the current iteration, as described in Section 9.3.2.

9.4.1.2 Results on 20K Image Dataset

The image dataset is more imbalanced than the UCI datasets. We first set aside 4K images to be used as the test subset; the remaining 16K images were used for training and validation. We compared five schemes: SVMs, BM (the boundary movement method by changing the parameter b in C-SVMs), BP (the biased penalty method of assigning different C to penalize different class in C-SVMs), ACT, and KBA. (The details of BM and BP has been presented in Section 9.2.) Notice that in this experiment, we used the L_1-norm RBF function for $D(\mathbf{x})$ in ACT, since the L_1-norm RBF works best for the image dataset [6].

Table 9.5 presents the prediction accuracy for twelve representative categories out of 116, sorted by their imbalance ratios. KBA improves the accuracy over SVMs by 5.3, 5.9, and 15.5 percentile points on the three subgroup datasets, respectively. KBA achieves the best prediction accuracy for seven out of twelve categories among all schemes (marked by bold font). BM is inferior to SVMs for almost all categories. Finally, BP outperforms SVMs, but only slightly. (We have predicted BP's ineffectiveness, due to the KKT conditions, in Section 9.2.)

Remark 9.1. From Table 9.5, we can see that on this challenging dataset of several diversified classes, the results of all algorithms, including KBA, are not stellar (class-predication accuracy is less than 50% for almost all classes). This low accuracy is caused partly by a large number of classes (116), and partly by not-so-perfect image-feature extraction. Nevertheless, a 50% prediction accuracy is far better than that of a random predication, which is $1/116 = 0.86\%$. ∎

Table 9.5: Image-dataset prediction accuracy

Category	Ratio	SVMs	BM	BP	ACT	KBA
Mountain	34 : 1	24.8	21.2	24.8	33.3	**34.5**
Snow	37 : 1	46.4	47.5	47.8	**54.6**	52.3
Desert	39 : 1	33.7	31.8	34.3	**39.1**	36.8
Dog	44 : 1	32.9	28.5	35.2	41.5	**42.7**
Woman	54 : 1	27.9	25.3	26.2	35.3	**39.1**
Church	66 : 1	**21.8**	19.4	**21.8**	20.0	20.6
Leaf	80 : 1	26.1	27.2	24.8	32.6	**37.2**
Lizard	101 : 1	13.9	11.8	15.1	22.2	**25.4**
Parrot	263 : 1	7.1	3.5	7.1	14.3	**18.4**
Horse	264 : 1	14.3	10.4	14.3	28.6	**32.9**
Leopard	283 : 1	7.7	5.6	7.7	**23.1**	**23.1**
Shark	1232 : 1	0.0	0.0	0.0	**16.6**	**16.6**

9.4.2 Non-Vector-Space Evaluation

For our multi-camera video-surveillance project, we recorded video data at a campus parking lot. We collected trajectories depicting five motion patterns: *circling* (30 instances), *zigzag-pattern* or *M-pattern* (22 instances), *back-forth* (40 instances), *go-straight* (200 instances), and *parking* (3, 161 instances). We divided these events into benign and suspicious categories and aimed to detect suspicious events with high accuracy. The benign-event category consists of patterns *go-straight* and *parking*, and the suspicious-event category consists of the other three patterns.

For each experiment, we chose 60% of the data as the training set, keeping the remaining 40% to use as our testing data. We employed a sequence-alignment kernel to compare similarity between two trajectories (see [4] for details). Figure 9.5(a) reports the sensitivities of using SVMs and three methods of improving the SVMs. All three methods — BM, BP, and KBA — improve sensitivity. Among the three, KBA achieves the largest magnitude of improvement over SVMs, around 30 percentile points. Figure 9.5(b) shows that all methods maintain high specificity. We note that BM method performs well for detecting *M-pattern* and *back-forth*; however, it does not do well consistently over all patterns. The performance of the BM method can be highly dependent on the data distribution. Overall, BP does not work effectively, which bears out our prediction in Section 9.2.

9.5 Concluding Remarks

We have presented the kernel-boundary-alignment algorithm for tackling the imbalanced training-data challenge. Through theoretical justifications and empirical studies, we show this method to be effective. We believe that kernel-boundary

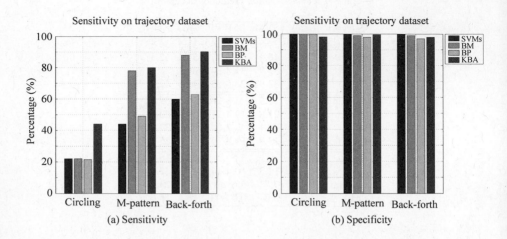

Fig. 9.5: Sensitivity vs. specificity on trajectory dataset (See color insert)

alignment is attractive, not only because of its accuracy, but also because it can be applied to learning both vector-data and sequence-data (e.g., DNA sequences and spatio-temporal patterns) through modifying the kernel matrix directly. Future research includes studies on formulating a robust way of incorporating prior knowledge of the imbalanced datasets to estimate the "ideal" boundary. Some prior work has been done in incorporating the prior knowledge into the optimization formulation of SVMs, such as [36]. However, the incorporation is usually not robust and very depends on the prediction rules of the prior knowledge. Researchers can look into a more robust way and apply it on KBA.

References

1. Wu, G., Chang, E.Y. KBA: Kenel boundary alignment considering imbalanced data distribution. *IEEE Trans. Knowledge and Data Engineering*, 17(6):786–795, 2005.
2. Weiss, G.M. Mining with rarity: A unifying framework. *SIGKDD Explorations*, (1):7–19, June 2004.
3. Fawcett, T., Provost, F. Adaptive fraud detection. In *Proceedings of ACM SIGKDD*, pages 291–316, 1997.
4. Wu, G., Wu, Y., Jiao, L., Wang, Y.F., Chang, E.Y. Multi-camera spatio-temporal fusion and biased sequence-data learning for security surveillance. In *Proceedings of ACM International Conference on Multimedia*, pages 528–538, November 2003.
5. Vapnik, V. *The Nature of Statistical Learning Theory*. Springer, New York, 1995.
6. Tong, S., Chang, E. Support vector machine active learning for image retrieval. In *Proceedings of ACM International Conference on Multimedia*, pages 107–118, October 2001.
7. Joachims, T. Text categorization with support vector machines: learning with many relevant features. In *Proceedings of ECML*, pages 137–142, 1998.

8. Fukunaga, K. *Introduction to Statistical Pattern Recognition*. Academic Press, Boston, MA, 2nd edition, 1990.
9. Provost, F. Learning with imbalanced data sets. *Invited paper for the AAAI'2000 Workshop on Imbalanced Data Sets*.
10. Kubat, M., Matwin, S. Addressing the curse of imbalanced training sets: One-sided selection. In *Proceedings of the fourteenth International Conference on Machine Learning (ICML)*, pages 179–186, 1997.
11. Chan, P., Stolfo, S. Learning with non-uniform class and cost distributions: Effects and a distributed multi-classifier approach. *Workshop Notes KDD Workshop on Distributed Data Mining*, pages 1–9, 1998.
12. Breiman, L. Bagging predictors. *Machine Learning*, 24:123–140, 1996.
13. Chawla, N., Bowyer, K., Hall, L., Kegelmeyer, W.P. Smote: synthetic minority over-sampling technique. *Journal of Artificial Intelligence Research (JAIR)*, 16:321–357, 2002.
14. Weiss, G.M., Provost, F. Learning when training data are costly: The effect of class distribution on tree induction. *Journal of Artificial Intelligence Research*, 19:315–354, 2003.
15. Nugroho, A., Kuroyanagi, S., Iwata, A. A solution for imbalanced training sets problem by combnet-ii and its application on fog forecasting. *IEICE Transactions on Information and Systems*, E85-D(7):1165–1174, July 2002.
16. Cardie, C., Howe, N. Improving minority class prediction using case-specific feature weights. In *Proceedings of the fourteenth International Conference on Machine Learning*, pages 57–65, 1997.
17. Drummond, C., Holte, R. Exploiting the cost (in)sensitivity of decision tree splitting criteria. In *Proceedings of the seventeenth International Conference on Machine Learning*, pages 239–246, 2000.
18. Ling, C., Li., C. Data mining for direct marketing - specific problems and solutions. In *Proceedings of ACM SIGKDD*, pages 73–79, 1998.
19. Amari, S., Wu, S. Improving support vector machine classifiers by modifying kernel functions. *Neural Networks*, 12(6):783–789, 1999.
20. Crammer, K., Keshet, J., Singer, Y. Kernel design using boosting. In *Proceedings of NIPS*, pages 537–544, 2002.
21. Karakoulas, G., Taylor, J.S. Optimizing classifiers for imbalanced training sets. In *Proceedings of NIPS*, pages 253–259, 1998.
22. Lin, Y., Lee, Y., Wahba, G. Support vector machines for classification in nonstandard situations. *Machine Learning*, 46:191–202, 2002.
23. Ong, C.S., Smola, A.J., Williamson, R.C. Hyperkernels. In *Proceedings of NIPS*, pages 478–485, 2003.
24. Veropoulos, K., Campbell, C., Cristianini, N. Controlling the sensitivity of support vector machines. In *Proceedings of the International Joint Conference on Artificial Intelligence*, pages 55–60, 1999.
25. Wu, X., Srihari, R. New v-support vector machines and their sequential minimal optimization. In *Proceedings of the twentieth International Conference on Machine Learning (ICML)*, pages 824–831, Washington, DC, August 2003.
26. Kuhn, H.W., Tucker, A.W. *Non-linear programming*. Proc. of Berkeley Syrup. on Mathematical Statistics and Probability, Univ. Calif. Press, 1961.
27. Cristianini, N., Shawe-Taylor, J., Elisseeff, A., Kandola, J. On kernel target alignment. In *Proceedings of NIPS*, pages 367–373, 2001.
28. Scholkopf, B., Smola, A. *Learning with Kernels: Support Vector Machines, Regularization, Optimization, and Beyond*. MIT Press, Cambridge, MA, 2002.
29. Wu, G., Chang, E. Adaptive feature-space conformal transformation for imbalanced data learning. In *Proceedings of the twentieth International Conference on Machine Learning (ICML)*, pages 816–823, Washington DC, August 2003.
30. Kandola, J., Shawe-Taylor, J. Refining kernels for regression and uneven classification problems. In *Proceedings of the ninth International Workshop on Artificial Intelligence and Statistics*, 2003.

31. Burges, C.   Geometry and invariance in kernel based methods. in adv. in kernel methods: Support vector learning.   In *Advances in kernel methods: support vector learning*, pages 89–116. MIT Press, 1999.
32. Hastie, T., Tibshirani, R., Friedman, J.   *The Elements of Statistical Learning: Data Mining, Inference, and Prediction*. Springer, New York, 2001.
33. Burges, C.   A tutorial on support vector machines for pattern recognition.   In *Proceedings of ACM SIGKDD*, pages 955–974, 1998.
34. Dietterich, T., Bakiri, G.   Solving multiclass learning problems via error-correcting output codes. *Journal of Artifical Intelligence Research*, 2:263–286, 1995.
35. Bradley, A.P.   The use of the area under the roc curve in the evaluation of machine learning algorithms. *Pattern Recognition*, 30:1145–1159, 1997.
36. Wu, X., Srihari, R.   Incorporating prior knowledge with weighted margin support vector machines. pages 326–333, Seattle, Washington, August 2004.

Chapter 10
PSVM: Parallelizing Support Vector Machines on Distributed Computers

Abstract Support Vector Machines (SVMs) suffer from a widely recognized scalability problem in both memory use and computational time. To improve scalability, we have developed a parallel SVM algorithm (PSVM), which reduces memory use through performing a row-based, approximate matrix factorization, and which loads only essential data to each machine to perform parallel computation. Let n denote the number of training instances, p the reduced matrix dimension after factorization (p is significantly smaller than n), and m the number of machines. PSVM reduces the memory requirement by the Interior Point Method (IPM) from $\mathcal{O}(n^2)$ to $\mathcal{O}(np/m)$, and improves computation time to $\mathcal{O}(np^2/m)$. Empirical studies show PSVM to be effective. This chapter[†] was first published in NIPS'07 [1] and the open-source code was made available at [2].
Keywords: Support Vector Machines, Interior Point Method, Incomplete Cholesky Factorization, MPI, distributed systems, matrix factorization.

10.1 Introduction

Support Vector Machines (SVMs) are a core machine learning technology. They enjoy strong theoretical foundations and excellent empirical successes in many pattern recognition applications. Unfortunately, SVMs do not scale well with respect to the size of training data. Given n training instances, the time to train an SVM model is about $\mathcal{O}(n^2)$ in the average case, and so is the memory required by the Interior Point Method (IPM) to solve the quadratic optimization problem. These excessive costs make SVMs impractical for large-scale applications.

Let us examine the resource bottlenecks of SVMs in a binary classification setting to explain our proposed solution. Given a set of training data $\mathscr{X} = \{(\mathbf{x}_i, y_i) | \mathbf{x}_i \in$

[†] ©NIPS, 2007. This chapter is a minor revision of the author's work with Kaihua Zhu, Hongjie Bai, Hao Wang, Zhihuan Qiu, Jian Li, and Hang Cui published in NIPS'07 and then in Scaling Up Machine Learning by Cambridge University Press. Permission to publish this chapter is granted by copyright agreements.

$\mathbf{R}^d\}_{i=1}^n$, where $\mathbf{x}_i$ is an observation vector, $y_i \in \{-1, 1\}$ is the class label of $\mathbf{x}_i$, and n is the size of $\mathscr{X}$, we apply SVMs on $\mathscr{X}$ to train a binary classifier. SVMs aim to search a hyperplane in the *Reproducing Kernel Hilbert Space* (RKHS) that maximizes the margin between the two classes of data in $\mathscr{X}$ with the smallest training error [3]. This problem can be formulated as the following quadratic optimization problem:

$$\min \quad \mathscr{P}(\mathbf{w}, b, \xi) = \frac{1}{2}\|\mathbf{w}\|_2^2 + C\sum_{i=1}^n \xi_i \qquad (10.1)$$
$$s.t. \quad 1 - y_i(\mathbf{w}^T\phi(\mathbf{x}_i) + b) \le \xi_i, \quad \xi_i > 0,$$

where $\mathbf{w}$ is a weighting vector, b is a threshold, C a regularization hyperparameter, and $\phi(\cdot)$ a basis function which maps $\mathbf{x}_i$ to an RKHS space. The decision function of SVMs is $f(\mathbf{x}) = \mathbf{w}^T\phi(\mathbf{x}) + b$, where $\mathbf{w}$ and b are attained by solving $\mathscr{P}$ in Eq.(10.1). The optimization problem in (10.1) is called the primal formulation of SVMs with L_1 loss. It is hard to solve $\mathscr{P}$ directly, partly because the explicit mapping via $\phi(\cdot)$ can make the problem intractable and partly because the mapping function $\phi(\cdot)$ is often unknown. The method of *Lagrangian multipliers* is thus introduced to transform the primal formulation into the dual one

$$\min \quad \mathscr{D}(\alpha) = \frac{1}{2}\alpha^T\mathbf{Q}\alpha - \alpha^T\mathbf{1} \qquad (10.2)$$
$$s.t. \quad \mathbf{0} \le \alpha \le \mathbf{C}, \, \mathbf{y}^T\alpha = 0,$$

where $[\mathbf{Q}]_{ij} = y_iy_j\phi^T(\mathbf{x}_i)\phi(\mathbf{x}_j)$, and $\alpha \in \mathbf{R}^n$ is the Lagrangian multiplier variable (or dual variable). The weighting vector $\mathbf{w}$ is related with α in $\mathbf{w} = \sum_{i=1}^n \alpha_i\phi(\mathbf{x}_i)$.

The dual formulation $\mathscr{D}(\alpha)$ requires an inner product of $\phi(\mathbf{x}_i)$ and $\phi(\mathbf{x}_j)$. SVMs utilize the *kernel trick* by specifying a kernel function to define the inner-product $K(\mathbf{x}_i, \mathbf{x}_j) = \phi^T(\mathbf{x}_i)\phi(\mathbf{x}_j)$. We thus can rewrite $[\mathbf{Q}]_{ij}$ as $y_iy_jK(\mathbf{x}_i, \mathbf{x}_j)$. When the given kernel function K is psd (positive semi-definite), the dual problem $\mathscr{D}(\alpha)$ is a convex Quadratic Programming (QP) problem with linear constraints, which can be solved via the *Interior-Point method* (IPM) [4]. Both the computational and memory bottlenecks of the SVM training is the IPM solver to the dual formulation of SVMs in Eq.(10.2).

Currently, the most effective IPM algorithm is the primal-dual IPM [4]. The principal idea of the primal-dual IPM is to remove inequality constraints using a barrier function and then resort to the iterative Newton's method to solve the KKT linear system related to the Hessian matrix $\mathbf{Q}$ in $\mathscr{D}(\alpha)$. The computational cost is $O(n^3)$ and the memory usage $O(n^2)$.

In this work, we propose a parallel SVM algorithm (PSVM) to reduce memory use and to parallelize both data loading and computation. Given n training instances each with d dimensions, PSVM first loads the training data in a round-robin fashion onto m machines. The memory requirement per machine is $\mathscr{O}(nd/m)$. Next, PSVM performs a parallel row-based Incomplete Cholesky Factorization (ICF) on the loaded data. At the end of parallel ICF, each machine stores only a fraction of the factorized matrix, which takes up space of $\mathcal{O}(np/m)$, where p is the column dimension of the factorized matrix. (Typically, p can be set to be about $\sqrt{n}$ without noticeably degrading training accuracy.) PSVM reduces memory use of IPM from $O(n^2)$ to $O(np/m)$, where p/m is much smaller than n. PSVM then performs parallel IPM to solve the quadratic optimization problem in Eq.(10.2). The computation time is improved from about $\mathcal{O}(n^2)$ of a decomposition-based algorithm (e.g., SVMLight [5], LIBSVM [6], SMO [7], and SimpleSVM [8]) to $\mathcal{O}(np^2/m)$. This work's main contributions are: (1) PSVM achieves memory reduction and computation speedup via a parallel ICF algorithm and parallel IPM. (2) PSVM handles kernels (in contrast to other algorithmic approaches [9, 10]). (3) We have implemented PSVM on our parallel computing infrastructures. PSVM effectively speeds up training time for large-scale tasks while maintaining high training accuracy.

PSVM is a practical, parallel approximate implementation to speed up SVM training on today's distributed computing infrastructures for dealing with Web-scale problems. What we do **not** claim are as follows: (1) We make no claim that PSVM is the sole solution to speed up SVMs. Algorithmic approaches such as [11, 12, 9, 10] can be more effective when memory is not a constraint or kernels are not used. (2) We do not claim that the algorithmic approach is the only avenue to speed up SVM training. Data-processing approaches such as [13] can divide a serial algorithm (e.g., LIBSVM) into subtasks on subsets of training data to achieve good speedup. (Data-processing and algorithmic approaches complement each other, and can be used together to handle large-scale training.)

10.2 Interior Point Method with Incomplete Cholesky Factorization

Interior Point Method (IPM) is one of the state-of-the-art algorithms to solve convex optimization problem with inequality constraints and the primal-dual IPM is one of the most efficient IPM methods. Whereas the detailed derivation could be found in ([14, 4]), this section briefly reviews primal-dual IPM.

First, we take (10.2) as a primal problem (it is the dual form of SVMs, however, it is treated as primal optimization problem here) and its dual form can be written as

$$\max_{v,\lambda,\xi} \quad \mathscr{D}'(\alpha,\lambda) = -\frac{1}{2}\alpha^T Q\alpha - C\sum_{i=1}^{n}\lambda_i \qquad (10.3)$$

$$s.t. \quad -Q\alpha - vy + \xi - \lambda = -\mathbf{1}$$

$$\xi \geq 0, \lambda \geq 0,$$

where λ, ξ and v are the dual variables in SVMs for constraints $\alpha \leq \mathbf{C}$, $\alpha \geq \mathbf{0}$ and $\mathbf{y}^T\alpha = 0$, respectively.

The basic idea of the primal-dual IPM is to optimize variables α, λ, ξ, and ν concurrently. The algorithm applies Newton's method on each variable iteratively to gradually reach the optimal solution. The basic flow is depicted in Algorithm 10.2, where μ is a tuning parameter and the *surrogate gap*

$$\hat{\eta} = C \sum_{i=1}^{n} \lambda_i - \alpha^T \lambda + \alpha^T \xi \qquad (10.4)$$

is used to compute t and check convergence. We omit how to compute s here as all the details could be found in [14].

$\alpha = 0$, $\nu = 0$, $\lambda \geq \mathbf{0}$, $\xi \geq \mathbf{0}$
repeat
    Determine $t = 2n\mu/\hat{\eta}$
    Compute $\triangle \mathbf{x}$, $\triangle \lambda$, $\triangle \xi$, and $\triangle \nu$ according to Eq.(10.5).
    Determine step length $s > 0$ through backtracking line search and update $\alpha = \alpha + s\triangle \mathbf{x}$,
    $\lambda = \lambda + s\triangle \lambda$, $\xi = \xi + s\triangle \xi$, $\nu = \nu + s\triangle \nu$.
until α is primal feasible and λ, ξ, ν is dual feasible and the surrogate gap $\hat{\eta}$ is smaller than a threshold

Fig. 10.1: Interior Point Method

Newton update, the core step of IPM, could be written as solving the following equation

$$\begin{pmatrix} \mathbf{Q}_{nn} & \mathbf{I}_{nn} & -\mathbf{I}_{nn} & \mathbf{y}_n \\ -diag(\lambda)_{nn} & diag(\mathbf{C}-\alpha)_{nn} & \mathbf{0}_{nn} & \mathbf{0}_n \\ diag(\xi)_{nn} & \mathbf{0}_{nn} & diag(\alpha)_{nn} & \mathbf{0}_n \\ \mathbf{y}^T & \mathbf{0}_n^T & \mathbf{0}_n^T & 0 \end{pmatrix} \begin{pmatrix} \triangle \mathbf{x} \\ \triangle \lambda \\ \triangle \xi \\ \triangle \nu \end{pmatrix} = \qquad (10.5)$$
$$- \begin{pmatrix} \mathbf{Q}\alpha - \mathbf{1}_n + \nu \mathbf{y} + \lambda - \xi \\ vec(\lambda_i(C-\alpha_i) - \frac{1}{t}) \\ vec(\xi_i \alpha_i - \frac{1}{t}) \\ \mathbf{y}^T \alpha \end{pmatrix},$$

where $diag(\mathbf{v})$ means generating an $n \times n$ square diagonal matrix whose diagonal element in the ith row is v_i; $vec(\alpha_i)$ means generating a vector with the ith component as α_i; I_{nn} is an identity matrix.

IPM boils down to solving the following equations in the Newton step iteratively.

$$\triangle \lambda = -\lambda + \text{vec}\left(\frac{1}{t(C - \alpha_i)}\right) + \text{diag}(\frac{\lambda_i}{C - \alpha_i})\triangle \mathbf{x} \tag{10.6}$$

$$\triangle \xi = -\xi + \text{vec}\left(\frac{1}{t\alpha_i}\right) - \text{diag}(\frac{\xi_i}{\alpha_i})\triangle \mathbf{x} \tag{10.7}$$

$$\triangle v = \frac{\mathbf{y}^T \Sigma^{-1} \mathbf{z} + \mathbf{y}^T \alpha}{\mathbf{y}^T \Sigma^{-1} \mathbf{y}} \tag{10.8}$$

$$\mathbf{D} = \text{diag}(\frac{\xi_i}{\alpha_i} + \frac{\lambda_i}{C - \alpha_i}) \tag{10.9}$$

$$\triangle \mathbf{x} = \Sigma^{-1}(\mathbf{z} - \mathbf{y}\triangle v), \tag{10.10}$$

where Σ and $\mathbf{z}$ depend only on $[\alpha, \lambda, \xi, v]$ from the last iteration as follows:

$$\Sigma = \mathbf{Q} + \text{diag}(\frac{\xi_i}{\alpha_i} + \frac{\lambda_i}{C - \alpha_i}) \tag{10.11}$$

$$\mathbf{z} = -\mathbf{Q}\alpha + \mathbf{1}_n - v\mathbf{y} + \frac{1}{t}\text{vec}(\frac{1}{\alpha_i} - \frac{1}{C - \alpha_i}). \tag{10.12}$$

The computation bottleneck is on matrix inverse, which takes place on Σ for solving $\triangle v$ in (10.8) and $\triangle \mathbf{x}$ in (10.10). We will mainly focus on this part as the other computations are trivial. Obviously, when the data set size is large, it is virtually infeasible to compute inversion of an $n \times n$ matrix due to resource and time constraints. It is beneficial to employ matrix factorization to factorize $\mathbf{Q}$. As $\mathbf{Q}$ is positive semi definite, there always exists an exact Cholesky factor: a lower-triangular matrix $\mathbf{G}$ that $\mathbf{G} \in \mathbb{R}^{n*n}$ and $\mathbf{Q} = \mathbf{GG}^T$. If we truncate $\mathbf{G}$ to $\mathbf{H}$ ($\mathbf{H} \in \mathbb{R}^{n*p}$ and $p \ll n$) by keeping only the most important p columns (i.e., minimizing $trace(\mathbf{Q} - \mathbf{HH}^T)$), this will become incomplete Cholesky factorization and $\mathbf{Q} \approx \mathbf{HH}^T$. In other words, $\mathbf{H}$ is somehow "close" to $\mathbf{Q}$'s exact Cholesky factor $\mathbf{G}$.

If we factorize $\mathbf{Q}$ this way and $\mathbf{D}$ is an identity matrix, according to SMW (the *Sherman-Morrison-Woodbury formula*) ([15]), we can write Σ^{-1} as

$$\Sigma^{-1} = (\mathbf{D} + \mathbf{Q})^{-1} \approx (\mathbf{D} + \mathbf{HH}^T)^{-1}$$
$$= \mathbf{D}^{-1} - \mathbf{D}^{-1}\mathbf{H}(\mathbf{I} + \mathbf{H}^T\mathbf{D}^{-1}\mathbf{H})^{-1}\mathbf{H}^T\mathbf{D}^{-1},$$

where $(\mathbf{I} + \mathbf{H}^T\mathbf{D}^{-1}\mathbf{H})$ is a $p \times p$ matrix. As p is usually small, it is practically feasible to compute it. In the following section, we will introduce how to parallelize the key steps of IPM to further speed it up.

10.3 PSVM Algorithm

The key step of PSVM is parallel ICF (PICF). Traditional column-based ICF [16, 17] can reduce computational cost, but the initial memory requirement is $O(np)$, and hence not practical for very large data set. PSVM devises parallel row-based ICF

Input: n training instances; p: rank of ICF matrix H; m: number of machines
Output: H distributed on m machines
Variables:
$\mathbf{v}$: fraction of the diagonal vector of Q that resides in local machine
k: iteration number
$\mathbf{x}_i$: the ith training instance
M: machine index set, $M = \{0, 1, \ldots, m-1\}$
I_c: row-index set on machine c ($c \in M$), $I_c = \{c, c+m, c+2m, \ldots\}$
1: **for** $i = 0$ to $n-1$ **do**
2:     Load $\mathbf{x}_i$ into machine $i\%m$.
3: **end for**
4: $k \leftarrow 0$; $H \leftarrow 0$; $\mathbf{v} \leftarrow$ the fraction of the diagonal vector of Q that resides in local machine.
   ($\mathbf{v}(i)(i \in I_m)$ can be obtained from $\mathbf{x}_i$)
5: Initialize *master* to be machine 0.
6: **while** $k < p$ **do**
7:     Each machine $c \in M$ selects its local pivot value, which is the largest element in $\mathbf{v}$:

$$\mathbf{lpv}_{k,c} = \max_{i \in I_c} \mathbf{v}(i),$$

and records the local pivot index, the row index corresponds to $\mathbf{lpv}_{k,c}$:

$$\mathbf{lpi}_{k,c} = \arg\max_{i \in I_c} \mathbf{v}(i)$$

8:     Gather $\mathbf{lpv}_{k,c}$'s and $\mathbf{lpi}_{k,c}$'s ($c \in M$) to *master*.
9:     The *master* selects the largest local pivot value as global pivot value $\mathbf{gpv}_k$ and records in i_k,
       row index corresponding to the global pivot value.

$$\mathbf{gpv}_k = \max_{c \in M} \mathbf{lpv}_{k,c}$$

10:    The *master* broadcasts $\mathbf{gpv}_k$ and i_k.
11:    Change *master* to machine $i_k\%m$.
12:    Calculate $H(i_k, k)$ according to Eq.(10.13) on *master*.
13:    The *master* broadcasts the pivot instance $\mathbf{x}_{i_k}$ and the pivot row $H(i_k, :)$. (Only the first $k+1$
       values of the pivot row need to be broadcast, since the remainder are zeros.)
14:    Each machine $c \in M$ calculates its part of the kth column of H according to Eq.(10.14).
15:    Each machine $c \in M$ updates $\mathbf{v}$ according to Eq.(10.15).
16:    $k \leftarrow k+1$
17: **end while**

Fig. 10.2: Row-based PICF

(PICF) as its initial step, which loads training instances onto parallel machines and performs factorization simultaneously on these machines. Once PICF has loaded n training data distributedly on m machines, and reduced the size of the kernel matrix through factorization, IPM can be solved on parallel machines simultaneously. We present PICF first, and then describe how IPM takes advantage of PICF.

10.3.1 Parallel ICF

ICF can approximate Q ($Q \in R^{n \times n}$) by a smaller matrix H ($H \in R^{n \times p}, p \ll n$), i.e., $Q \approx HH^T$. ICF, together with SMW (the *Sherman-Morrison-Woodbury formula*), can greatly reduce the computational complexity in solving an $n \times n$ linear system. The work of [16] provides a theoretical analysis of how ICF influences the optimization problem in Eq.(10.2). They proved that the error of the optimal objective value introduced by ICF is bounded by $C^2 l \varepsilon / 2$, where C is the hyperparameter of SVM, l is the number of support vectors, and ε is the bound of ICF approximation (i.e. $tr(Q - HH^T) < \varepsilon$). Experimental results in Section 10.4 show that when p is set to $\sqrt{n}$, the error can be negligible.

Our row-based parallel ICF (PICF) works as follows: Let vector $\mathbf{v}$ be the diagonal of Q and suppose the pivots (the largest diagonal values) are $\{i_1, i_2, \ldots, i_k\}$, the kth iteration of ICF computes three equations:

$$H(i_k, k) = \sqrt{\mathbf{v}(i_k)}, \qquad (10.13)$$

$$H(J_k, k) = (Q(J_k, k) - \sum_{j=1}^{k-1} H(J_k, j)H(i_k, j))/H(i_k, k), \qquad (10.14)$$

$$\mathbf{v}(J_k) = \mathbf{v}(J_k) - H(J_k, k)^2, \qquad (10.15)$$

where J_k denotes the complement of $\{i_1, i_2, \ldots, i_k\}$. The algorithm iterates until the approximation of Q by $H_k H_k^T$ (measured by $trace(Q - H_k H_k^T)$) is satisfactory, or the predefined maximum iterations (or say, the desired rank of the ICF matrix) p is reached.

As suggested by G. Golub, a parallelized ICF algorithm can be obtained by constraining the parallelized Cholesky Factorization algorithm, iterating at most p times. However, in the proposed algorithm [15], matrix H is distributed by columns in a round-robin way on m machines (hence we call it column-based parallelized ICF). Such column-based approach is optimal for the single-machine setting, but cannot gain full benefit from parallelization for two major reasons:

1. Large memory requirement. All training data are needed for each machine to calculate $Q(J_k, k)$. Therefore, each machine must be able to store a local copy of the training data.
2. Limited parallelizable computation. Only the inner product calculation

$$\sum_{j=1}^{k-1} H(J_k, j)H(i_k, j)$$

in Eq.(10.14) can be parallelized. The calculation of pivot selection, the summation of local inner product result, column calculation in Eq.(10.14), and the vector update in Eq.(10.15) must be performed on one single machine.

To remedy these shortcomings of the column-based approach, we propose a row-based approach to parallelize ICF, which we summarize in Figure 10.3. Our row-based approach starts by initializing variables and loading training data onto m machines in a round-robin fashion (Steps 1 to 5). The algorithm then performs the ICF main loop until the termination criteria are satisfied (e.g., the rank of matrix H reaches p). In the main loop, PICF performs five tasks in each iteration k:

1. Distributedly find a pivot, which is the largest value in the diagonal $\mathbf{v}$ of matrix Q (steps 7 to 10). Notice that PICF computes only needed elements in Q from training data, and it does not store Q.
2. Set the machine where the pivot resides as the *master* (step 11).
3. On the *master*, PICF calculates $H(i_k, k)$ according to Eq.(10.13) (step 12).
4. The *master* then broadcasts the pivot instance $\mathbf{x}_{i_k}$ and the pivot row $H(i_k, :)$ (step 13).
5. Distributedly compute Eqs.(10.14) and (10.15) (steps 14 and 15).

At the end of the algorithm, H is stored distributedly on m machines, ready for parallel IPM (presented in the next section). PICF enjoys three advantages: parallel memory use ($\mathscr{O}(np/m)$), parallel computation ($\mathscr{O}(p^2 n/m)$), and low communication overhead ($\mathscr{O}(p^2 \log(m))$). Particularly on the communication overhead, its fraction of the entire computation time shrinks as the problem size grows. We will verify this in the experimental section. This pattern permits a larger problem to be solved on more machines to take advantage of parallel memory use and computation.

Example

We use a simple example to explain how PICF works. Suppose we have three machines (or processors) and eight data instances, PICF first loads the data in a round-robin fashion on the three machines (numbered as #0, #1, and #2).

processor	data (*label id* : *value* [*id* : *value* $\cdots$])	row index
#0	-1 1:0.943578 2:0.397088	0
#1	-1 1:0.397835 2:0.097548	1
#2	1 1:0.821040 2:0.197176	2
#0	1 1:0.592864 2:0.452824	3
#1	1 1:0.743459 2:0.605765	4
#2	-1 1:0.406734 2:0.687923	5
#0	-1 1:0.398752 2:0.820476	6
#1	-1 1:0.592647 2:0.224432	7

Suppose the Laplacian kernel is used:

$$K(\mathbf{x}_i, \mathbf{x}_j) = e^{-\gamma \|\mathbf{x}_i - \mathbf{x}_j\|}, \tag{10.16}$$

and we set $\gamma = 1.000$. The first five columns of $Q_{ij} = y_i y_j K(\mathbf{x}_i, \mathbf{x}_j)$ is

$$Q = \begin{pmatrix} 1.000000 & 0.429436 & -0.724372 & -0.666010 & -0.664450 \\ 0.429436 & 1.000000 & -0.592839 & -0.576774 & -0.425776 \\ -0.724372 & -0.592839 & 1.000000 & 0.616422 & 0.614977 \\ -0.666010 & -0.576774 & 0.616422 & 1.000000 & 0.738203 \\ -0.664450 & -0.425776 & 0.614977 & 0.738203 & 1.000000 \\ 0.437063 & 0.549210 & -0.404520 & -0.656240 & -0.657781 \\ 0.379761 & 0.484884 & -0.351485 & -0.570202 & -0.571542 \\ 0.592392 & 0.724919 & -0.774414 & -0.795640 & -0.587344 \end{pmatrix}_{8 \times 5}$$

Note that the kernel matrix doesn't reside in memory, it is computed on-demand according to Eq.(10.16).

Iteration $k = 0$

PICF initializes v, whose elements are all one at the start. The elements of v are stored on the same machines as their corresponding x_i.

processor local diagonal vector v row index

$$\#0 \qquad v = \begin{pmatrix} 1.000000 \\ 1.000000 \\ 1.000000 \end{pmatrix} \qquad \begin{matrix} 0 \\ 3 \\ 6 \end{matrix}$$

processor local diagonal vector v row index

$$\#1 \qquad v = \begin{pmatrix} 1.000000 \\ 1.000000 \\ 1.000000 \end{pmatrix} \qquad \begin{matrix} 1 \\ 4 \\ 7 \end{matrix}$$

processor local diagonal vector v row index

$$\#2 \qquad v = \begin{pmatrix} 1.000000 \\ 1.000000 \end{pmatrix} \qquad \begin{matrix} 2 \\ 5 \end{matrix}$$

PICF next chooses the pivot. Each machine finds the maximum pivot and its index, and then broadcasts to the rest of the machines. Each machine then finds the largest value, and its corresponding index is the index of the global pivot. PICF sets the machine where the pivot resides as the *master* machine. In the first iteration, since all elements of v are one, the *master* can be set to machine #0. The global pivot value is 1, and its index 0.

Once the global pivot has been identified, PICF follows Eq.(10.13) to compute $H(i_0, 0) = H(0,0) = \sqrt{v(i_0)} = \sqrt{1} = 1$. The *master* broadcasts the pivot instance and the first $k+1$ value (in iteration $k = 0$, the *master* broadcasts only one value) of the pivot row of H (the i_0^{th} row of H). That is, the *master* broadcasts pivot instance $x_0 = -1 \ 1 : 0.943578 \ 2 : 0.397088$ and 1.

Next, each machine can compute rows of the first column of H according to Eq.(10.14). Take $H(4,0)$ as an example, which is located at machine #1. $Q(4,0)$ can be computed by the Laplacian kernel function using the broadcast pivot instance x_0 and x_4 on machine #1:

$$Q(4,0) = y_4 y_0 K(x_4, x_0) = y_4 y_0 exp(-\gamma \|x_4 - x_0\|) = -0.664450.$$

$H(0,0)$ can be obtained from the pivot row of H, which has been broadcast in the previous step. We thus get

$$H(4,0) = (Q(4,0) - \sum_{j=0}^{-1} H(4,j)H(0,j))/H(0,0) = Q(4,0)/H(0,0) = -0.664450.$$

Similarly, the other elements of the first column of H can be calculated on their machines. The result on machine #0 is as follows:

$$H_0 = \begin{pmatrix} 1.000000 & 0.000000 & 0.000000 & 0.000000 \\ 0.000000 & 0.000000 & 0.000000 & 0.000000 \\ 0.000000 & 0.000000 & 0.000000 & 0.000000 \end{pmatrix}$$

$$\downarrow$$

$$\begin{pmatrix} 1.000000 & 0.000000 & 0.000000 & 0.000000 \\ -0.666010 & 0.000000 & 0.000000 & 0.000000 \\ 0.379761 & 0.000000 & 0.000000 & 0.000000 \end{pmatrix}$$

The final step of the first iteration updates $\mathbf{v}$ distributedly according to Eq.(10.15).

$$v = \begin{pmatrix} v(0) - H(0,0)^2 \\ v(3) - H(3,0)^2 \\ v(6) - H(6,0)^2 \end{pmatrix} = \begin{pmatrix} 1.000000 - 1.000000^2 \\ 1.000000 - (-0.666010)^2 \\ 1.000000 - 0.379761^2 \end{pmatrix} = \begin{pmatrix} 0.000000 \\ 0.556430 \\ 0.855782 \end{pmatrix}$$

Iteration $k = 1$

PICF again, obtains local pivot values (the largest element of $\mathbf{v}$ on each machine, and their indexes.

#0 $localPivotValue_{1,0} = 0.855782$ $localPivotIndex_{1,0} = 6$
#1 $localPivotValue_{1,1} = 0.815585$ $localPivotIndex_{1,1} = 3$
#2 $localPivotValue_{1,2} = 0.808976$ $localPivotIndex_{1,2} = 5$

After the above information has been broadcast and received, the global pivot value is identified as 0.855782, and the global pivot index $i_1 = 6$. The id of the *master* machine is $6\%3 = 0$. Next, PICF calculates $H(i_1, 1)$ on the *master* according to Eq.(10.13). $H(6,1) = \sqrt{v(i_6)} = \sqrt{0.855782} = 0.925085$. PICF then broadcasts the pivot instance x_6, and the first $k+1$ elements on the pivot row of H, which are 0.379761 and 0.925085. Each machine then computes the second column of H according to Eq.(10.14). The result on machine #0 is as follows:

$$H_0 = \begin{pmatrix} 1.000000 & 0.000000 & 0.000000 & 0.000000 \\ -0.666010 & 0.000000 & 0.000000 & 0.000000 \\ 0.379761 & 0.000000 & 0.000000 & 0.000000 \end{pmatrix}$$

$$\downarrow$$

$$\begin{pmatrix} 1.000000 & 0.000000 & 0.000000 & 0.000000 \\ -0.666010 & -0.342972 & 0.000000 & 0.000000 \\ 0.379761 & 0.925085 & 0.000000 & 0.000000 \end{pmatrix}$$

In the final step of the second iteration, PICF updates v distributedly according to Eq.(10.15).

$$v = \begin{pmatrix} v(0) - H(0,1)^2 \\ v(3) - H(3,1)^2 \\ v(6) - H(6,1)^2 \end{pmatrix} = \begin{pmatrix} 0.000000 - 0.000000^2 \\ 0.556430 - (-0.342972)^2 \\ 0.855782 - 0.925085^2 \end{pmatrix} = \begin{pmatrix} 0.000000 \\ 0.438801 \\ 0.000000 \end{pmatrix}$$

Iteration $k = 3$

We fast-forward to show the end result of the fourth and final iteration of this example. The ICF matrix is obtained as follows:

computer	ICF matrix H				row index
#0	1.000000	0.000000	0.000000	0.000000	0
#1	0.429436	0.347862	0.833413	0.000000	1
#2	-0.724372	-0.082584	-0.303618	0.147541	2
#0	-0.666010	-0.342972	-0.205731	0.260080	3
#1	-0.664450	-0.345060	-0.024483	0.662451	4
#2	0.437063	0.759837	0.116631	-0.154472	5
#0	0.379761	0.925085	0.000000	0.000000	6
#1	0.592392	0.247443	0.461294	-0.146505	7

10.3.2 Parallel IPM

Solving IPM can be both memory and computation intensive. The computation bottleneck is on matrix inverse, which takes place on Σ for solving $\triangle v$ in (10.8) and $\triangle x$ in Eq.(10.10). Eq.(10.11) shows that Σ depends on Q, and we have shown that Q can be approximated through PICF by HH^T. Therefore, the bottleneck of the Newton step can be sped up from $\mathcal{O}(n^3)$ to $\mathcal{O}(p^2 n)$, and be parallelized to $\mathcal{O}(p^2 n/m)$.

Parallel Data Loading

To minimize both storage and communication cost, PIPM stores data distributedly as follows:

• *Distribute matrix data. H* is distributedly stored at the end of PICF.

- *Distribute $n \times 1$ vector data*. All $n \times 1$ vectors are distributed in a round-robin fashion on m machines. These vectors are $\mathbf{z}$, α, ξ, λ, $\Delta\mathbf{z}$, $\Delta\alpha$, $\Delta\xi$, and $\Delta\lambda$.
- *Replicate global scalar data*. Every machine caches a copy of global data including ν, t, n, and $\Delta\nu$. Whenever a scalar is changed, a broadcast is required to maintain global consistency.

Parallel Computation of $\Delta\nu$

Rather than walking through all equations, we describe how PIPM solves Eq.(10.8), where Σ^{-1} appears twice. An interesting observation is that parallelizing $\Sigma^{-1}z$ (or $\Sigma^{-1}y$) is simpler than parallelizing Σ^{-1}. Let us explain how parallelizing $\Sigma^{-1}z$ works, and parallelizing $\Sigma^{-1}y$ can follow suit.

According to SMW (the *Sherman-Morrison-Woodbury formula*), we can write $\Sigma^{-1}z$ as

$$
\begin{aligned}
\Sigma^{-1}z &= (D+Q)^{-1}z \approx (D+HH^T)^{-1}z \\
&= D^{-1}z - D^{-1}H(I+H^TD^{-1}H)^{-1}H^TD^{-1}z \\
&= D^{-1}z - D^{-1}H(GG^T)^{-1}H^TD^{-1}z.
\end{aligned}
$$

$\Sigma^{-1}z$ can be computed in four steps:

1. Compute $D^{-1}z$.
   D can be derived from locally stored vectors, following Eq.(10.9). $D^{-1}z$ is an $n \times 1$ vector, and can be computed locally on each of the m machines.
2. Compute $t_1 = H^TD^{-1}z$.
   Every machine stores some rows of H and their corresponding part of $D^{-1}z$. This step can be computed locally on each machine. The results are sent to the *master* (which can be a randomly picked machine for all PIPM iterations) to aggregate into t_1 for the next step.
3. Compute $(GG^T)^{-1}t_1$.
   This step is completed on the *master*, since it has all the required data. G can be obtained from $I + H^TD^{-1}H$ by Cholesky Factorization. Computing $t_2 = (GG^T)^{-1}t_1$ is equivalent to solving the linear equation system $t_1 = (GG^T)t_2$. PIPM first solves $t_1 = Gy_0$, then $y_0 = G^Tt_2$. Once it has obtained y_0, PIPM can solve $G^Tt_2 = y_0$ to obtain t_2. The *master* then broadcasts t_2 to all machines.
4. Compute $D^{-1}Ht_2$.
   All machines have a copy of t_2, and can compute $D^{-1}Ht_2$ locally to solve for $\Sigma^{-1}z$.

Similarly, $\Sigma^{-1}y$ can be computed at the same time. Once we have obtained both, we can solve $\Delta\nu$ according to Eq.(10.8).

10.3.3 Computing Parameter b and Writing Back

When the IPM iteration stops, we have the value of α and hence the classification function

$$f(x) = \sum_{i=1}^{N_s} \alpha_i y_i \mathbf{k}(s_i, x) + b.$$

Here N_s is the number of support vectors and s_i are support vectors. In order to complete this classification function, b must be computed. According to the SVM model, given a support vector s, we obtain one of the two results for $f(s)$: $f(s) = +1$, if $y_s = +1$, or $f(s) = -1$, if $y_s = -1$.

In practice, we can select M, say $1,000$, support vectors and compute the average of the bs:

$$b = \frac{1}{M} \sum_{j=1}^{M} \left(y_{s_j} - \sum_{i=1}^{N_s} \alpha_i y_i \mathbf{k}(s_i, s_j) \right).$$

Since the support vectors are distributed on m machines, PSVM collects them in parallel to compute b. For this purpose, we transform the above formula into the following:

$$b = \frac{1}{M} \sum_{j=1}^{M} y_{s_j} - \frac{1}{M} \sum_{i=1}^{N_s} \alpha_i y_i \sum_{j=1}^{M} \mathbf{k}(s_i, s_j).$$

The M support vectors and their labels ys are first broadcast to all machines. All m machines then compute their local results. Finally, the local results are summed up by a reduce operation [18]. When b has been computed, the last task of PSVM is to store the model file on GFS [19] for later classification use.

10.4 Experiments

We conducted experiments on PSVM to evaluate its: (1) class-prediction accuracy, (2) scalability on large datasets, and (3) overheads. The experiments were conducted on up to 500 machines in our data center. Not all machines are identically configured; however, each machine is configured with a CPU faster than 2GHz and memory larger than 4GB.

10.4.1 Class-Prediction Accuracy

PSVM employs PICF to approximate an $n \times n$ kernel matrix Q with an $n \times p$ matrix H. This experiment evaluated how the choice of p affects class-prediction accuracy. We set p of PSVM to n^t, where t ranges from 0.1 to 0.5 incremented by 0.1, and compared its class-prediction accuracy with that achieved by LIBSVM. The first two columns of Table 10.1 enumerate the datasets[1] and their sizes with which we

---

[1] *RCV* is located at http://jmlr.csail.mit.edu/papers/volume5/lewis04a/lyrl2004_rcv1v2_README.htm. The *image* set is a binary-class image dataset consisting of 144 perceptual features. The others are

Table 10.1: Class-prediction accuracy with different p settings

dataset	samples (train/test)	LIBSVM	$p = n^{0.1}$	$p = n^{0.2}$	$p = n^{0.3}$	$p = n^{0.4}$	$p = n^{0.5}$
svmguide1	3,089/4,000	0.9608	0.6563	0.9	0.917	0.9495	0.9593
mushrooms	7,500/624	1	0.9904	0.9920	1	1	1
news20	18,000/1,996	0.7835	0.6949	0.6949	0.6969	0.7806	0.7811
Image	199,957/84,507	0.849	0.7293	0.7210	0.8041	0.8121	0.8258
CoverType	522,910/58,102	0.9769	0.9764	0.9762	0.9766	0.9761	0.9766
RCV	781,265/23,149	0.9575	0.8527	0.8586	0.8616	0.9065	0.9264

experimented. We use Gaussian kernel, and select the best C and σ for LIBSVM and PSVM, respectively. For *CoverType* and *RCV*, we loosed the terminate condition (set -e 1, default 0.001) and used shrink heuristics (set -h 1) to make LIBSVM terminate within several days. The table shows that when t is set to 0.5 (or $p = \sqrt{n}$), the class-prediction accuracy of PSVM approaches that of LIBSVM.

We compared only with LIBSVM because it is arguably the best open-source SVM implementation in both accuracy and speed. Another possible candidate is CVM [12]. Our experimental result on the *CoverType* dataset outperforms the result reported by CVM on the same dataset in both accuracy and speed. Moreover, CVM's training time has been shown unpredictable by [20], since the training time is sensitive to the selection of stop criteria and hyper-parameters. For how we position PSVM with respect to other related work, please refer to our disclaimer in the end of Section 10.1.

10.4.2 Scalability

For scalability experiments, we used three large datasets. Table 10.2 reports the speedup of PSVM on up to $m = 500$ machines. Since when a dataset size is large, a single machine cannot store the factorized matrix H in its local memory, we cannot obtain the running time of PSVM on one machine. We thus used 10 machines as the baseline to measure the speedup of using more than 10 machines. To quantify speedup, we made an assumption that the speedup of using 10 machines is 10, compared to using one machine. This assumption is reasonable for our experiments, since PSVM does enjoy linear speedup when the number of machines is up to 30.

We trained PSVM three times for each dataset-m combination. The speedup reported in the table is the average of three runs with standard deviation provided in brackets. The observed variance in speedup was caused by the variance of machine loads, as all machines were shared with other tasks running on our data centers. We can observe in Table 10.2 that the larger is the dataset, the better is the speedup.

---

obtained from http://www.csie.ntu.edu.tw/~cjlin/libsvmtools/datasets/. We separated the datasets into training/testing (see Table 10.1 for the splits) and performed cross validation.

Table 10.2: Speedup (p is set to $\sqrt{n}$); LIBSVM training time is reported on the last row for reference

Machines	Image (200k)		CoverType (500k)		RCV (800k)	
	Time (s)	Speedup	Time (s)	Speedup	Time (s)	Speedup
10	1,958 (9)	10*	16,818 (442)	10*	45,135 (1373)	10*
30	572 (8)	34.2	5,591 (10)	30.1	12,289 (98)	36.7
50	473 (14)	41.4	3,598 (60)	46.8	7,695 (92)	58.7
100	330 (47)	59.4	2,082 (29)	80.8	4,992 (34)	90.4
150	274 (40)	71.4	1,865 (93)	90.2	3,313 (59)	136.3
200	294 (41)	66.7	1,416 (24)	118.7	3,163 (69)	142.7
250	397 (78)	49.4	1,405 (115)	119.7	2,719 (203)	166.0
500	814 (123)	24.1	1,655 (34)	101.6	2,671 (193)	169.0
LIBSVM	4,334 NA	NA	28,149 NA	NA	184,199 NA	NA

Figures 10.3(a), (b) and (c) plot the speedup of *Image*, *CoverType*, and *RCV*, respectively. All datasets enjoy a linear speedup[2] when the number of machines is moderate. For instance, PSVM achieves linear speedup on *RCV* when running on up to around 100 machines. PSVM scales well till around 250 machines. After that, adding more machines receives diminishing returns. This result led to our examination on the overheads of PSVM, presented next.

10.4.3 Overheads

PSVM cannot achieve linear speedup when the number of machines continues to increase beyond a data-size-dependent threshold. This is expected due to communication and synchronization overheads. Communication time is incurred when message passing takes place between machines. Synchronization overhead is incurred when the *master* machine waits for task completion on the slowest machine. (The *master* could wait forever if a child machine fails. We have implemented a check-point scheme to deal with this issue.)

The running time consists of three parts: computation (Comp), communication (Comm), and synchronization (Sync). Figures 10.3(d), (e) and (f) show how Comm and Sync overheads influence the speedup curves. In the figures, we draw on the top the computation only line (Comp), which approaches the linear speedup line. Computation speedup can become sublinear when adding machines beyond a threshold. This is because the computation bottleneck of the unparallelizable step 12 in Fig-

---

[2] We observed super-linear speedup when 30 machines were used for training *Image* and when up to 50 machines were used for *RCV*. We believe that this super-linear speedup resulted from performance gain in the memory management system when the physical memory was not in contention with other processes running at the data center. This benefit was cancelled by other overheads (explained in Section 10.4.3 when more machines were employed.

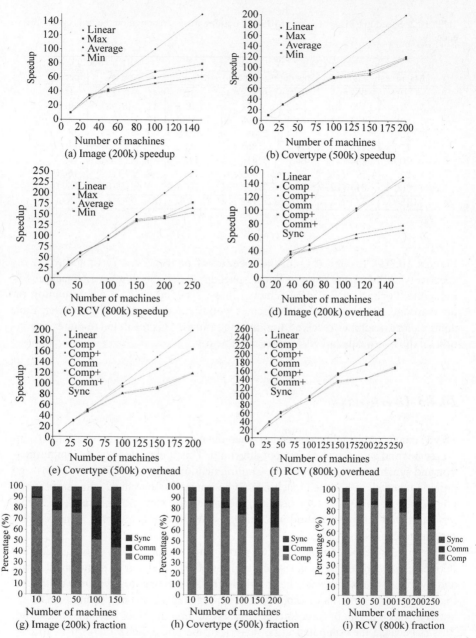

Fig. 10.3: Speedup and overheads of three datasets (See color insert)

ure 10.3 (which computation time is $\mathcal{O}(p^2)$). When m is small, this bottleneck is insignificant in the total computation time. According to the Amdahl's law; however, even a small fraction of unparallelizable computation can cap speedup. Fortunately, the larger the dataset is, the smaller is this unparallelizable fraction, which is $\mathscr{O}(m/n)$. Therefore, more machines (larger m) can be employed for larger datasets (larger n) to gain speedup.

When communication overhead or synchronization overhead is accounted for (the Comp + Comm line and the Comp + Comm + Sync line), the speedup deteriorates. Between the two overheads, the synchronization overhead does not impact speedup as much as the communication overhead does. Figures 10.3(g), (h), and (i) present the percentage of Comp, Comm, and Sync in total running time. The synchronization overhead maintains about the same percentage when m increases, whereas the percentage of communication overhead grows with m. As mentioned in Section 10.3.1, the communication overhead is $\mathscr{O}(p^2 \log(m))$, growing sub-linearly with m. But since the computation time per node decreases as m increases, the fraction of the communication overhead grows with m. Therefore, PSVM must select a proper m for a training task to maximize the benefit of parallelization.

10.5 Concluding Remarks

In this chapter, we have shown how SVMs can be parallelized to achieve scalable performance. PSVM distributively loads training data on parallel machines, reducing memory requirement through approximate factorization on the kernel matrix. PSVM solves IPM in parallel by cleverly arranging computation order. Through empirical studies, we have shown that PSVM does not sacrifice class-prediction accuracy significantly for scalability, and it scales well with training data size. Open source code of PSVM was made available at [2].

References

1. Chang, E.Y., Zhu, K., Wang, H., Bai, H., Li, J., Qiu, Z., Cui, H. Parallelizing support vector machines on distributed computers. In *Proceedings of NIPS*, 2007.
2. Chang, E.Y., Bai, H., Zhu, K., Wang, H., Li, J., Qiu, Z. Google PSVM Open Source. *http://code.google.com/p/psvm/*.
3. Vapnik, V. *The Nature of Statistical Learning Theory*. Springer, New York, 1995.
4. Mehrotra, S. On the implementation of a primal-dual interior point method. *SIAM J. Optimization*, 2:575–601, 1992.
5. Joachims, T. Making large-scale svm learning practical. In *Advances in Kernel Methods - Support Vector Learning*, 1998.
6. Chang, C.C., Lin, C.J. LIBSVM: a library for support vector machines. 2001.
7. Platt, J. Sequential minimal optimization: A fast algorithm for training support vector machines. Technical Report MSR-TR-98-14, Microsoft Research, 1998.
8. Vishwanathan, S., Smola, A.J., Murty, M.N. Simplesvm. In *Proceedings of ICML*, pages 760–767, 2003.
9. Joachims, T. Training linear svms in linear time. In *Proceedings of ACM KDD*, pages 217–226, 2006.
10. Chu, C.T., Kim, S.K., Lin, Y.A., Yu, Y., Bradski, G., Ng, A.Y., Olukotun, K. Map reduce for machine learning on multicore. In *Proceedings of NIPS*, pages 281–288, 2006.

11. Lee, Y.J., Huang, S.Y. Reduced support vector machines: A statistical theory. *IEEE Transactions on Neural Networks*, 18(1):1–13, 2007.
12. Tsang, I.W., Kwok, J.T., Cheung, P.M. Core vector machines: Fast svm training on very large data sets. *Journal of Machine Learning Research*, 6:363–392, 2005.
13. Graf, H.P., Cosatto, E., Bottou, L., Dourdanovic, I., Vapnik, V. Parallel support vector machines: The cascade svm. In *Proceedings of NIPS*, pages 521–528, 2005.
14. Boyd, S. *Convex Optimization*. Cambridge University Press, 2004.
15. Golub, G.H., Loan, C.F.V. *Matrix Computations*. Johns Hopkins University Press, 1996.
16. Fine, S., Scheinberg, K. Efficient svm training using low-rank kernel representations. *Journal of Machine Learning Research*, 2:243–264, December 2001.
17. Bach, F.R., Jordan, M.I. Predictive low-rank decomposition for kernel methods. In *Proceedings of International Conference on Machine Learning (ICML)*, pages 33–40, 2005.
18. Dean, J., Ghemawat, S. Mapreduce: Simplified data processing on large clusters. In *Proceedings of OSDI: Symposium on Operating System Design and Implementation*, pages 137–150, 2004.
19. Ghemawat, S., Gobioff, H., Leung, S.T. The Google file system. In *Proceedings of 19th ACM Symposium on Operating Systems Principles*, pages 29–43, 2003.
20. Loosli, G., Canu, S. Comments on the core vector machines: Fast svm training on very large data sets. *Journal of Machine Learning Research*, 8:291–301, 2007.

Chapter 11
Approximate High-Dimensional Indexing with Kernel

Abstract Indexing high-dimensional data for efficient nearest-neighbor searches poses challenges. It is well known that when data dimension is very high, the search time can exceed the time required for performing a linear scan on the entire dataset. To alleviate this *dimensionality curse*, indexing schemes such as *locality sensitive hashing* (LSH) and *M-trees* were proposed to perform *approximate* searches. In this chapter[†], we present a *hypersphere indexer*, named SphereDex, to perform such searches. SphereDex partitions the data space using concentric hyperspheres. By exploiting geometric properties, SphereDex can perform effective pruning. Our empirical study shows that SphereDex enjoys three advantages over competing schemes for achieving the same level of search accuracy. First, SphereDex requires fewer disk-seek operations. Second, SphereDex can maintain disk accesses sequential most of the time. And third, it requires fewer distance computations. More importantly, SphereDex can be extended to support *hyperplane queries* for Support Vector Machines (SVMs) or the kernel methods. In classification problems using SVMs, the data instances closest to the hyperplane are considered to be most ambiguous, and the ones farthest away from the hyperplane to be most certain (or most confident) regarding their class membership. Hyperplane queries, rather than point queries, are essential to supporting fast retrieval of applications using SVMs. In the end of this chapter, we illustrate how SphereDex can be extended to support both nearest and farthest neighbor hyperplane query processing.

Keywords: Dimensionality curse, high-dimensional indexing, nearest neighbor query, approximate nearest neighbor query.

11.1 Introduction

Nearest neighbor search has generated a great deal of interest because of a wide range of applications such as text, image/video, and bio-data retrieval. These applications represent objects (text documents, images, or bio-data) as *feature vectors* in very high-dimensional spaces. A user submits a query object to a search engine, which returns objects that are similar to the query object. The degree of similarity between two objects is measured by some distance function (e.g., Euclidean) over their feature vectors. The search is performed by returning the objects that are *nearest* to the query object in the high-dimensional vector space.

With the vast volume of data available for search, indexing is essential to provide scalable search performance. However, when data dimension is high (higher than 20 or so), no nearest-neighbor algorithm can be significantly faster than a linear scan of the entire dataset. Let n denote the size of a dataset and d the dimension of data, the theoretical studies of [2, 3, 4, 5] show that when $d \gg \log n$, a linear search will outperform classic search structures such as k-d-trees [6], SR-trees [7], and SS-trees [8]. Several recent studies (e.g., [3, 4, 9]) provide empirical evidence, all confirming this phenomenon of *dimensionality curse*.

The prohibitive nature of exact nearest-neighbor search has led to the development of *approximate nearest-neighbor search* that returns instances approximately similar to the query instance [2, 10]. The first justification behind approximate search is that a feature vector is often an approximate characterization of an object, so we are already dealing with approximations [11]. Second, an approximate set of answers suffices if the answers are relatively close to the query concept. Of late, two approximate indexing schemes, *locality sensitive hashing* (LSH) [12] and M-trees [13], have been employed in applications such as image-copy detection [14] and bio-sequence-data matching [15]. These approximate indexing schemes speed up similarity search significantly (over a sequential scan) by slightly lowering the bar for accuracy.

This chapter presents a *hypersphere indexer*, named SphereDex, to perform approximate nearest-neighbor searches. First, the indexer finds a roughly central instance among a given set of instances. Next, the instances are partitioned based on their distances from the central instance. SphereDex builds an *intra-partition* (or local) index within each partition to efficiently prune out irrelevant instances. It also builds an *inter-partition* index to help a query to identify a good starting location in a neighboring partition to search for nearest neighbors. A search is conducted by first finding the partition to which the query instance belongs. (The query instance does not need to be an existing instance in the database.) SphereDex then searches in this and the neighboring partitions to locate nearest neighbors of the query. Notice that since each partition has just two neighboring partitions, and neighboring partitions can largely be sequentially laid out on disks, SphereDex can enjoy sequential IO performance (with a tradeoff of transferring more data) to retrieve candidate partitions into memory. Even in situations (e.g., after a large batch of insertions) when one sequential access might not be feasible for retrieving all candidate partitions, SphereDex can keep the number of non-sequential disk accesses low. Once a partition has been retrieved from the disk, SphereDex exploits geometric properties to perform intelligent intra-partition pruning so as to minimize the computational cost for finding the top-k approximate nearest neighbors. Through empirical studies on two very large, high-dimensional datasets, we show that SphereDex significantly outperforms both LSH and M-trees in both IO and CPU time.

In summary, the main benefits of SphereDex are as follows:

1. SphereDex is a hypersphere partitioning scheme for indexing high-dimensional data. By the nature of the partitioning, it can maintain sequential disk-IO performance under reasonable volumes of insertions.
2. SphereDex employs an effective pruning algorithm, which performs actual distance computations on only a small fraction of instances in a partition.
3. SphereDex can work with data in the projected kernel space. SphereDex is designed to support both point queries and hyperplane queries.

The rest of the chapter is organized into four sections. In Section 11.2, we discuss related work. Section 11.3 describes in details SphereDex's operations: creation, search, insertion, and deletion. Section 11.4 presents experimental results. Finally, we offer our closing remarks in Section 11.5.

11.2 Related Reading

For over a decade, indexing high-dimensional data has been an active area of research, and a wide variety of approaches have been attempted. We present some representative methods in this section, but our discussion is by no means exhaustive. (For a comprehensive survey, please consult [16, 17].)

Existing indexers can be divided into two categories: *coordinate-based* and *distance-based*. The coordinate-based methods work on data instances residing in a vector space by partitioning the space into *minimum bounding regions* (MBRs). A top-k nearest-neighbor query can be treated as a range query, and ideally, only a small number of MBRs need to be examined for finding the best matches. Example coordinate-based methods are the X-tree [18], the R^*-tree [19], the TV-tree [20], the SR-tree [7], and the M-tree [21], to name a few. A big disadvantage of these tree-structures is the exponential decay in performance that accompanies an increase in the dimensionality of data instances [5]. This phenomenon has been reported for the R^*-tree, the X-tree, and the SR-tree, among others. The decay in performance can be attributed to the almost exponential number of rectanglar MBRs (in the cases of the X-tree and the R^*-tree) or spherical MBRs (in the cases of the SR-tree, the TV-tree, and the M-tree)[1] that need to be examined for finding nearest neighbors. In con-

---

[1] Each dimension of a rectangular MBR has two neighboring rectangular MBRs. To ensure retrieval of exact set of top-k nearest neighbors, a search needs to examine all 3^d neighboring MBRs [11]. For the tree-structures using spherical MBRs, [5] reports that their performance decays almost as rapid as the tree-structures using rectangluar MBRs.

trast to these tree-structures, each partition of SphereDex has only two neighboring partitions, independent of data dimension.

The distance-based methods do not require an explicit vector space, but rely only on the existence of a pairwise distance metric. The M-tree [21] is a representative scheme that uses the distances between instances to build an indexing structure. (Some other distance-based methods are the multi-vantage-point trees [22], geometric near-neighbor access trees [23], and spatial approximation trees [24].) The M-tree has been shown to be a more effective method for indexing high-dimensional data [21]. Given a query point, it prunes out instances based on distance. The M-tree performs IOs to retrieve relevant data blocks into memory for processing; however, the disk accesses cannot be sequential since the data blocks can be randomly scattered on disks. In contrast to the M-tree, SphereDex can largely maintain contiguous layout of neighboring partitions on disks, and hence its IOs can be sequential and efficient.

As mentioned in Section 11.1, when data dimension is very high, the cost of supporting exact queries can be higher than that of a linear scan. The work of [12] proposes an approximate indexing strategy using locality sensitive hashing (LSH). This approach attempts to hash nearest-neighbor instances into the same bucket, with high probability. A top-k approximate query can be supported by retrieving the bucket into which the query point has been hashed. To improve search accuracy, multiple hash functions must be used. Theoretically, the number of hash functions to be used is given by $(\frac{n}{B})^\rho$, where n is the number of instances in the dataset, B is the number of instances that can be accommodated in a single bucket, and ρ a tunable parameter. Another approximate approach is clustering for indexing [11, 25] (ClinDex and DynDex). Queries are handled by ranking the clusters with respect to their distances to the query and then processing the clusters in the order of their closeness to the query.

As we will show in Section 11.4, SphereDex outperforms both LSH and an approximate version of M-trees[2]. First, SphereDex requires fewer IOs compared to competing approaches to achieve the same level of search accuracy. Second, when a dataset is relatively static (without a large number of insertions), SphereDex can largely maintain its IOs sequential. Furthermore, as depicted in greater details in [1], SphereDex can support hyperplane queries to work with kernel methods discussed in Chapters 2, 3, 5, and 6, whereas traditional schemes cannot.

11.3 Algorithm SphereDex

Most high-dimensional indexing methods require a large number of *buckets* (data blocks or data pages) to be retrieved for answering a query. Additionally, these buckets may be located at random locations on disks, and hence a seek overhead

---

[2] The M-tree also provides an approximate version of the algorithm [13]. For a fair comparison, we use the approximate version of the code provided by M. Patella to conduct experiments. The code is available on request, but is not part of the basic download available at [26].

is required for reading each bucket. The large number of IOs and the high seek overhead of each IO can make query processing exceedingly expensive.

SphereDex aims to improve the performance of nearest-neighbor searches using a three-pronged approach. First, we attempt to reduce the number of IOs. Second, when multiple disk blocks must be retrieved, we make the IO sequential as much as possible. Third, when data is finally staged in main memory for processing, we minimize the amount of data that needs to be examined to obtain the top-k approximate nearest neighbors. Before we begin, let us present our problem statement formally in the following definition.

Definition 11.1. (ε-Nearest Neighbor Search) [27] Given a set P of points in a normed space l_p^d, preprocess P so as to efficiently return a point $p \in P$ for any query point q, such that $d(q,p) \leq (1+\varepsilon)d(q,P)$, where $d(q,P)$ is the distance of q to its closest point in P.

Generalizing this to $k > 1$, we wish to find the k points $p_1, p_2, \cdots, p_k$ such that the distance of p_i to the query q is at most $(1 + \varepsilon)$ times the distance from the ith nearest point to q. Hereafter, we refer to the normed space l_p^d as *data space*. Vectors in this data space are referred to as *data instances*. In the remainder of this section, we present the three operations of SphereDex: *creation*, *search*, and *update*.

11.3.1 Create — Building the Index

The indexer is created in four steps.

1. Finding the instance $\mathbf{x}_c$ that is approximately centrally located in the data space,
2. Separating the instances into partitions based on their distances from the central instance $\mathbf{x}_c$,
3. Constructing for each partition a local indexing structure (*intra-partition indexer*), and
4. Creating an *inter-partition* index.

11.3.1.1 Choosing the Center Instance

Input:
  1. Number of instances: n.
  2. Dataset instances: $\mathbf{x}_1, \mathbf{x}_2, \cdots, \mathbf{x}_n$.
  Output: Approximate center instance $\mathbf{x}_c$.

We use an approximately central instance in the dataset to serve as the center of the concentric hyperspherical partitions. This center is found by finding the instance at the smallest distance from the centroid of the data instances. This is done to ensure that the instances are "roughly" uniformly distributed in all directions around the reference point.

The centroid of the available instances can be computed in $O(nd)$ time, and finding the instance at the smallest distance from the centroid takes $O(nd)$ time. Therefore, overall this step can be accomplished in $O(nd)$ time and $O(d)$ space. When the size of the dataset is large, we can use sampling to lower this cost, since we are only interested in the approximate center of the distribution of data instances.

11.3.1.2 Partitioning the Instances

Input
  1. Number of instances: n.
  2. Dataset instances: $\mathbf{x}_1, \mathbf{x}_2, \cdots, \mathbf{x}_n$.
  3. Approximate center instance: $\mathbf{x}_c$.
  4. Number of instances per partition: g.
Output:
  1. Number of partitions: n_p.
  2. Partitions: $P[1], P[2], \cdots, P[n_p]$.

Once the center instance has been determined, our next step is to partition the instances in the dataset based on their distances from the center instance. We sort the distances of the instances from the central instance. To separate the instances, we need to partition this sorted array. This step requires $O(n \log n + nd)$ time and $O(n)$ space. The created partitions are then placed on the disk. Our placement policy aims to achieve two goals. First, we would like to place adjacent partitions contiguously on disk to achieve sequential disk accesses. Second, we need to reserve space within partitions to anticipate insertions of new data instances. Since instances within a partition can be placed in any order, in-partition free-space management is straightforward. However, tradeoff exists between the amount of free space reserved for insertions and IO efficiency. At one extreme, no free space is reserved and sequential IOs can be ensured, but the dataset cannot permit insertions. At the other extreme, abundant space is reserved for insertions, but IO resolution (the amount of useful data retrieved over the amount of data retrieved) can be low. We discuss the placement policy in greater detail when we discussion how SphereDex handles the insertion operation in Section 11.3.3. In Section 11.4.5 we quantify the tradeoff between free space and IO performance.

11.3.1.3 Intra-Partition Index

Input
  1. Partitions: $P[1], P[2], \cdots, P[n_p]$.
  2. Number of instances per partition: g.
Output: Local indices for each partition: $S[1], S[2], \cdots, S[n_p]$.

An intra-partition (or local) index is created for each partition and stored on disk along with the instances in the partition. For each instance, this local index stores a sorted list of instances ranked according to their distances from this instance. Maintaining this data structure for all pairs would take up $O(g^2)$ space. Instead, we store two reduced sets of information.

1. The distances of only $O(\log g)$ instances for every instance $\mathbf{x}$ in the partition. Let $L = (l_1, l_2, \cdots, l_g)$ be the ordering of all instances in the partition based on their distances from $\mathbf{x}$. Let $L_1 = (l_1, l_2, \cdots, l_{\frac{g}{2}})$ and $L_2 = (l_{\frac{g}{2}+1}, l_{\frac{g}{2}+2}, \cdots, l_g)$ be two equal halves of L. In our local index we store the distances of $4 \times \log \frac{g}{2}$ instances with $\log \frac{g}{2}$ distances from each end of L_1 and L_2, thus selecting

$$l_1, l_2, l_4, \cdots, l_{2^{\lfloor \log \frac{g}{2} \rfloor}};$$
$$l_{\frac{g}{2}}, l_{\frac{g}{2}-1}, l_{\frac{g}{2}-3}, \cdots, l_{\frac{g}{2}-2^{\lfloor \log \frac{g}{2} \rfloor}+1};$$
$$l_{\frac{g}{2}+1}, l_{\frac{g}{2}+2}, l_{\frac{g}{2}+4}, \cdots, l_{\frac{g}{2}+2^{\lfloor \log \frac{g}{2} \rfloor}+1};$$
$$l_g, l_{g-1}, l_{g-3}, \cdots, l_{g-2^{\lfloor \log \frac{g}{2} \rfloor}+1}. \tag{11.1}$$

2. The distances of the r closest instances. Out of g instances, we store just a small fraction of the distances. We show in our experiments that a typical g is on the order of thousands, and setting $r = 30$, we can save tremendous amount of space without losing search accuracy. We will discuss the detailed reason in Section 11.3.2.3.

Combining the above two, the local index of SphereDex uses $O(g \log g)$ space per partition. The total space required for the local index would therefore be $O(n \log g)$. Constructing this local index for all partitions requires $O(ng \log g + ngd)$ time. Operations on this local index are explained in Section 11.3.2 when we discuss how query-processing takes place.

11.3.1.4 Inter-Partition Index

Input
  1. Number of partitions: n_p.
  2. Partitions: $P[1], P[2], \cdots, P[n_p]$.
  3. Number of instances per partition: g.
Output: Inter-partition indices: I.

We also maintain a data structure which contains neighborhood information across adjacent partitions. This index gives SphereDex a good location to start searching k-NN when a search transitions from one partition to the next. Intuitively, a good starting point to continue searching is an instance that is very close to the current set of top-k results. This index stores, for each instance in a partition, its nearest neighbor(s) in the adjacent partition(s). Building this index requires $O(ngd)$ time. The storage required to maintain this index is of the order $O(n)$.

11.3.2 Search — Querying the Index

Given a query instance, query processing begins by finding approximate nearest neighbors of the query in the partition to which it belongs. Thereafter, adjacent partitions are retrieved and searched for better nearest neighbors. The set of nearest neighbors improves as we continue to retrieve and process adjacent partitions. SphereDex terminates its search for top-k when the constituents of the top-k set do not change over the evaluation of multiple partitions, or the query time expires. SphereDex achieves speedup over the naive linear scan method in two ways. First, SphereDex does not examine all partitions for a query, and the partitions it examines could be sequentially retrieved from disks. Second, SphereDex examines only a small fraction of the instances in a partition. The remainder of this section details these steps, explaining how SphereDex effectively approximates the top-k result for achieving significant speedup.

11.3.2.1 Identification of Starting Partition

Input
  1. Query instance: $\mathbf{x}_q$.
  2. Center instance: $\mathbf{x}_c$.
  3. Delimiter distances: *Delim*.
Output: Partition number: α.

First, we identify the partition to which the query instance $\mathbf{x}_q$ belongs. Recall that the partitions were constructed by separating the sorted list of instances in the dataset, based on their distances to the center instance $\mathbf{x}_c$. Based on the distance between $\mathbf{x}_q$ and $\mathbf{x}_c$, we can find exactly where the query instance should be on the sorted list via a binary search. (The query instance $\mathbf{x}_q$ does not have to be an existing query instance.) Instead of maintaining the distances of all the instances from the center, we keep another sorted list, *Delim*, of the distances of only the starting instance of each partition. The cost of this binary search is $O(\log n_p)$ (n_p denotes the number of partitions).

11.3.2.2 Intra-Partition Pruning

Input
  1. Query instance: $\mathbf{x}_q$.
  2. Starting partition: $P[\alpha]$.

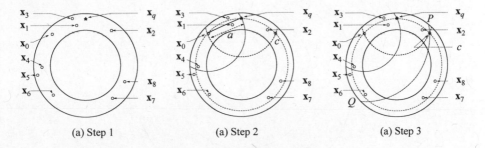

(a) Step 1   (a) Step 2   (a) Step 3

Fig. 11.1: Arrangement of instances

3. *Intra-partition index*: $S[\alpha]$.

4. *Inter-partition index*: $I[\alpha]$.

Output: Approximate k-NNs of x_q.

Once the starting partition $P[\alpha]$ has been identified, we retrieve the instances in that partition and the associated local index $S[\alpha]$ from the disk (if not yet in main memory). We then find the one single nearest neighbor of x_q among the instances in $P[\alpha]$. From that one single nearest neighbor, we can use the intra-partition index $S[\alpha]$ to harvest the approximate k-NNs of x_q.

The exact nearest neighbor of the query instance in the partition can be identified by computing distances of all instances in $P[\alpha]$ to x_q, but this is an expensive operation. Instead, we use an iterative procedure to find the nearest neighbor. The process starts by selecting an arbitrary instance x_0 in $P[\alpha]$, and then iteratively finds instances closer to x_q (than x_0) until no improvement is possible. At first glance, this procedure might sound expensive, but it is not, as we explain next through an example in Figure 11.1.

Figure 11.1(a) depicts an example partition $P[\alpha]$. The local index of x_0 contains nine data instances: $x_0 \cdots x_8$. The query instance x_q is located at the top of the figure (we assume x_q was not in the dataset). We use x_0 as an anchor to find instances closer to x_q. Let the distance between x_0 and x_q be a. Starting at x_0, we seek to find an instance as close to x_q as possible. The intra-partition index of x_0 contains an ordered list of instances based on their distances from x_0. (Note that, we have, in fact, stored $O(\log g)$ instances in the ordered list. We discuss its impact in Section 11.3.2.3). For the example in Figure 11.1, the neighboring points of x_0 appear in the order of x_3, x_1, x_4, x_5, x_2, x_6, x_7, and x_8 on the sorted list of x_0.

To find an instance closer than x_0 to x_q, we search this list for instances with a separation of about a from x_0. Pictorially, Figure 11.1(b) shows the region where a better nearest neighbor can reside is in the radius of a centered at x_q. Since we as-

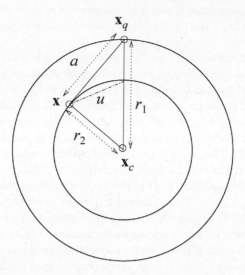

Fig. 11.2: Identifying relevant instances in other partitions

sume that $\mathbf{x}_q$ is an external data instance and we do not have an intra-partition index for it, we must solely rely on the intra-partition index of $\mathbf{x}_0$ to find a better nearest neighbor. Our pruning is performed in two steps. First, we prune out instances that cannot possibly be closer to $\mathbf{x}_q$ than $\mathbf{x}_0$. Second, we find an instance whose distance from $\mathbf{x}_0$ is closest to a and which is nearer to $\mathbf{x}_q$.

1. *Pruning away the impossible instances.* We have only the intra-partition index of $\mathbf{x}_0$ to work with. We are only interested in the instances that are a from $\mathbf{x}_q$. Figure 11.1(c) shows that the instances lying outside of the PQ arc cannot be closer to $\mathbf{x}_q$ than $\mathbf{x}_0$. This step allows us to prune instances $\mathbf{x}_7$ and $\mathbf{x}_8$.
2. *Finding an instance nearer to $\mathbf{x}_q$.* Next, we would like to re-sort the instances remaining on the list of $\mathbf{x}_0$, based on their likelihood of being close to $\mathbf{x}_q$. To quantify this likelihood for instance $\mathbf{x}_i$, we compute how close the distance between $\mathbf{x}_i$ and $\mathbf{x}_0$ is to the distance between $\mathbf{x}_q$ and $\mathbf{x}_0$ (which is a). The list does not need to be explicitly constructed since we have the distance between $\mathbf{x}_i$ and $\mathbf{x}_0$ sorted and stored in the intra-partition index. We find the position of the instance, $\mathbf{x}_r$, whose distance from $\mathbf{x}_0$ is closest to a. The rest of the instances on the re-sorted list can be obtained by looking up the adjacent instances of $\mathbf{x}_r$ in the local index. In our example, this re-sorted list is $\mathbf{x}_4$, $\mathbf{x}_5$, $\mathbf{x}_1$, $\mathbf{x}_3$, $\mathbf{x}_2$ and $\mathbf{x}_6$.
   Now, it may be surprising to see that $\mathbf{x}_5$ and $\mathbf{x}_4$ appear before $\mathbf{x}_1$ on the re-sorted list. This is because we know only their distances from $\mathbf{x}_0$. Fortunately, pruning out $\mathbf{x}_5$ and $\mathbf{x}_4$ is simple — we need to remove instances that are farther from $\mathbf{x}_q$ than $\mathbf{x}_0$ as we go down the list. In this case, $\mathbf{x}_5$ and $\mathbf{x}_4$ are farther from $\mathbf{x}_q$ than $\mathbf{x}_0$. After removing them from the re-sorted list, we harvest $\mathbf{x}_1$ and use it as the anchor instance for the next pruning iteration.

The iterations continue till no nearer instance than the current anchor can be found. The approximate top-k instances nearest to x_q are the k nearest neighbors on the current anchor's intra-partition sorted list.

11.3.2.3 Lowering the Size of the Local Index

The size of the local index, if we store the pairwise distances of all instances in the partition would be equal to $O(g^2)$. Since there are n/g partitions in total assuming that the instances are equally divided among the partitions, the total size of the index would be $O(ng)$. Here, we explain how this size can be reduced to $O(n \log g)$ with a low probability of loss of accuracy.

To understand why a reduced index can be used, we revisit how the query processing works. At each iteration, we are looking for instances based on their perceived nearness to the query point. Suppose we were examining the local index of instance x. Instances appearing in the local index of x can either be closer to the query instance or farther away from it.

Now, instead of storing the distances of all the instances from x in the local index of x, if we stored just the distances of $\log g$ items, the probability of not finding an instance closer to the query point after evaluating all these instances would be $\frac{1}{2^{\log g}}$, which is essentially $\frac{1}{g}$. This probability can be reduced further by storing a larger number of instances ($> \log g$). How many do we need to store to maintain high search accuracy?

Let $L = (l_1, l_2, l_3, \cdots, l_g)$ be the list representing the ordering of instances in the partition based on their distances from x. Let $L_1 = (l_1, l_2, \cdots, l_{\frac{g}{2}})$ and $L_2 = (l_{\frac{g}{2}+1}, l_{\frac{g}{2}+2}, \cdots, l_g)$ be two equal halves of L. In our local index we store distances of $4 \times \log \frac{g}{2}$ instances with $\log \frac{g}{2}$ distances from each end of both the lists L_1 and L_2 (shown in Eq.(11.1) in Section 11.3.1.)

For $g = 1,000$, the probability of not finding an instance closer to the query instance, after evaluating all the instances in the local index of x, is less than 10^{-11}.

As we get closer to the query instance, the number of instances on the list that are examined becomes small and the possibility of not finding an instance closer to x_q becomes more likely. This situation can easily be addressed by also adding the instances closest to x in the partition to our index. That is, we add the distances of the instances, say $l_1, l_2, \cdots, l_{30}$ to our index. The figure of 30 was chosen because $\frac{1}{2^{30}}$ is approximately equal to 10^{-9}. Thus, even when we do get close to the query instance, the probability of not finding an instance is very low. An added advantage of picking instances as explained above, is that the instances are picked from all regions of the original list without excessively large gaps between the chosen instances.

Therefore, the number of distances stored in the local index of instance x is bounded by $4 \times \log \frac{g}{2} + 30$. (However, since some of the chosen positions overlap, the number of instances chosen is lower than this number.) For $g = 1,000$, the number of instances stored is about 60. That is, the size of the local index for each partition is of the order $O(g \log g)$, and the size of the index for all the partitions is $O(n \log g)$.

11.3.2.4 Search within Adjacent Partition

As SphereDex progresses its search from the initial partition to which $\mathbf{x}_q$ belongs to the other partitions, identifying instances close to $\mathbf{x}_q$ in a partition to which $\mathbf{x}_q$ does not belong requires a little more work. Considering the example in Figure 11.2, we see two partitions, with the query point $\mathbf{x}_q$ belonging to one partition and the instance being evaluated $\mathbf{x}$ belonging to another partition. The radii of the two partitions are given by r_1 and r_2. The values of r_1 and r_2 are determined by the distances of the instances from the center instance $\mathbf{x}_c$. In addition, we also know the distance a between $\mathbf{x}$ and $\mathbf{x}_q$.

In order to identify suitable instances in the partition to which $\mathbf{x}$ belongs, we need to compute the distance u shown in the figure. The instances of interest in this partition lie close to the point of intersection of the line joining $\mathbf{x}_q$ and $\mathbf{x}_c$ with the hypersphere of radius r_2. Since we know the lengths r_1, r_2 and a, we know the angle formed at the center $\mathbf{x}_c$ by the side a of the triangle with vertices $\mathbf{x}_q, \mathbf{x}$ and $\mathbf{x}_c$. Now to find u we note that we have an isosceles triangle with two sides of length r_2. Since we know the angles between these sides we can find the length of the third side which is actually u. We note that the computations outlined in this section would be applicable for any instance $\mathbf{x}$ not lying on the surface of the hypersphere $\mathbf{x}_q$. That is, even in the first partition examined, if the instance $\mathbf{x}$ does not lie on the surface of the hypersphere corresponding to $\mathbf{x}_q$, we would need to perform the above computations.

11.3.2.5 Inter-Partition Co-Operation

As the algorithm proceeds, we advance from the partition to which the query instance belongs, to partitions adjacent to it. Having identified the instance $\mathbf{x}$ in a partition close to the query instance, we use the inter-partition index to select a good starting instance in the adjacent partition. Since the inter-partition index contains the instance from the adjacent partition closest to the instance $\mathbf{x}$, there is a high probability that the inter-partition index gives us an instance very close to the query instance. This has the effect of lowering the number of iterations needed to converge to the approximate best instance in the adjacent partition.

11.3.2.6 Stopping Criteria

When query time is the criterion, and the time is limited, query processing terminates when the time expires. SphereDex in such circumstances may not be able to find all top-k results, but it can still provide reasonable results depending on the

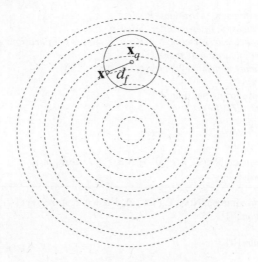

Fig. 11.3: Stopping criteria

amount of time available. In Section 11.4, we show the tradeoff between time and accuracy. One can select a termination time when the expected accuracy will have a high probability of being achieved.

When quality is the criterion, the selected stopping criterion can ensure that the best approximate results are obtained. In our quest for the top-k instances closest to the given query instances, we start evaluation in the partition containing the query point. Having found the instance in the partition that is closest to the query point, we select its k nearest neighbors as our results from the partition. Let the distance of the instance among these k farthest from the query point be represented by d_f (Figure 11.3). Then d_f represents an upper bound on the farthest distance among the *best* set of instances. A partition farther than d_f from the partition containing the query instance can safely be excluded from consideration, since it cannot contain an instance with a distance less than d_f from x_q. We need to evaluate partitions which are within a radius of d_f from the current partition. Since d_f is continually revised as the partitions are processed, the stopping criteria can only be revised in the direction of reducing the number of partitions that need to be processed, and hence SphereDex will converge. Figure 11.4 documents the entire algorithm.

11.3.3 Update — Insertion and Deletion

SphereDex in general does not need to process a large number of partitions to achieve high retrieval accuracy. In the experiment section, we will show that SphereDex incurs less number of IOs compared to competing methods. In addition to keeping the number of IOs small, SphereDex intends take advantage of the

Input: *Query instance*:$\mathbf{x}_q$; *Partitions*: P;
           Local indices: S; *Inter-partition index*: I;
           Partition delimiters: *Delim*; *Number of instances per partition*: g;
           Number of nearest neighbors: k;
Output: *Approximate nearest neighbors* :$\mathbf{z}_1, \cdots , \mathbf{z}_k$;
$r_1 = distance(\mathbf{x}_q, \mathbf{x}_c)$
/* Determination of starting partition*/
$\alpha = $ Binary_search(r_1, *Delim*)
$i = 0$
$start1 = 0$
while True **do**
    $num1 = $ Approximate_NN($\mathbf{x}_q$, k, $P[\alpha - i]$, $S[\alpha - i]$, I, $start1$, r_1)
    $start1 = I[num1][2]$
    if $i \neq 0$ **then**
        $num2 = $ Approximate_NN($\mathbf{x}_q$, k, $P[\alpha + i]$, $S[\alpha + i]$, I, $start2$, r_1)
        $start2 = I[num2][1]$
    else
        $start2 = I[num1][1]$
    end if
end while

Procedure Approximate_NN($\mathbf{x}_q$, k, $P[\alpha]$, $S[\alpha]$, I, p, r_1)
    while True **do**
        $a = distance(\mathbf{x}_q, \mathbf{x}_{P[\alpha][p]})$
        $r_2 = distance(\mathbf{x}_{P[\alpha][p]}, \mathbf{x}_c)$
        $u = $ Compute_u(a, r_1, r_2)
        $u_{delim} = $ Compute_u$_{delim}$(u, r_2)
        $pos = $ Binary_search($S[\alpha][1]$, u)
        $pos_{lim} = $ Binary_search($S[\alpha][1]$, u_{delim})
        $p_{new} = p$
        for $i = 0$ to g **do**
            if $pos - i \geq 0$ **then**
                $t = distance(\mathbf{x}_q, \mathbf{x}_{P[\alpha]S[\alpha][p][pos-i]})$
                if $t > a$ **then**
                    $p_{new} = pos - i$; break
                end if
            end if
            if $pos + i < pos_{lim}$ **then**
                $t = distance(\mathbf{x}_q, \mathbf{x}_{P[\alpha]S[\alpha][p][pos+i]})$
                if $t > a$ **then**
                    $p_{new} = pos + i$; break
                end if
            end if
        end for
        if $p_{new} == p$ **then**
            break
        end if
        $p = p_{new}$
    end while
    for $i = 1$ to k **do**
        Update_topk($P[\alpha]S[\alpha][p][i]$)
    end for
    return $P[\alpha][p]$

Fig. 11.4: Nearest neighbor search algorithm

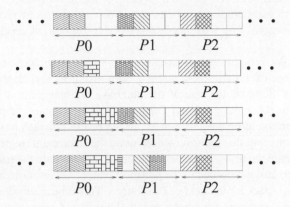

Fig. 11.5: Insertion of instances

geometric property of the hyperspherical partitions, and keep the IOs largely sequential.

When the dataset is static, i.e., no insertion is allowed, SphereDex can ensure sequential IOs by placing adjacent partitions contiguously on disks. If insertions do not take place often, one can keep a pool of inserted instances on the side, without perturbing the indexer. The indexer can be periodically rebuilt. However, to make SphereDex a general indexing method, we need to deal with online insertions. We propose a placement scheme, which trades disk space for ensuring almost contiguous placement of adjacent partitions. Of course, one can always hypothesize a pathological scenario to break a contiguous placement scheme. However, we believe that our scheme can work reasonably well most of time.

Our placement scheme is *space reservation and migration*. In each partition, we allocate free space anticipating insertions. When a new instance is inserted, we can find the partition to which it belongs, and then place the instance in the free space. Since we have the flexibility to place the instances in a partition in any order, as long as the space is available, we can handle insertion without breaking the contiguous partition placement. When the free space in the inserted partition is full, we can migrate data instances to adjacent partitions to continue maintaining contiguous placement.

The storage of instances with free space set aside in each partition can be visualized as in Figure 11.5. The shaded areas are already populated with instances and the unshaded areas can accept new instances. As new instances are inserted into a partition, the free space can be filled up. Suppose each partition stores g instances, and sets aside g free slots for insertions. After g new instances have been inserted into the same partition, the free space is used up. When another new instance is to be inserted in the same partition, we need to migrate some instances to the adjacent partitions to maintain balanced free space. Suppose free space is available in $P1$. We can move instances to $P1$ to free up space for the new instance. The placement scheme needs to maintain balanced free space among partitions proactively to accommodate insertions. (We will shortly discuss the situation when all partitions are full.)

[Case Study] We now present a case study to estimate the cost of a worst-case scenario. Suppose each partition contains g data instances and allocates free space for g insertions. Suppose the size of the database doubles, and all new instances are inserted into the same partition. Further, the worst choice of partition where this might occur would be the partition Pc, in the center between both ends. This is because insertions into this partition would need the maximum migration of data.

In order to store the first g instances, we need to access only the partition they belong to. The next instance added to the partition causes a migration of $g/2$ instances in $P1$, freeing up space for the next $g/2$ instances. This migration occurs in $P1$ again after another $g/2$ instances have been added. The second migration is of size $g/4$. Therefore, for the addition of the next g instances the number of migrations is log g. The third set of g instances would cause a migration in $P2$ followed by a migration in $P1$. The number of instances which cause migrations in total is $\frac{n}{g}$ log g. Since the average number of partitions accessed by a migration is $\frac{n}{2g}$, the total cost of the seeks is given by

$$\frac{1}{2}(\frac{n}{g})^2 \log g.$$

The rest of the instances $n - \frac{n}{g}$ log g, need a single disk access. Thus, the total number of disk accesses is given by

$$\frac{1}{2}(\frac{n}{g})^2 \log g + n - \frac{n}{g} \log g.$$

Since this is the cost for the insertion of n new instances, the cost per instance is given by

$$\frac{1}{2}\frac{n}{g^2} \log g + 1 - \frac{1}{g} \log g.$$

Even if a million new instances were added into the same partition, with $g = 1,000$, this is fairly small cost. We further wish to stress that these are not random seeks, since all the seeks would have been done in sequence.

In the above scenario, we assume additional instances do not cause a partition to split. In case partitions do split, the above cost is essentially doubled and the cost of splitting is added. Since the ordering of instances within a partition is unknown, when a split does occur, we read in the partition and rearrange the instances in it before writing that partition back to the disk as two partitions. From the computational aspect, splitting a partition requires rebuilding a local index each for the old and the new partition.

The above example presents the worst-case scenario for insertions. In Section 11.4 we present experimental data to show that in the usual course of operation, the number of disk accesses and the number of partition-splitting cases are very minor even in the face of insertions causing a doubling of the dataset.

When it becomes impossible to accommodate any further instances (all the partitions are filled to capacity), two different approaches can be taken. The first approach borrows from the above and migrates instances in each partition, starting with the outermost partition and moving inwards, creating additional space in every partition. The second approach is to leave the currently full structure untouched and build a new indexing structure with the fresh instances on the same lines as the previously constructed indexing structure. That is, now we have a fresh set of empty partitions which can accommodate new instances based on their distances from the center. In this case, a query would be handled by querying both the original structure and the new one. If multiple disks are available, it would be advantageous to place the second structure on a different disk, allowing parallel seeks to be conducted. If they are placed on the same disk, the number of seeks doubles.

Insertion of instances requires an update of the local index. As explained in Section 11.3.2.3, the instances in the partition maintain their closest 30 instances in their local index to conserve storage. If an inserted instance changes this closest set for a given instance in the partition, then the instance in the partition will necessitate an update of the local index. Before the inserted instance creates its local index, we compute its distances from the instances in the partition. If the inserted instance does not lie within the neighborhood of the closest 30 instances in the partition, we then choose the instance closest to the inserted instance and replace the appropriate entry.

Deletion of instances is helpful since it frees up space to store new instances. Deletion is handled by setting a flag denoting that the instance has been deleted. When a substantial number of the instances in the local index of an instance have been deleted, that instance requires to reconstruct its local index by computing afresh distances from the other instances in the partition.

11.4 Experiments

In this section, we present an evaluation of SphereDex and make comparisons with existing techniques. We were interested in the following key questions:

- Using random IOs, how does SphereDex compare with LSH/M-tree?
- When can SphereDex maintain sequential IOs, what is the additional gain of SphereDex?
- What is the percentage of data SphereDex needs to process compared to competing schemes?
- How do insertions affect sequential access?

11.4.1 Setup

We chose to compare SphereDex with two representative schemes: LSH algorithm and the M-tree. LSH was chosen because it outperforms many traditional schemes, and has been used in real applications. The M-tree was chosen partly because of its popularity and partly because it is a distance-based indexing scheme. It should be noted that the original formulation of the M-tree algorithm [21] was designed to perform exact nearest-neighbor searches. An approximate version of the algorithm was presented in [13]. For a fair comparison, we chose the algorithm presented in [13] and used the code provided by M. Patella. The code is available on request, but is not part of the basic download available at http://www-db.deis.unibo.it/Mtree/download.html.

11.4.1.1 Datasets

We used two datasets for conducting our experiments. The first dataset contains 275, 465 feature vectors. Each of these is a 60-dimensional vector representing texture information of blocks of large aerial photographs. This dataset was obtained from B.S. Manjunath [28]. The size and dimensionality of the dataset provides challenging problems in high dimensional indexing [29]. These features were obtained by applying Gabor filters to the image tiles. The Gabor filter bank consists of 5 scales and 6 orientations of filters. Therefore, the total number of filters is 30. The mean and standard deviation of each filtered output are used to construct the feature vector, and hence we have a dimensionality of 60 (30×2). The texture features are extracted from 40 large aerial photos. Before the feature extraction, each airphoto is first partitioned into non-overlapping tiles, each sized 64×64. The feature vectors are extracted from these tiles.

The second dataset contains 314, 499 feature vectors. Each of these is a 144-dimensional vector representing color and texture information in the image. Research [30] has shown that colors that can be named by all cultures ("culture colors") are generally limited to eleven — black, white, red, yellow, green, blue, brown, purple, pink, orange and gray. Color is first divided into 12 color bins, 11 for culture colors and one bin for outliers. Each color is associated with eight additional features. These are the color histograms: *color means* (in H, S and V channels), *color variances* (in H, S and V channels) and two shape characteristics: *elongation* and *spreadness*. Color elongation characterizes the shape of a color, and spreadness characterizes how the color scatters within the image [31]. Thus, the feature vector contains 96 color features. The texture features are obtained using a discrete wavelet transform (DWT) using quadrature mirror filters [32]. The wavelet decomposition of a 2-D image yields twelve features for each of four sub-images. Hence, we have 48 texture features within a total of 144 features. (We use this feature set throughout this book.)

In our experiments on each dataset, we chose 2, 000 instances randomly from the dataset and created an index using the rest of the dataset. As in [12] our experiments aimed at finding the performance for top-10 approximate NN search. The results are averaged over all the approximate 10-NN searches.

11.4.1.2 Comparison Metric

The objective of our experiment was to ascertain how quickly the approximate results could be obtained using the index. To quantify the degree of approximation, we used the same error metric as used by LSH. Following [27] the effective error for the 1-nearest neighbor search problem is defined as

$$E = \frac{1}{|Q|} \sum_{query \mathbf{x}_q \in Q} (\frac{d_{index}}{d^*} - 1),$$

where d_{index} denotes the distance from a query point $\mathbf{x}_q$ to a point found by the indexing approach, d^* is the distance from $\mathbf{x}_q$ to the closest point, and the sum is taken over all queries. For the approximate k nearest neighbor problem, as in [12] we measure the distance ratios between the closest point to the nearest neighbor, the second closest one to the second nearest neighbor and so on, finally averaging the ratios. In the case of LSH, the sum is calculated over all queries, each of which returns more than k results. Under LSH, queries returning less than k results are defined as misses.

Indexing strategies are usually compared using the number of disk IOs performed to achieve a specified level of accuracy, the reasoning being that the more the number of IOs performed the slower the retrieval. However, disk IO has the twin characteristics of seek and transfer. While the transfer is usually fast, seeks are much slower. Our method attempts to minimize the number of seeks by maximizing sequential access. Thus, the number of seek operations carried out by our method is essentially 2. Directly comparing the number of IOs performed by our method with that performed by other methods would be unfair, since the transfer time with our method is not the same as the transfer time in the others.

To understand the relationship between seek time and data transfer time, we look at the time the disk takes to transfer 1 MB of data. Current disk technology can maintain a sustained throughput in the range 50 to 70 MB/s. Therefore, transferring 1 MB of data takes roughly roughly 14 to 20 ms. Average seek times are usually in the range of 10 ms. Thus, it takes roughly twice the time to transfer 1 MB of data as it takes for a seek operation. Since the exact time varies from disk to disk, we use the ratio of the times in our calculations. Note that a higher value of the ratio has been selected to allow the competing index structures the maximum benefit. Further, since the other index structures transfer only small chunks of data in each access to the disk, the time required by them to transfer data has been assumed to be zero. We also present graphs with different ratios between seek and transfer time for the sake of completeness.

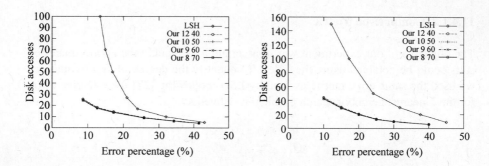

Fig. 11.6: SphereDex *vs.* LSH under random access for both datasets

11.4.2 Performance with Disk IOs

We also compared the performance of the two approaches in the cases when the index needs to be accessed from the disk. In such cases, the cost of distance computations becomes insignificant when compared to the cost of the disk access. As mentioned above, we do not count the cost of data transfer for LSH. We count only the total number of seeks performed by LSH. The number of seeks performed by LSH is controlled by parameter l.

11.4.2.1 Performance under Random Access

In the case of the first dataset, LSH used $l \approx 70$ hash functions to achieve acceptable error rates. Hence the total time taken by LSH would be given by

$$t_{LSH} = l \times t_{seek} = 70 \times t_{seek}.$$

Here t_{seek} is the average seek time. Compared to this, our method requires to retrieve whole partitions. Thus, we need to compute the space required to store the retrieved partitions. In the case of the first dataset, where each instance has 60 features, each partition takes up space

$$g \times d + g \times a \times \log g.$$

But $a \times \log g$ is equal to 60×3.25 when $g = 1,000$. Hence, the total space consumed is equal to

$$1,000 \times 60 + 1,000 \times 60 \times 3.25,$$

where, as in LSH, each dimension can be stored in 1 byte. Therefore, the total space required to store a partition is given by 0.255 MB. The total time consumed is therefore given by

$$r \times t_{seek} + r \times 0.255 \times t_{transfer},$$

where, $t_{transfer}$ is the time taken to transfer 1MB of data from the disk and r is the number of partitions transferred. Since $t_{transfer}$ takes less than twice the average seek time, the equation simplifies to

$$r \times t_{seek}(1 + 0.51).$$

Taking the ratio of the times taken by LSH and SphereDex we get

$$Speedup = \frac{70 \times t_{seek}}{r \times t_{seek}(1 + 0.51)} = \frac{70}{r \times (1.51)}.$$

Therefore, if our approach is able to retrieve the approximate set of instances in less than $\frac{70}{1.51}$ partitions, then we have speedup. This value is roughly equal to 46. We show in the left-hand side of Figure 11.6 that the average number of partitions that had to be evaluated (the number of disk accesses) before achieving the same level of error was 12. The speedup is hence on the order of ≈ 4 times.

Similar analysis for the second dataset can be done as follows. The space required for each partition of the second dataset is given by $1,000 \times (144 + 3.25 \times 60)$ bytes (0.339 MB). For the second dataset, LSH required $l > 130$ to achieve 15% error rate. Therefore,

$$Speedup = \frac{130 \times t_{seek}}{r \times t_{seek}(1 + 2 \times 0.339)} = \frac{130}{r \times (1 + 0.678)}.$$

To achieve parity with the time taken by LSH we would need to evaluate $r = 77$ partitions. We show in the right-hand side of Figure 11.6 that in this case the average value of r is 21. The speedup is approximately 3.5.

M-tree is another indexing structure which uses distances between instances. Our experiments with M-tree show that the number of IO operations performed by the index are much higher on both the data sets. This makes using M-trees for searching prohibitively expensive. Figure 11.7 presents comparisons between our indexing strategy and M-trees on the both datasets.

11.4.2.2 Performance under sequential access

Next, we examined performance with sequential access of partitions. In this case, we would need at most two seek operations and the rest of the time would be spent in transferring the partitions. Here, we assume that space has been reserved for another g instances in each partition to take care of insertions.

The total time consumed for the first dataset is therefore given by

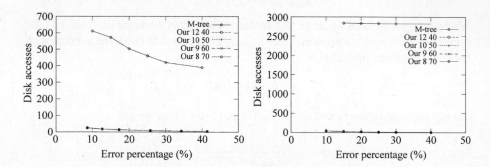

Fig. 11.7: SphereDex *vs.* M-tree under random access for both datasets

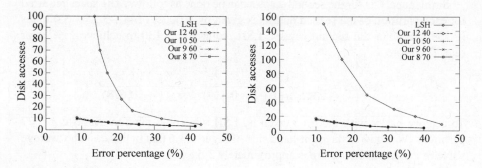

Fig. 11.8: SphereDex *vs.* LSH under sequential access for both datasets

$$2 \times t_{seek} + 2r \times 0.255 \times t_{transfer}.$$

Since $t_{transfer}$ is less than twice the average seek time, the equation simplifies to

$$2 \times t_{seek}(1 + 2r \times 0.255).$$

Taking the ratio of the times taken by LSH and our method, we get

$$Speedup = \frac{70 \times t_{seek}}{2 \times t_{seek}(1 + r \times 0.255)} = \frac{35}{1 + r \times 0.255}.$$

Therefore, if our approach can retrieve the approximate set of instances in fewer than $\frac{34}{2 \times 0.255}$ partitions then we achieve speedup. This value is roughly equal to 66. Since the average number of partitions required to be evaluated before achieving the same level of error is 12, the speedup is on the order of 5.5 times.

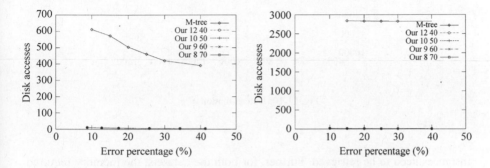

Fig. 11.9: SphereDex *vs.* M-tree under sequential access for both datasets

Similar analysis for the second dataset can be done as follows.

$$Speedup = \frac{130 \times t_{seek}}{2 \times t_{seek}(1 + 2r \times 0.339)} = \frac{65}{1 + 2r \times 0.339}.$$

To achieve parity with the time taken by LSH we would need to evaluate $r = 94$ partitions. The average value of r is 21 for 15% error. Hence, the speedup is approximately 4.5 times.

Our experiments with M-tree show that the number of IO operations performed by the index are much higher on both the data sets. Figure 11.9 presents comparisons between our indexing strategy and M-trees on the both datasets.

11.4.3 Choice of Parameter g

These set of experiments focused on finding a good value of g. In Figure 11.10, the x-axis represents the value of g and the y-axis represents the number of instances that needed to be retrieved to obtain error levels of $< 15\%$. The values for each g were obtained by averaging over 200 query evaluations. We see that in both the cases, there are a range of values over which g can be varied with approximately the same number of instances being evaluated. From the cpu-performance point of view, we would like to set g to a value such that, multiple partitions and their associated index structures can be place in the $L2$ cache at the same time. We notice that, for both the datasets, a g value of $\approx 1,000$ presents a threshold after which larger number of

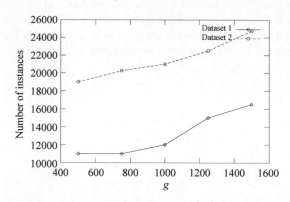

Fig. 11.10: Determination of g

instances need to be retrieved. Further, for both the datasets, the memory required (0.255MB and 0.339 MB) to store each partition with $g = 1000$ allows us to place multiple partitions in the $L2$ cache.

11.4.4 Impact of Insertions

Since insertions play such an important role in our indexing structure, we also examined the effect of inserting instances into the data set. To observe this effect, we randomly picked a subset of instances in the original dataset to construct the index, and then we inserted the rest of the instances into the index one at a time. We found that for insertions of instances numbering less than 80% of the original dataset, no extra disk seeks were necessary for either dataset. When we inserted approximately the same number of instances as the original dataset contained, the average number of seeks that had to be performed in the case of the first dataset was 1.00376. Only 391 instances (out of nearly 137,000) needed 519 extra disk accesses because of full partitions. This translates to roughly 1.327 extra accesses for each of these instances.

For the second dataset, the average number of disk accesses over all inserted instances was 1.00285, with only 400 instances (out of nearly 157,000) needing a total of 449 extra disk accesses because of full partitions. This translates to roughly 1.12 extra disk accesses for this subset of instances. Therefore, even with a doubling of the dataset we are able to maintain an average insertion cost of almost one per instance. Since we need at least one disk access for inserting an instance, this is very close to the best possible.

11.4.5 Sequential vs. Random

To evaluate the impact of the volume of space reservation for both the datasets, we examined the equations governing the time taken under random access and sequential access. The time taken under random access is given by

$$r(t_{seek} + t_{transfer} \times size),$$

where r is the number of partitions evaluated, t_{seek} is the seek time, $t_{transfer}$ is the transfer time per MB of data and $size$ is the size each partition.

Under sequential access the total time taken is given by

$$2t_{seek} + r \times t_{transfer} \times size \times \alpha,$$

where the additional variable $\alpha (\geq 1)$ indicates the volume of space reservation.

Equality is achieved when

$$r(t_{seek} + t_{transfer} \times size) = 2t_{seek} + r \times t_{transfer} \times size \times \alpha.$$

Assuming a ratio of 2 between $t_{transfer}$ and t_{seek}, we can now obtain the values of α for each of the datasets based on the values of r and $size$. For the first dataset ($r = 12$, $size = 0.255$ MB), we get $\alpha = 2.63$. Therefore, we could potentially reserve 1.63 times the space originally occupied in a partition for future instances to achieve the same level of performance as random access. For the second dataset ($r = 21$, $size = 0.339$MB), we get $\alpha = 2.5$. Thus, in the case of both the datasets space reservation for an equal number of instances would still allow us to outperform random access.

The above analysis helps us decide between random placement of partitions vs. space reservation based on the expected volume of insertion. If the volume of insertion is low, we can suitably lower the space reservation to get higher benefits from sequential access of instances. Analytically, keeping $size$ fixed we can obtain the speedup of sequential over random access as

$$\frac{r \times (1 + 2 \times size)}{2r \times \alpha \times size + 2} \approx \frac{r \times (1 + 2 \times size)}{2r \times \alpha \times size} = \frac{1}{\alpha} + \frac{1}{2\alpha \times size}.$$

This implies that if $size \times \alpha < 0.5$ we are guaranteed speedup.

11.4.6 Percentage of Data Processed

Real-world datasets are frequently too large in size to be stored in the main memory. Therefore, we focus on datasets which require to perform IO operations on disks. However, we do show that even if the instances could all be stored in memory, our method uses fewer distance comparisons than those made by rival schemes.

LSH was initially proposed as a main memory algorithm [3]. However, newer versions of LSH allow disk accesses also. The main advantage of having a main memory algorithm is that the original LSH does not need to maintain multiple copies of the entire dataset, only pointers to the data items. Thus, if we use l hash functions in LSH, then each instance is hashed to l different buckets. This means that in addition to the d features associated with each instance, we have another l pointers. In the case of disk-based LSH however, l copies of the dataset must be maintained.

To compare the performance of our approach with that of LSH in the main memory setting, we examined the space used by LSH for the two datasets under consideration. For the results reported in [12], the authors needed $l \approx 70$ to achieve an error rate less than 15%. Thus, for each instance an additional 70 pointers were maintained. Considering that the space used by a pointer is 4 bytes on our machine, the total memory used by LSH for the pointers is of the order 4×70 bytes for each data instance. Compared to this, the space used by our indexing structure is $O(\log g)$, g being the number of instances assigned to a partition. As explained in Section 11.3.2.3, when $g = 1,000$, we store approximately 60 distances for each instance. If we limit ourselves to 2 bytes for each instance, we can get 4 digits of precision. Also, since there are 1000 instances in the partition, to maintain unique identifiers for them we need 10 bits only. The total storage space is therefore $60 \times (2 + \frac{10}{8})$, which is equal to 60×3.25. Since the memory required by the index is comparable in both cases, we can compare the number of distance computations to determine which method is faster.

In the case of the second dataset, LSH requires at least $l = 120$ to achieve reasonable error rates ($< 15\%$). However, the size of our index remains unchanged at 60 distances per instance. Therefore, as the size of the dataset increases, the size of the index per instance (and hence per partition) remains the same, while the size of the index per instance for LSH grows. Again, in the memory-based scenario, we can compare the results for the two approaches purely on the basis of the number of distance computations they perform.

Remark: We analytically examine the performance under sustained use. By sustained use we mean that the indexing structure has already been queried multiple times. Here we notice an important distinction between LSH and our indexing approach. In LSH the query instance maps to a specific bucket. Unless the bucket is in memory, it must be retrieved from the disk. Neighboring buckets do not contain any information that can be useful for that query. In contrast, under our indexing scheme, even if the exact partition is not available in main memory, processing can continue with the closest partition available while the target partition and its neighbors are being retrieved from the disk. The quality of results will vary according to the distance of the available partition from the target partition. That is, if the closest available partition is far away from the query instance's target partition, the top-k results will not be very good. But if there is a partition close to the target partition, the approximate top-k results have a high probability of being good. Therefore, the presence of partitions in memory allows us to address queries and generate intermediate results while the missing partitions are being recovered from the disk.

The above procedure also provides a method of seeding the memory, since now we can actually keep every r^{th} partition in main memory, where r is the ratio of the total number of partitions to the number of partitions that can fit into main memory. As new partitions are retrieved, the least used partitions are paged out of memory.

11.4.7 Summary

We enumerate the advantages of SphereDex as compared to LSH.

1. The average time needed to address an approximate nearest neighbor query is about a magnitude shorter.
2. In the case of LSH there is the possibility of turning up with no results (or fewer than k results). This is defined as a miss. In our case there is no chance of a miss.
3. The size of the index in our method does not increase with the number of dimensions in the dataset.
4. As discussed in [25] that it is possible to continue processing to obtain better results if the user wishes to do so.
5. The number of hash functions, l, chosen in LSH depends on the size of the dataset. Hence, the duplication of instances also grows with the size of the dataset in the case of LSH.

11.5 Concluding Remarks

SphereDex can be extended to support *hyperplane queries* for Support Vector Machines (SVMs) or the kernel methods. In classification problems using SVMs, the data instances closest to the hyperplane are considered to be most ambiguous, and the ones farthest away from the hyperplane to be most certain (or most confident) regarding their class membership. Hyperplane queries, rather than point queries, are essential to supporting fast retrieval of applications using SVMs. In the end of this chapter, we illustrate how SphereDex can be extended to support both nearest and farthest neighbor hyperplane query processing.

11.5.1 Range Queries

Range queries are specified by the query instance and a distance. These queries are interested in determining nearest neighbors of the query instance which lie within the given distance. Using our indexing structure handling range queries is straight-forward because the specified distance helps us determine the number of partitions that need to be processed. This is because we now already know d_f in Figure 11.3.

Thus, the query processing proceeds exactly as before and terminates when the partitions beyond the specified distance is encountered.

11.5.2 Farthest Neighbor Queries

Finding the farthest neighbors given the query point can also be accomplished with minor modification. In Figure 11.11, we are interested in finding the farthest instance from $\mathbf{x}_q$ in the partition. We already know the distance s between $\mathbf{x}_0$) and $\mathbf{x}_q$ and need to find the distance s'. We can also determine the diameter $2r$. The Pythagorus theorem lets us find the distance s' given s and $2r$. It is important to note that we need to start from the outermost ring when we look for farthest instances.

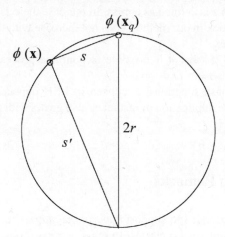

Fig. 11.11: Farthest instances

References

1. Panda, N., Chang, E.Y. Efficient top-k hyperplane query processing for multimedia information retrieval. In *Proceedings of ACM Multimedia*, pages 317–326, 2006.
2. Arya, S., Mount, D., Netanyahu, N., Silverman, R., Wu, A. An optimal algorithm for approximate nearest neighbor searching in fixed dimensions. In *Proceedings of ACM SODA*, pages 573–82, 1994.
3. Indyk, P., Motwani, R. Approximate nearest neighbors: towards removing the curse of dimensionality. In *Proceedings of VLDB*, pages 604–613, 1998.
4. Kleinberg, J.M. Two algorithms for nearest-neighbor search in high dimensions. In *Proceedings of ACM STOC*, pages 599–608, 1997.

5. Weber, R., Schek, H.J., Blott, S. A quantitative analysis and performance study for similarity-search methods in high-dimensional spaces. In *Proc. 24th Int. Conf. Very Large Data Bases VLDB*, pages 194–205, 1998.

6. Bentley, J. Multidimensional binary search trees used for associative binary searching. *Communications of ACM*, 18(9):509–517, 1975.

7. Katayama, N., Satoh, S. The SR-tree: an index structure for high-dimensional nearest neighbor queries. In *Proceedings of ACM SIGMOD Int. Conf. on Management of Data*, pages 369–380, 1997.

8. White, D.A., Jain, R. Similarity indexing with the SS-Tree. In *Proceedings of IEEE ICDE*, pages 516–523, 1996.

9. Kushilevitz, E., Ostrovsky, R., Rabani, Y. Efficient search for approximate nearest neighbor in high dimensional spaces. In *Proceedings of the 30th STOC*, pages 614–23, 1998.

10. Clarkson, K. An algorithm for approximate closest-point queries. In *Proceedings of SCG*, pages 160–164, 1994.

11. Li, C., Chang, E., Garcia-Molina, H., Wilderhold, G. Clindex: Approximate similarity queries in high-dimensional spaces. *IEEE Transactions on Knowledge and Data Engineering (TKDE)*, 14(4):792–808, July 2002.

12. Gionis, A., Indyk, P., Motwani, R. Similarity search in high dimensions via hashing. *VLDB Journal*, pages 518–529, 1999.

13. Ciaccia, P., Patella, M. Pac nearest neighbor queries: Approximate and controlled search in high-dimensional and metric spaces. In *Proceedings of IEEE ICDE*, pages 244–255, 2000.

14. Qamra, A., Meng, Y., Chang, E.Y. Enhanced perceptual distance functions and indexing for image replica recognition. *IEEE Transactions on Pattern Analysis and Machine Intelligence (PAMI)*, 27(3), 2005.

15. Buhler, J. Efficient large-scale sequence comparison by locality-sensitive hashing. *Bioinformatics*, 17:419–428, 2001.

16. Keim, D.A. Tutorial on high-dimensional index structures: Database support for next decade's applications. In *Proceedings of the ACM SIGMOD*, page 501, 1998.

17. Moule, M.E., Sakuma, J. Fast approximate similarity search in extremely high-dimensional data sets. In *Proceedings of IEEE ICDE*, pages 619–630, 2005.

18. Berchtold, S., Keim, D., Kriegel, H. The X-tree: An index structure for high-dimensional data. In *Proceedings of the 22nd Conference on Very Large Databases VLDB*, pages 28–39, 1996.

19. Beckmann, N., Kriegel, H., Schneider, R., Seeger, B. The R^* tree: An efficient and robust access method for points and rectangles. In *Proceedings of ACM SIGMOD Int. Conf. on Management of Data*, pages 322–331, 1990.

20. Lin, K.I., Jagadish, H.V., Faloutsos, C. The TV-tree: An index structure for high-dimensional data. *VLDB Journal*, 3(4):517–542, 1994.

21. Ciaccia, P., Patella, M., Zezula, P. M-tree: An efficient access method for similarity search in metric spaces. In *Proc. 23rd Int. Conf. on Very Large Databases VLDB*, pages 426–435, 1997.

22. Bozkaya, T., Ozsoyoglu, M. Indexing large metric spaces for similarity search queries. *ACM Transactions on Database Systems*, 24(3):361–404, 1999.

23. Brin, S. Near neighbor search in large metric spaces. *VLDB Journal*, pages 574–584, 1995.

24. Navarro, G. Searching in metric spaces by spatial approximation. In *SPIRE/CRIWG*, pages 141–148, 1999.

25. Ferhatosmanoglu, H., Tuncel, E., Agrawal, D., Abbadi, A.E. Approximate nearest neighbor searching in multimedia databases. In *Proceedings if IEEE ICDE*, pages 503–511, 2001.

26. Patella, M. http://www-db.deis.unibo.it/mtree/download.html.

27. Arya, S., Mount, D.M., Netanyahu, N.S., Silverman, R., Wu, A.Y. An optimal algorithm for approximate nearest neighbor searching fixed dimensions. *Journal of the ACM*, 45(6):891–923, 1998.

28. Manjunath, B.S. Airphoto dataset. *http://vision.ece.ucsb.edu/download.html*.

29. Manjunath, B.S., Ma, W.Y. Texture features for browsing and retrieval of image data. *IEEE Trans. Pattern Anal. Mach. Intell.*, 18(8):837–842, 1996.

30. Goldstein, E.B. *Sensation and Perception (5th Edition)*. Brooks/Cole, 1999.

31. Leu, J.G. Computing a shape's moments from its boundary. *Pattern Recognition*, 24:949–957, 1991.
32. Smith, J., Chang, S.F. Automated image retrieval using color and texture. *IEEE Transactions on Pattern Analysis and Machine Intelligence*, November 1996.

Chapter 12
Speeding Up Latent Dirichlet Allocation with Parallelization and Pipeline Strategies

Abstract Previous methods of distributed Gibbs Sampling for LDA run into either memory or communication bottleneck. To improve scalability, this chapter[†] presents two strategies: (1) parallelization — carefully assigning documents among processors based on word locality, and (2) pipelining — masking communication behind computation through a pipeline scheme. In addition, we employ a scheduling algorithm to ensure load balancing both spatially (among machines) and temporally. Experiments show that our strategies can significantly reduce the unparallelizable communication bottleneck and achieve good load balancing, and hence improve LDA's scalability.

Keywords: Latent Dirichlet Allocation, pipeline processing, data placement, distributed systems.

12.1 Introduction

Latent Dirichlet Allocation (LDA) was first proposed by Blei, Ng and Jordan to model documents [2]. Each document is modeled as a mixture of K latent topics, where each topic, k, is a multinomial distribution $V\phi_k$ over a W-word vocabulary. For any document d_j, its topic mixture $V\theta_j$ is a probability distribution drawn from a Dirichlet prior with parameter α. For each ith word x_{ij} in d_j, a topic $z_{ij} = k$ is drawn from $V\theta_j$, and x_{ij} is drawn from $V\phi_k$. The generative process for LDA is thus given by

$$\theta_j \sim Dir(\alpha), \phi_k \sim Dir(\beta), z_{ij} = k \sim \theta_j, x_{ij} \sim \phi_k, \tag{12.1}$$

where $Dir(*)$ denotes Dirichlet distribution. The graphical model for LDA is illustrated in Figure 12.1, where the observed variables, i.e., words x_{ij} and hyper parameters α and β, are shaded.

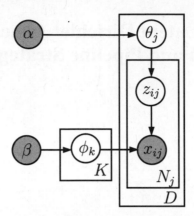

Fig. 12.1: The graphical model for LDA

The computation complexity of Gibbs sampling is K multiplied by the total number of word occurrences in the training corpus. Prior work has explored multiple alternatives for speeding up LDA, including both parallelizing LDA across multiple machines and reducing the total amount of work required to build an LDA model. Three representative distributed LDA algorithms are Dirichlet Compound Multinomial LDA (DCM-LDA) [3], Approximate Distributed LDA (AD-LDA) [4], and Asynchronous Distributed LDA (AS-LDA) [5], which all parallelize Gibbs Sampling on distributed machines. These algorithms suffer from either high communication cost or long convergence time (an approximate method reduces communication time but increases number of Gibbs Sampling iterations). In this chapter, we present PLDA+[1], which uses distributed data-placement and pipeline strategies to reduce the communication bottleneck. The distributed data placement strategy aims to first separate CPU-bound tasks and communication-bound tasks onto two sets of machines. It then ensures that both computation and communication loads can be balanced among parallel machines. The pipeline strategy aims to mask communication time by computation time; and hence the communication bottleneck can be reduced. Experiments show that the strategies of PLDA+ can significantly improve scalability of LDA over our initial attempt at Google [6].

The rest of the chapter is organized as follows: We first present LDA and related distributed algorithms in Section 12.2. In Section 12.3 we present Approximate Distributed LDA (AD-LDA) and explain how it works via a simple example. In Section 12.4 we analyze the bottleneck of AD-LDA. Sections 12.4.3 and 12.4.4 depict PLDA+ in details. Section 12.5 demonstrates that the speedup of PLDA+ on large-scale document collections significantly outperforms AD-LDA. In Section 12.6 we introduce two large-scale applications of distributed LDA. Finally, we discuss future research plans in Section 12.7. For the convenience of readers, we summarize the notation used in this chapter in Table 12.1.

Table 12.1: Symbols associated with LDA used in this chapter

D	Number of documents
T	Number of topics
W	Vocabulary size
N	Number of words in the corpus
x_{ij}	The ith word in d_j document
z_{ij}	Topic assignment for word x_{ij}
C_{kj}	Number of topic k assigned to d_j document
C_{wk}	Number of word w assigned to topic k
C_k	Number of topic k in corpus
C^{doc}	Document-topic count matrix
C^{word}	Word-topic count matrix
C^{topic}	Topic count matrix
$\mathbf{V}\theta_j$	Probability of topics given document d_j
$\mathbf{V}\phi_k$	Probability of words given topic k
α	Dirichlet prior
β	Dirichlet prior
P	Number of processors
P_w	Number of P_w processors
P_d	Number of P_d processors
p_i	The ith processor

12.2 Related Reading

According to the generative process of LDA shown in Eq.(12.1), the full joint distribution over all parameters and variables is

$$p(\mathbf{V}x, \mathbf{V}z, \mathbf{V}\theta, \mathbf{V}\phi | \alpha, \beta) = p(\mathbf{V}\phi | \beta) \prod_{j=1}^{D} p(\mathbf{V}\theta_j | \alpha) \prod_{i=1}^{N_j} p(x_{ij} | \mathbf{V}\phi, z_{ij}) p(z_{ij} | \mathbf{V}\theta_j),$$

(12.2)

where $\mathbf{V}x = \{x_{ij}\}$ is the observed word occurrences in D documents, $\mathbf{V}z = \{z_{ij}\}$ is the assigned latent topics to words $\mathbf{V}x$, and N_j the number of word occurrences in document d_j. Similar to most previous work, we use symmetric Dirichlet priors in LDA for simplicity. Given the observed words $\mathbf{V}x$, the task of inference for LDA is to compute the posterior distribution of the latent topic assignments $\mathbf{V}z$, the topic mixtures of documents $\mathbf{V}\theta$, and the topics $\mathbf{V}\phi$.

Blei, Ng and Jordan [2] proposed using a Variational Expectation Maximization (VEM) algorithm for obtaining maximum-likelihood estimate of Φ from V. This algorithm iteratively executes an E-step and an M-step, where the E-step infers the topic distribution of each training document, and the M-step updates model parameters using the inference result. Unfortunately, this inference is intractable, so variational Bayes is used in the E-step for approximate inference. Minka and Lafferty proposed a comparable algorithm [7], which uses another approximate inference method, Expectation Propagation (EP), in the E-step.

Griffiths and Steyvers [8] proposed using Gibbs sampling, a Markov-chain Monte Carlo method, to perform inference for LDA. By assuming a Dirichlet prior,

β, $\mathbf{V}\phi$ can be integrated (hence removed from the equation) using the Dirichlet-multinomial conjugacy. MCMC is widely used as an inference method for latent topic models, e.g., Author-Topic Model [9], Pachinko Allocation [10], and Special Words with Background Model [11]. Moreover, since the memory requirement of VEM is not nearly as scalable as that of MCMC [12], most existing distributed methods for LDA use Gibbs Sampling for inference, e.g., Dirichlet Compound Multinomial LDA [3], Approximate Distributed LDA [4], and Asynchronous Distributed LDA [5]. In this chapter, we thus focus on Gibbs Sampling for approximate inference. In Gibbs Sampling, it is usual to integrate out the mixtures θ and topics ϕ and just sample the latent variables $\mathbf{z}$. The process is called *collapsing*. When performing Gibbs Sampling for LDA, we maintain two matrices: word-topic count matrix C^{word} in which each element C_{wk} is the number of word w assigned to topic k, and document-topic count matrix C^{doc} in which each element C_{kj} is the number of topic k assigned to d_j document. Moreover, we maintain a topic count vector C^{topic} in which each element C_k is the number of topic k assignments in document collection. Given the current state of all but one variable z_{ij}, the conditional probability of z_{ij} is

$$p(z_{ij} = k | \mathbf{z}^{\neg ij}, \mathbf{x}^{\neg ij}, x_{ij} = w, \alpha, \beta) \propto \frac{C_{wk}^{\neg ij} + \beta}{C_k^{\neg ij} + W\beta} \left(C_{kj}^{\neg ij} + \alpha \right), \qquad (12.3)$$

where $\neg ij$ means that the corresponding word is excluded in the counts. Whenever z_{ij} is assigned to a new topic drawn from Eq.(12.3), C^{word}, C^{doc} and C^{topic} are updated. After enough sampling iterations to burn in the Markov chain, $\mathbf{V}\theta$ and $\mathbf{V}\phi$ can be estimated by

$$\theta_{kj} = \frac{C_{kj} + \alpha}{\sum_{k=1}^{T} C_{kj} + T\alpha}, \quad \text{and} \qquad (12.4)$$

$$\phi_{wk} = \frac{C_{wk} + \beta}{\sum_{w=1}^{W} C_{wk} + W\beta}, \qquad (12.5)$$

where θ_{kj} indicates the probability of topic k given document j, and ϕ_{wk} indicates the probability of word w given topic k. Griffiths and Steyvers conducted an empirical study of VEM, EP and Gibbs sampling and the comparison shows that Gibbs sampling converges to a known ground-truth model more rapidly than either VEM or EP [8].

LDA Performance Enhancement

The computation complexity of Gibbs sampling is K multiplied by the total number of word occurrences in the document collection. Prior work has explored multiple alternatives for speeding up LDA, including both parallelizing LDA across multiple processors and reducing the total amount of work required to build an LDA model. Relevant distributed methods for LDA include:

- Nallapati, et al. [13] and Wolfe, et al. [14] both reported distributed computing of the VEM algorithm for LDA [2].
- Mimno and McCallum proposed Dirichlet Compound Multinomial LDA (DCM-LDA) [3], where the data sets are distributed to processors, Gibbs Sampling is performed in each processor independently without any communication between processors, and finally a global clustering of the topics is performed.
- Newman, et al. [4] proposed Approximate Distributed LDA (AD-LDA), where each processor performs a local Gibbs Sampling iteration followed by a global update using a reduce-scatter operation. Since the Gibbs Sampling in each processor is performed with the local word-topic matrix, which is only updated at the end of each iteration, it is named with *approximate* distributed LDA.
- An asynchronous distributed learning algorithm of LDA was proposed in [5], where no global synchronization step like that in [4] is required. Each processor performs a local Gibbs Sampling step followed by a step of communicating with other *random* processors. We name this method as AS-LDA.

In addition to these parallelization techniques, the following optimizations can reduce LDA model learning times by reducing the total computational cost:

- Gomes, et al. [15] presented an enhancement of the VEM algorithm using a bounded amount of memory.
- Porteous, et al. [16] proposed a method to accelerate the computation of Eq.(12.3). The acceleration is achieved by no approximations but using the property that the topic probability vectors for document d_j, $\mathbf{V}\theta_j$, are sparse in most cases.

12.3 AD-LDA: Approximate Distributed LDA

Before introducing PLDA+, let us review our prior implementation [6] of the AD-LDA algorithm [4]. We present the algorithm's dependency on the collective communication operation, *AllReduce*, and how to express the AD-LDA algorithm in the model of MPI. AD-LDA serves as the performance yardstick of PLDA+.

12.3.1 Parallel Gibbs Sampling and AllReduce

AD-LDA distributes D training documents over P processors, with $D_p = D/P$ documents on each processor. AD-LDA partitions document content $\mathbf{V}x = \{\mathbf{V}x_d\}_{d=1}^{D}$ into $\{\mathbf{V}x_{|1}, \ldots, \mathbf{V}x_{|P}\}$ and the corresponding topic assignments $\mathbf{V}z = \{\mathbf{V}z_d\}_{d=1}^{D}$ into $\{\mathbf{V}z_{|1}, \ldots, \mathbf{V}z_{|P}\}$, where $\mathbf{V}x_{|p}$ and $\mathbf{V}z_{|p}$ exist only on processor p. Document-topic count matrix, C^{doc}, are likewise distributed. We denote the document-topic count matrix on processor p as $C^{doc}_{|p}$. Each processor maintains its own copy of word-topic count matrix, C^{word}. Moreover, AD-LDA uses $C^{word}_{|p}$ to temporarily store word-topic counts accumulated from local documents' topic assignments on each processor.

In each Gibbs sampling iteration, each processor p updates $\mathbf{V}z_{|p}$ by sampling every $z_{ij|p} \in \mathbf{V}z_{|p}$ from the approximate posterior distribution:

$$p(z_{ij|p} = k \mid \mathbf{V}z^{\neg ij}, \mathbf{V}x^{\neg ij}, x_{ij|p} = w) \propto \frac{C_{wk}^{\neg ij} + \beta}{C_k^{\neg ij} + W\beta} \left(C_{jk|p}^{\neg ij} + \alpha \right), \quad (12.6)$$

and updates $C_{|p}^{doc}$ and $C_{|p}^{word}$ according to the new topic assignments. After each iteration, each processor recomputes word-topic counts of its local documents $C_{|p}^{word}$ and uses the AllReduce operation to reduce and broadcast the new C^{word} to all processors.

12.3.2 MPI Implementation of AD-LDA

Our AD-LDA implementation [6] uses MPI [17] to parallelize LDA learning. The MPI model supports `AllReduce` via an API function:

```
int MPI_Allreduce(void *sendbuf, void *recvbuf, int
                  count, MPI_Datatype datatype, MPI_Op op);
```

When a *worker*, meaning a thread or a process that executes part of the parallel computing job, finishes sampling, it shares topic assignments and waits for AllReduce by invoking `MPI_Allreduce`, where `sendbuf` points to word-topic counts of its local documents: a vector of `count` elements with type `datatype`. The worker sleeps until the MPI implementation finishes AllReduce and the results are in each worker's buffer `recvbuf`. During the reduction process, word-topic counts vectors are aggregated element-wise by the addition operation `op` explained in Section 12.3.1.

Figure 12.2 presents the detail of MPI implementation for AD-LDA. The algorithm first attempts to load checkpoints $\mathbf{V}z_{|p}$ if a machine failure took place and the computation is in the recovery mode. The procedure then performs initialization (lines 5 to 9), where for each word, its topic is sampled from a uniform distribution. Next, $C_{|p}^{doc}$ and $C_{|p}^{word}$ can be computed from the histogram of $\mathbf{V}z_{|p}$ (line 11). To obtain C^{word}, the algorithm invokes `MPI_Allreduce` (line 12). In the Gibbs sampling iterations, each word's topic is sampled from the approximate posterior distribution (Eq.(12.6)) and $C_{|p}^{word}$ and $C_{|p}^{doc}$ is updated accordingly (lines 14 to 18). At the end of each iteration, the algorithm checkpoints $\mathbf{V}z_{|p}$ (line 20) and recomputes $C_{|p}^{word}$ (line 21). Using $C_{|p}^{word}$, the algorithm perform global `MPI_AllReduce` to obtain up-to-date C^{word} for the next iteration (line 22). After a sufficient number of iterations, the "converged" LDA model is outputted by the master (line 24).

Different MPI implementations may use different AllReduce algorithms. The state-of-the-art is the recursive doubling and halving (RDH) algorithm presented in [17], which was used by many MPI implementations including the well known MPICH2. RDH includes two phases: *Reduce-scatter* and *All-gather*. Each phase

1: **if** there is a checkpoint **then**
2:      $t \leftarrow$ The number of iterations already done
3:      Load $\mathbf{V}z_{|p}$ from the checkpoint
4: **end if**
5: $t \leftarrow 0$
6: Load documents on current processor p into $\mathbf{V}x_{|p}$
7: **for** each word $x_{ij|p} \in \mathbf{V}x_{|p}$ **do**
8:      Draw a sample k from uniform distribution $U(1, K)$
9:      $z_{ij|p} \leftarrow k$
10: **end for**
11: Compute $C_{|p}^{doc}$ and $C_{|p}^{word}$
12: MPI_AllReduce($C_{|p}^{word}$, C^{word}, $W \times T$, ``float-number'', ``sum'')
13: **for** ; $t <$ iteration-num; $t \leftarrow t + 1$ **do**
14:      **for** each word $x_{ij|p} \in \mathbf{V}x_{|p}$ **do**
15:           $C_{j,z_{ij}|p}^{doc} \leftarrow C_{j,z_{ij}|p}^{doc} - 1, C_{x_{ij},z_{ij}}^{word} \leftarrow C_{x_{ij},z_{ij}}^{word} - 1$
16:           $z_{ij} \leftarrow$ draw new sample from (12.6), given C^{word} and $C_{j|p}^{doc}$
17:           $C_{j,z_{ij}|p}^{doc} \leftarrow C_{j,z_{ij}|p}^{doc} + 1, C_{x_{ij},z_{ij}}^{word} \leftarrow C_{x_{ij},z_{ij}}^{word} + 1$
18:      **end for**
19: **end for**
20: Checkpoint $\mathbf{V}z_{|p}$
21: Recompute $C_{|p}^{word}$
22: MPI_AllReduce($C_{|p}^{word}$, C^{word}, $W \times T$, ``float-number'', ``sum'')
23: **if** this is the master worker **then**
24:      Output C^{word}
25: **end if**

Fig. 12.2: The MPI implementation of AD-LDA

runs a recursive algorithm, and in each recursion level, workers are grouped into pairs and exchange data in both directions. This algorithm is particularly efficient when the number of workers is a power of 2, because no worker would be idle during communication.

RDH provides no facilities for fault recovery. In order to provide fault-recovery capability in AD-LDA, the worker state can be check-pointed before AllReduce. This ensures that when one or more processors fail in an iteration, the algorithm can roll back all workers to the end of the most recent succeeded iteration, and restart the failed iteration. The checkpointing code is executed immediately before the invocation of MPI_Allreduce in AD-LDA. In practice, only $\mathbf{V}z_{|p}$ is flushed onto the disk, because $\mathbf{V}x_{|p}$ can be reloaded from data set, $C_{|p}^{doc}$ and C^{word} can also be recovered from the histogram of $\mathbf{V}z_{|p}$. The recovery code is at the beginning of AD-LDA: if there is a checkpoint on the disk, load it; otherwise perform random initialization.

12.4 PLDA+

To further speed up AD-LDA [4], PLDA+ algorithm employs distributed data placement and pipeline processing strategies.

12.4.1 Reduce Bottleneck of AD-LDA

As presented in the previous section, in our AD-LDA implementation [6], D documents are distributed over P processors with approximately D/P documents on each processor. This is shown with a $D/P\text{-}W$ matrix in Figure 12.3(a), where W indicates the vocabulary of the collection of documents. The word-topic count matrix is also distributed, with each processor keeping a local copy, which is the $W\text{-}K$ matrix in the figure.

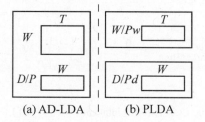

(a) AD-LDA   (b) PLDA

Fig. 12.3: The assignments of documents and word-topic count matrix for AD-LDA and PLDA+

In AD-LDA, after each iteration of Gibbs Sampling, local word-topic counts on each machine are globally synchronized. This synchronization process is expensive partly because a large amount of data is sent and partly because the synchronization starts only when the slowest machine has completed its work. To avoid unnecessary wait, AS-LDA does not perform global synchronization like AD-LDA. AS-LDA only synchronizes word-topic counts with its neighbors. However, since word-topic counts can be outdated, the sampling process may take a larger number of iterations than that AD-LDA takes to converge. Figure 12.4 illustrates the spread patterns of the updated topic distribution of a word from one processor to the others. AD-LDA has to synchronize all word updates after one full Gibbs Sampling iteration, whereas AS-LDA performs updates only with a small subset of processors. The memory requirement of both AD-LDA and AS-LDA is $O(KW)$, since the whole word-topic matrix is maintained on all machines.

Although having different strategies for model combination, existing distributed methods share two characteristics:

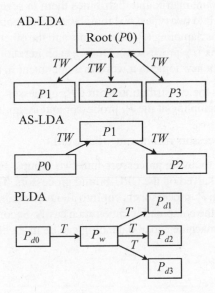

Fig. 12.4: The spread patterns of the updated topic distribution of a word from one processor for AD-LDA, AS-LDA and PLDA+

- These methods have to maintain all word-topic counts in memory of each processor; and
- These methods have to send and receive the entire word-topic matrix between processors for updates.

For the former characteristic, suppose we want to estimate a $\mathbf{V}\phi$ with W words and K topics from a large-scale data set. When either W or K is large to a certain extent, the memory requirement will exceed that available on a typical processor. Due to the bottleneck of latency and transfer-rate of hard disks, it is not practical to maintain the word-topic counts on hard disks. This characteristic makes the existing distributed methods face a significant challenge in terms of memory scalability. For the latter characteristic, the communication bottleneck caps the room for speeding up the algorithm. This communication bottleneck will only acerbate over years as a study of high performance computing [18] shows that floating-point instructions improve speed historically at 59% a year, but inter-processor bandwidth improves 26% a year, and inter-processor latency improves only 15% a year.

12.4.2 Framework of PLDA+

To address the increasing communication bottleneck, PLDA+ uses an enhanced distributed method for LDA. In addition to partitioning documents, PLDA+ also partitions the word-topic count matrix and distributes them to several processors. Thus, processors are divided into two types: one maintains documents and document-topic matrix to perform Gibbs Sampling (P_d processors, and the other stores and maintains word-topic count matrix (P_w processors). During each iteration of Gibbs Sampling, a P_d processor assigns a new topic to a word in a document in three steps:

1. Fetching the word's topic distribution from a P_w processor,
2. Performing Gibbs sampling at the P_d processor and assigning a new topic to the word, and
3. Updating all P_w processors maintaining that word.

There are two reasons to divide processors into two groups. First, the communication bottleneck can be halved on the CPU-bound processors. This way, not only the communication time on P_w processors is cut into about one half, the reduced IO time can also be masked by the computation time much easily. Second, by separating two tasks onto two sets of machines, load balancing can be more flexibly performed.

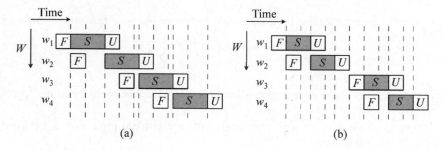

Fig. 12.5: Pipeline-based Gibbs Sampling in PLDA+ (Top: $t_s > t_f + t_u$. Bottom: $t_s < t_f + t_u$)

Besides improving parallelization, PLDA+ employs pipeline processing. The pipeline technique has been used in many applications to increase throughput, such as the instruction pipeline in modern CPUs [19] and in graphics processors [20]. Although pipeline does not decrease the time for a job to be processed, it can efficiently improve throughput by overlapping IOs with computation. Figure 12.5 illustrates the Pipeline-based Gibbs Sampling for four words, i.e., w_1, w_2, w_3 and w_4, where F indicates the fetching operation, U indicates the updating operation, and S the Gibbs Sampling operation. In this figure, the top chart demonstrates the case when $t_s > t_f + t_u$, and the bottom chart the case when $t_s < t_f + t_u$, where t_s, t_f and t_u denote the time of Gibbs Sampling, fetching topic distribution, and updating topic distribution, respectively.

On the top chart of Figure 12.5, PLDA+ begins by fetching the topic distribution of w_1. Then it begins Gibbs Sampling on w_1, and at the same time, it fetches the topic distribution of w_2. After it has finished Gibbs Sampling for w_1, it updates the topic distribution of w_1 on P_w processors. When $t_s > t_f + t_u$, PLDA+ can begin Gibbs Sampling on w_2 immediately after it has completed that for w_1. Total ideal time for PLDA+ to process W words is $Wt_s + t_f + t_u$.

The bottom chart of Figure 12.5 shows a suboptimal scenario where the IO time cannot be entirely masked. PLDA+ is not able to begin Gibbs Sampling for w_3 until after some communication delay. The example shows that in order to mask communication, the tasks must be scheduled to ensure as much as possible that $t_s > t_f + t_u$. There are two important scheduling considerations:

1. Word bundling. To ensure t_s to be sufficiently long to mask IOs, Gibbs Sampling can be performed on a group of words.
2. Low latency IO scheduling. IOs must be scheduled in such a way that a CPU-bound task is minimally delayed by a fetch operation.

Since each round of Gibbs Sampling can be performed in any word order, it makes word bundling flexible. First, rather than processing one document after another, PLDA+ performs Gibbs Sampling according to a word order. A word that occurs several times on the documents at a node can be process in a loop. Moreover, for words that do not occur frequently, they can be bundled with frequently-occurred words to ensure that t_s is sufficiently long. In fact, if one can estimate $t_f + t_u$, one can decide how many word-occurrences to process in each Gibb Sampling batch. The remaining challenge is that one ought to ensure that $t_f + t_u$ can indeed be shorter than t_s. If a fetch cannot be completed by the time when the last Gibbs Sampling task has completed, the wait time adds to the bottleneck, and hence hampers speedup.

To perform Gibbs Sampling word by word, PLDA+ builds word indexes to documents on each P_d processor. Words are organized in a *circular queue* as shown on the top of Figure 12.6. Gibbs Sampling is performed by going around the circular queue. To avoid concurrent access to the same words, different processes are scheduled to begin at a different position of the queue. For example, Figure 12.6 shows four P_d processors, P_{d1}, P_{d2}, P_{d3} and P_{d4} start their first word from w_1, w_3, w_5 and w_7, respectively. To ensure that this scheduling algorithm works, PLDA+ must distribute the word-topic matrix also in a circular fashion on P_w machines. This static allocation scheme enjoys two benefits. First, the workload among P_w processors can be relatively balanced. Second, avoiding two P_d nodes from concurrently updating the same word can roughly maintain serializability of the word-topic matrix on P_w nodes. This makes PLDA+ more advantageous over an asynchronous scheme such as AS-LDA [5], which may miss updates. The detailed description of word placement is presented in Section 12.4.3.1.

12.4.3 Algorithm for P_w Processors

The task of the P_w processors is to process fetch and update queries from P_d processors. PLDA+ distributes the word-topic matrix to P_w machines according to words. After allocation, each P_w processor keeps approximately W/P_w words with their

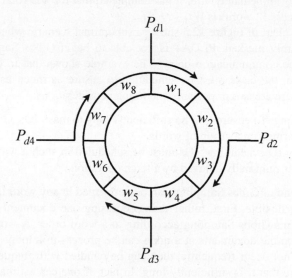

Fig. 12.6: Vocabulary circular queue in PLDA+

topic distributions. Figure 12.7 depicts the word-topic matrix distribution process to P_w machines.

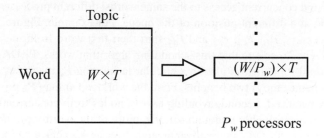

Fig. 12.7: The construction of word-topic matrix in P_w processors

12.4.3.1 Word Placement over P_w Processors

The goal of word allocation is to ensure *spatial* load balancing. To balance load, one would like to make sure that all nodes receive about the same number of work requests in a round of Gibbs Sampling.

For bookkeeping, PLDA+ maintains two data structures. First, for each word, it records how many P_d processors on which that word resides. Form W words, PLDA+ maintains a P_d vector $\mathbf{m} = (m_1, m_2, \ldots, m_W)$. The second data structure keeps track of each P_w processor's workload, or the number of word occurrences on that processor. The workload vector is denoted as $\mathbf{l} = (l_1, l_2, \ldots, l_{P_w})$.

A simple placement method is to place words independently and uniformly at random onto P_w processors. This method is referred to as *Random Word Allocation*. Unfortunately, this random placement method may cause load imbalance among P_w processors in high probability. To balance workload, PLDA+ uses the *Weighted Round-Robin* method for word placement. It first sorts words in *decreasing* order by their weights, and then picks the word with the largest weight from the vocabulary and assigns to a processor in a round-robin fashion. This placement process is repeated until all words have been placed. Weighted Round-Robin has been empirically shown to achieve balanced load with high probability ([21]).

12.4.3.2 Processing Requests from P_d Processors

Each P_w processor handles all requests related to the words it is responsible for maintaining. After allocating words with their topic distributions over P_w processors, P_w processors begin to receive and respond the requests from P_d processors. A P_w processor pw first builds its responsible word-topic count matrix $C^{word}_{|pw}$ by receiving initial word-topic counts from all P_d processors. Then, that P_w processor pw begins to process requests from P_d processors. PLDA+ defines three types of requests (communications):

- $fetch(w, pw, pd)$. Node pw requests for fetching topic distribution of word w from a P_d processor pd. For the request, the P_w processor pw retrieves the topic distribution $\phi^{(pw)}_w$, which will be used by the pd node as $n^{\neg ij}_{wk}$ in Eq.(12.3) for performing Gibbs Sampling.
- $update(w, \mathbf{u}, pw, pd)$. Node pw updates topic distribution for word w using $\mathbf{u}$ after receiving the information from node pd.
- $fetch(pw, pd)$. Node pw requests for all topic counts from node pd. The P_w Processor pw requires the data from pd to sum up the topic distributions of all words on pw in vector $\mathbf{n}^{(pw)} = (n^{(pw)}_k, k = 1, \cdots, T)$, which will be used as $n^{\neg ij}_k$ in Eq.(12.3) for performing Gibbs Sampling.

12.4.4 Algorithm for P_d Processors

The algorithm for P_d processors executes according to the following steps:

1. At the beginning, it allocates documents over P_d processors and then builds inverted index for documents on each P_d processor.

2. It groups the words in vocabulary into *bundles* for Gibbs Sampling and IO requests.
3. It schedules word bundles to minimize communication bottleneck.
4. Finally, it performs pipelined Gibbs Sampling iteratively until the terminate condition is met.

In the following, we present these four steps in details.

12.4.4.1 Document Allocation and Building Inverted Index

Before performing Gibbs Sampling, D documents must be distributed onto P_d processors. The goal of document allocation is to achieve good CPU load balance among P_d processors. AD-LDA may suffer from imbalanced load problem since it has a global synchronization phase at the end of each Gibbs Sampling iteration, which may force fast processors to wait for the slowest processor. In contrast, Gibbs Sampling in PLDA+ is performed without the synchronization requirement. In other words, a processor that completes its work early can start its next round of sampling without having to wait for stragglers. Dealing with stragglers is a critical issue in distributed computing. PLDA+ tackles this problem through both static allocation and dynamic migration. PLDA+ first allocates words to nodes in a balanced fashion. Each P_d processor hosts approximate D/P_d documents. The time complexity of this allocation step is $O(D)$. After documents have been distributed, we build inverted index for documents on each P_d processor. The construction process is demonstrated in Figure 12.8. If a node is always a straggler due to run-time load imbalance or hardware configuration, the data on that node can be split and migrated onto additional nodes to eliminate stragglers.

Using inverted index, each time after a P_d processor has fetched the topic distribution of a word w, that processor performs Gibbs Sampling for all instances of w on that node. After that, the processor (or node) sends back the updated information to the corresponding P_w processor. The clear benefit is that for multiple occurrences of a word on a node, PLDA+ requires to perform only two communications: one fetch and one update, and substantially reducing communication cost. The index structure for each word w is:

$$w \rightarrow \{(d_1, z_1), (d_1, z_2), (d_2, z_1) \ldots\}, \tag{12.7}$$

in which, w occurs in document d_1 twice and there are 2 entries. In implementation, to save memory, all occurrences of w in d_1 can be recorded in one entry, $(d_1, \{z_1, z_2\})$.

12.4.4.2 Word bundle

Bundling words is to prevent the situation that too short the duration of Gibbs Samplings cannot mask a communication IO. Use an extreme example: a word appears

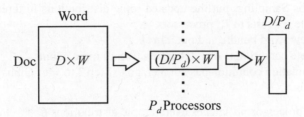

Fig. 12.8: The construction of data structure in P_d processors

only once in one document on a node. Performing Gibbs Sampling on that word takes a much shorter time than the time required to fetch and update the word-topic matrix. The remedy is intuitive: combining a few words into a bundle so that the IO time can be masked by the longer duration of Gibbs Sampling time.

To bundle words, each P_d processor groups words in sets, each matches words on a P_w processor. For each word set, words are sorted into a list according to their occurrence times in descending order. Then, words are picked from both ends of the list to form bundles. Each time a P_d node sends a request to a P_w node to fetch topic distributions for words in a bundle. The size of a bundle should be large enough so that the time to perform Gibbs sampling on a bundle is longer than the time to fetch the bundle from a P_w node.

12.4.4.3 Pipelined Gibbs Sampling

The core step of PLDA+ is the pipelined Gibbs Sampling. As shown in Eq.(12.3), to compute and assign a new topic for a given word $x_{ij} = w$ in a document d_j, we have to obtain C_w^{word}, C^{topic} and C_j^{doc}. The topic distribution of document j is maintained by P_d processors. While the up-to-date topic distribution C_w^{word} is maintained by a P_w processor, global topic count C^{topic} should be collected over all P_w processors. Therefore, before assigning a new topic for w in a document, a P_d processor has to request C_w^{word} and C^{topic} from P_w processors. After fetching C_w^{word} and C^{topic}, the P_d processor computes and assigns new topics for occurrences of w. Then the P_d processor returns the updated topic distribution of word w to the responsible P_w processor.

For a P_d processor pd, pipeline processing is performed according to the following steps:

1. Fetch overall topic counts for Gibbs Sampling.
2. Select F word bundles and put them in thread pool tp to fetch words' topic distributions. Once a request is responded from P_w processors, the returned topic distributions are put in a wait queue Q_{pd}.

3. Pick words' topic distributions from Q_{pd} to perform Gibbs Sampling.
4. After Gibbs Sampling, put the updated topic distributions in thread pool tp to send update requests to P_w processors.
5. Select a new word bundle and put it in tp.
6. If the update condition is met, fetch new overall topic counts.
7. If the termination condition has not met, go to Step 3 to start Gibbs Sampling for other words.

In Step 1, processor pd fetches overall topic distributions C^{topic}. In this step, pd sends requests $fetch(pw, pd)$ to each P_w processor $pw = 1, 2, \ldots, P_w$. The requests are returned with $(C^{topic}_{|pw}, pw = 1, 2, \ldots, P_w)$, and pd thus gets C^{topic} by sum overall topic counts from each P_w processors:

$$C^{topic} \leftarrow \sum_{pw} C^{topic}_{|pw}. \tag{12.8}$$

Since thread pool tp can send requests and process the returned results in parallel, in Step 2 it puts a number of requests to fetch topic distributions simultaneously in case some requests are responded with latency. Thus, once a response is returned, it can start Gibbs Sampling immediately. Here, we mention the pre-fetch number of requests as F. In PLDA+, F should be properly set to make sure that the wait queue Q_{pd} always has returned topic distributions of words waiting for Gibbs Sampling. If not, Gibbs Sampling is stalled by communication, which is considered a part of communication time of PLDA+. To make best use of threads in the thread pool, F should be larger than the number of threads in the pool.

It is expensive for P_w processors to process the request for overall topic counts because the operation has to access topic distributions of each word on each P_w processor. Fortunately, as indicated by the results of AD-LDA, topic assignments in Gibbs Sampling is not sensitive to the values of overall topic counts. Thus PLDA+ reduces the frequency of fetching overall topic counts to improve the efficiency of P_w processors. Therefore, in Step 6, PLDA+ does not fetch overall topic counts frequently. Experimental results show that fetching new overall topic counts only after performing one pass of Gibbs Sampling can obtain the same learning quality compared to LDA and AD-LDA.

Figure 12.9 summarizes a P_d node's interprocess communication with multiple P_w nodes. The figure shows a key reason for PLDA+ to reduce communication bottleneck: that a P_d node of PLDA+ commuicates with multiple P_w nodes, rather than that multiple P_d nodes of AD-LDA communicate with one master P_w node. Furthermore, the thread pool on P_d nodes enables pre-fetching, and thereby allows communication to be masked by computation working on completed requests.

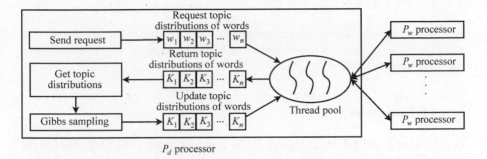

Fig. 12.9: PLDA+ Gibbs Sampling

12.4.5 Straggler Handling

So far, both presented data placement and scheduling schemes of PLDA+ are *static*. Static placement and scheduling cannot guarantee run-time load balancing. Run-time imbalanced workload can be caused by at least three reasons:

1. *Uneven hardware configuration.* Not all nodes are equally configured. In a realistic distributed environment, not all computer nodes are equipped with exactly the same class of processors, memory, and disks. Also, not all nodes are equally distanced. Computation on and communication with different nodes can thus take different amount of time to complete.
2. *Resource contention.* Distributed data centers must deal with a large number of simultaneous computation tasks. It is impossible to ask all nodes to be in a quiesce mode when PLDA+ is being executed. Therefore, PLDA+ can be slowed down by tass competing for resources.
3. *Failures.* When a large number of nodes are involved, the probability of failure becomes non-negligible. When a processor or a router fails, no static scheme can continue ensuring balanced workload among all nodes.

PLDA+ deals with run-time dynamics by employing two simple approaches. First, PLDA+ uses a reset and timeout scheme. When a P_w node notices that the number of requests in its work queue has reached a threshold, it informs all P_d nodes to reset their pointers into the circular queue depicted in Figure 12.6. In each request, the P_d node also registers a deadline. When the deadline has expired, the P_w node discards that request and proceeds to processing the next request. Occasionally missing a round of Gibbs Sampling does not affect overall performance due to the stochastic nature of Gibbs Sampling.

If a P_w node has missed too many request deadlines, then PLDA+ replicates that node to balance workload. For the details of a data replication scheme that can guarantee balanced workload in probability, please consult our previous work in [22].

12.4.6 Parameters and Complexity

In this section, we discuss the parameters that may influence the performance of PLDA+. We also analyze the complexity of PLDA+ compared to other distributed methods represented by AD-LDA.

12.4.6.1 Parameters

Given the total number of processors P, the first parameter is the proportion of the number of P_w processors to P_d processors, $\gamma = \frac{P_w}{P_d}$. The larger the value of γ, the more processors serve as P_w, and hence the average time of communication at P_d processors decreases. At the same time, the average time of Gibbs Sampling will increase due to less processors are used to perform that CPU-bound task. A good system design must balance the number of P_w and P_d processors to (1) make both computation and communication time low, and (2) ensure that communication is short enough to be masked by computation. This parameter can be derived once the average time for Gibbs Sampling and communication of the word-topic matrix is known. Suppose the total time of Gibbs Sampling for the whole data set is T_s, the communication time of transferring the topic distributions of all words from one processor to another processor is T_t. For P_d processors, the sampling time will be T_s/P_d. Suppose topic distributions of words can be simultaneously transferred to P_w processors, and thus transfer time will be T_t/P_w. To make sure the sampling time can overlap the fetching and updating process, PLDA+ thus must make sure that

$$\frac{T_s}{P_d} > \frac{2T_t}{P_w}. \tag{12.9}$$

Suppose $T_s = W\bar{t}_s$ where $\bar{t}_s$ is the average sampling time for all instances of a word, and $T_t = W\bar{t}_f = W\bar{t}_u$, where $\bar{t}_f$ and $\bar{t}_u$ are the average fetching and update time for a word, we can get

$$\gamma = \frac{P_w}{P_d} > \frac{\bar{t}_f + \bar{t}_u}{\bar{t}_s}, \tag{12.10}$$

where $\bar{t}_f$, $\bar{t}_u$ and $\bar{t}_s$ can be obtained by performing PLDA+ on a small data set and then empirically set a appropriate γ value. Under the computing environment of our experiments, we empirically set $\gamma = 3/5$.

The second parameter is the number of threads in thread pool R, which caps the number of parallel IO requests. Since thread pool is used to prevent from being blocked by some busy P_w processors and thus R is determined by the network environment. The setting of R can be empirically tuned during Gibbs Sampling. That is, when the waiting time during the previous iteration is large, the thread pool size is increased.

The third parameter is the number of requests F for pre-fetching topic distributions before performing Gibbs Sampling on P_d processors. This parameter is dependent on R.

The last parameter is the maximum interval $inter_{max}$ for fetching overall topic counts from all P_w processors during Gibbs Sampling of P_d processors. This parameter influences the quality of PLDA+. Experiments show that in order to learn LDA models with similar quality to AD-LDA and LDA, $inter_{max}$ should be set to W.

It should be noted that the optimal values of the parameters of PLDA+ are highly related to the distributed environment including network bandwidth and processor speed.

12.4.6.2 Complexity

Table 12.2 summarizes the complexity of P_d processors and P_w processors in both time and space. For comparison, the table also lists the complexity of LDA and AD-LDA. We assume $P = P_w + P_d$ when comparing PLDA+ with AD-LDA.

Table 12.2: Algorithm complexity

Method	Time complexity	Space complexity
LDA	NT	$T(D+W)+N$
AD-LDA	$\frac{NT}{P} + TW\log P$	$\frac{(N+TD)}{P} + TW$
PLDA+ - P_d	$\frac{NT}{P_d} + \delta$	$\frac{(N+TD)}{P_d}$
PLDA+ - P_w	-	$\frac{TW}{P_w}$

Finally, let us analyze the speedup efficiency of PLDA+. Suppose $\delta \rightarrow 0$ and $\gamma = \frac{P_w}{P_d}$ for PLDA+, the ideal parallel efficiency will be always:

$$\text{speedup efficiency} = \frac{S/P}{S/P_d} = \frac{P_d}{P} = \frac{1}{1+\gamma}, \tag{12.11}$$

where S denotes the running time of LDA on a single machine, S/P is the ideal time cost using P processors, and S/P_d is the ideal time achieved by PLDA+ with communication completely masked by Gibbs Sampling.

12.5 Experimental Results

This section compares the performance of PLDA+ and AD-LDA. The comparisons help understand benefits of data placement and pipeline processing strategies.

12.5.1 Datasets and Experiment Environment

We used four datasets shown in Table 12.3 to conduct experiments. The NIPS dataset consists of scientific articles appeared at NIPS conferences. Dianping dataset consists of restaurant reviews from dianping.com. NIPS and Dianping datasets are both relatively small, and we used them to carry out training-quality assessment. Two Wikipedia datasets were collected from English Wikipedia articles of the March 2008 snapshot from en.wikipedia.org. By setting the size of vocabulary to 20,000 and 200,000, respectively, the two Wikipedia datasets are named Wiki-20T and Wiki-200T. These two large datasets were used for testing scalability of PLDA+. The experiment environment was run on distributed machines with 2,048 processors, each with a 2HZ CPU, 3GB memory, and disk allocation of 100GB.

Table 12.3: Detailed information of data sets

	NIPS	Dianping	Wiki-20T	Wiki-200T
D_{train}	1,540	113,754	2,122,618	2,122,618
W	11,909	27,752	20,000	200,000
N	1,260,732	3,625,275	447,004,756	486,904,674
D_{test}	200	1,000	-	-

12.5.2 Perplexity

We used *test set perplexity* to measure the quality of LDA models learned by various distributed methods for LDA. Perplexity is a common way of evaluating language models in natural language processing, computed as:

$$Perp(\mathbf{x}^{\text{test}}) = \exp\Big(- \frac{1}{N^{\text{test}}} \log p(\mathbf{x}^{\text{test}}) \Big), \qquad (12.12)$$

where $\mathbf{x}^{\text{test}}$ denotes test set. A lower test perplexity value indicates a better quality. For every test document in the test set, we randomly designated half the words for fold-in, and the remaining words were used for testing. The document mixture θ_j was learned using the fold-in part, and the log probability of the test words was computed using this mixture. This ensures the test words were not used in estimating model parameters. The perplexity computation follows the standard way of averaging over multiple chains when making predictions with LDA models trained via Gibbs Sampling as shown in [8]. For PLDA+ and LDA, the test perplexity was computed using $S = 40$ samples from the posteriors of 40 independent chains using:

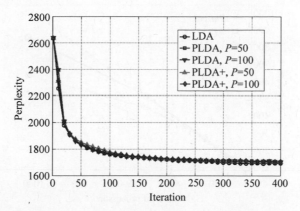

Fig. 12.12: Convergence of test perplexity versus iteration on NIPS with $T = 80$ (See color insert)

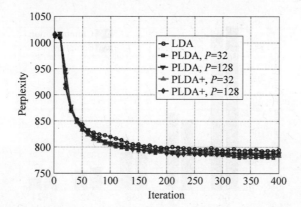

Fig. 12.13: Convergence of test perplexity versus iteration on Dianping with $T = 64$ (See color insert)

On the larger Wiki-200T dataset, the speedup of AD-LDA starts to flat out at $P = 512$, whereas PLDA continues to gain in speed[1]. For this dataset, we also list the sampling and communication time ratio of AD-LDA and PLDA+ in Figure 12.17. As shown in this figure, PLDA+ keeps communication time to quite low values from $P = 64$ to $P = 2,048$. While for AD-LDA, the communication time finally became a bottleneck to prevent it from speedup as the number of processors grows. Though eventually the Amdahl's law would kick in to cap speedup, it is evident that the reduced overhead of PLDA+ permits it to achieve much better speedup for training on larger datasets.

---

[1] For PLDA+, the parameter of pre-fetch number and thread pool size was set to $F = 100$ and $R = 50$. With $W = 200,000$ and $W = 1,000$, the matrix is 1.6GB, which is large for communication.

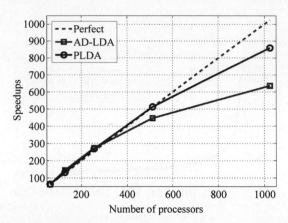

Fig. 12.14: Parallel speedup results for 64 to 1,024 processors on Wiki-20T (See color insert)

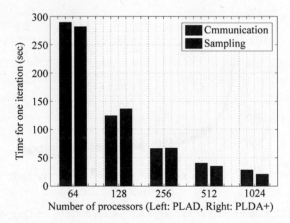

Fig. 12.15: Communication and sampling time for 64 to 1,024 processors on Wiki-20T (See color insert)

12.6 Large-Scale Applications

LDA has been shown effective in many tasks (e.g.,[23, 24, 25]). In this section, we use two large-scale applications, *community recommendation* of Google Orkut and *label suggestion* of Google Confucius [26], to demonstrate the usefulness of PLDA+.

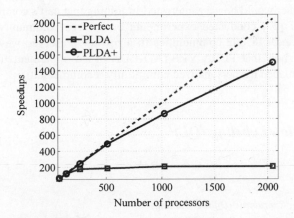

Fig. 12.16: Parallel speedup results for 64 to 2,048 processors on Wiki-200T (See color insert)

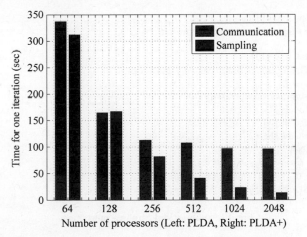

Fig. 12.17: Communication and sampling time for 64 to 2,048 processors on Wiki-200T (See color insert)

12.6.1 Mining Social-Network User Latent Behavior

Users of social networking services (e.g., Orkut, Facebook, and MySpace) can connect to each other explicitly by adding friends, or implicitly by joining communities. When the number of communities grows over time, finding an interesting community to join can be time consuming. We use LDA to model users' community membership [27]. On a matrix formed by users as rows and communities as columns, all values in user-community cells are initially unknown. When a user joins a community, the corresponding user-community cell is set to one. We apply LDA on the matrix to assign a probability value between zero and one to the unknown cells.

When LDA assigns a high probability to a cell, this can be interpreted as a prediction that that cell's user would be very interested in joining that cell's community.

The work of [27] conducted experiments on a large community data set of 492,104 users and 118,002 communities in a privacy-preserved way. The experimental results show that PLDA V1.0 (AD-LDA based implementation) achieves effective performance for personalized community recommendation.

12.6.2 Question Labeling (QL)

Question Labeling Subroutine (Offline Training)
Input: Questions $Q = \{q_1, \ldots, q_n\}$, in which $q_i = \{w_{i,1}, \ldots, w_{i,n}\}$, Labels $L = \{l_1, \ldots, l_m\}$ and their relationship with labels $R \in Q \times L$
Output: LDA models $M = \{(\theta, \phi, k)\}$
1: **for** $l \in L$
2:      $d_l = \{w|w \in d, \forall(d,l) \in R\}$
3: Remove stop words and rare words from all $d_l, l \in L$
4: $M = \{\}$
5: **for** $k \in \{32, 64, 128, 512\}$
6:      Train LDA model (θ, ϕ, k), with $\{d_l\}$ and k topics
7:      $M \leftarrow (\theta, \phi, k)$
Question Labeling Subroutine (Online)
Given: LDA Models M $= \{(\theta, \phi, k)\}$
Input: Question $q = \{w_1, \ldots, w_{|q|}\}$
Output: Suggested Labels $L_q = \{l_1, \ldots, l_n\}$
1: Infer $\theta_{q,k}$ with M
2: $S_k(\theta_q, \theta_{d_l}) = CosSim(\theta_{q,k}, \theta_{d_l}), \forall l \in L$
3: $S(q,l) = \overline{\{S_k(\theta_{q,k}, \theta_{d_l,k})\}}$ /* Mean similarity */
4: $L_q = \{l||\{l'|S(q,l') > S(q,l)\}| < N\}$ /*Top N labels*/

Fig. 12.18: QuestionLabeling subroutine

Confucius is a Q&A system developed by my team at Google Beijing, and has been launched in more than sixty countries [26]. The goal of QuestionLabeling is to help organize and route questions with automatically recommended labels. The QuestionLabeling subroutine takes a question as the input and outputs an ordered list of labels that best describe the question. Labels consist of a set of words or phrases that best describe the topic or type of the question. Confucius allows at most 5 labels per question, but puts no limit on the size of the global label vocabulary. Confucius organizes the most important category labels into a two-layer hierarchy, in order to provide a better browsing experience. QuestionLabeling is used by two other subroutines: UserRank and QuestionRouting . When ranking users, UserRank uses popular labels to compute the topic-dependent rank scores. QuestionRouting assigns questions to users via either subscription or expert identification, during which labels generated by QuestionLabeling are used for matching.

$$\log p(\mathbf{x}^{\text{test}}) = \sum_{j,w} n_{jw}^{\text{test}} \log \frac{1}{S} \sum_{k} \theta_{kj}^{S} \phi_{wk}^{S}, \tag{12.13}$$

where

$$\theta_{kj} = \frac{C_{kj}^{S} + \alpha}{\sum_{k=1}^{T} C_{kj}^{S} + T\alpha}, \qquad \phi_{wk} = \frac{C_{wk}^{S} + \beta}{\sum_{w=1}^{W} C_{wk}^{S} + W\beta}. \tag{12.14}$$

To compare the quality of PLDA+ to single-machine LDA and distributed AD-LDA, we computed the test perplexity for all methods after each iteration of Gibbs Sampling going through a round of whole vocabulary. The test perplexities on NIPS with the number of topics $K = 10, 20, 40, 80$, and Dianping with $K = 8, 16, 32, 64$ are shown in Figures 12.10 and 12.11, respectively. (Since we concerned only about training quality, the number of machines used in this experiment may not be relevant.)

From both figures we can see that the quality of PLDA+ is similar to single-machine LDA and distributed AD-LDA. Thus, we can conclude that PLDA+ can train as good a model as traditional LDA methods.

Figures 12.12 and 12.13 show the convergence of test perplexity versus # of iteration for LDA, AD-LDA and PLDA+ on NIPS and Dianping with different number of processors. (The parameters were set as depicted in Section 12.5.2.) The figures show the convergence rate of PLDA+ is virtually identical to LDA and AD-LDA.

12.5.3 Speedups and Scalability

The primary motivation for developing distributed algorithms for LDA is to achieve a good speedup. In this section, we report the speedup of PLDA+ comparing to AD-LDA. We used Wiki-20T and Wiki-200T for speedup experiments. By setting the number of topics $T = 1,000$, we ran PLDA+ and AD-LDA on Wiki-20T using $P = 64, 128, 256, 512$ and $1,024$ processors, and on Wiki-200T using $P = 64, 128, 256, 512, 1,024$ and $2,048$ processors. For PLDA+, the ratio of $P_w P_d$ was empirically set to $\gamma = 0.6$ according to the unit sampling time and transferring time. The number of threads in a thread pool is 50, which is sufficient to handle the peak load. As analyzed in Section 12.4.6.2, the ideal speedup efficiency of PLDA+ is $\frac{1}{1+\gamma} = 0.625$.

Figure 12.14 compares speedup performance. The speedup was computed relative to the time per iteration when using $P = 64$ processors, because it was not possible to run the algorithms on a smaller number of processors due to memory limitations. We assumed that the speedup on $P = 64$ to be 64, and then extrapolated on that basis. From the figure, we can observe that when P increases, PLDA+ simply achieves much better speedup than AD-LDA, thanks to the much reduced communication bottleneck of PLDA+.

Figure 12.15 compares the ratio of communication time over computation time when $P = 1,024$. The communication time of AD-LDA is 13.38 seconds, much

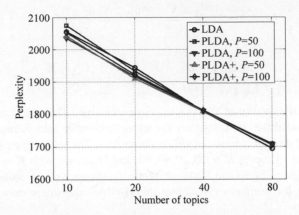

Fig. 12.10: Test perplexity on NIPS versus # topics T when the number of iterations is 400 (See color insert)

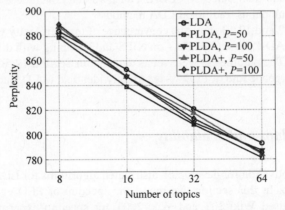

Fig. 12.11: Test perplexity on Dianping versus # topics T when the number of iterations is 400 (See color insert)

longer than that of PLDA+'s 3.68 seconds. The communication time of AD-LDA is about the same as its computation time at $P = 512$.

From the results, we can conclude that: (1) when word-topic matrix is not large, PLDA+ performs similarly to AD-LDA, and when the number of processors increases to large enough (e.g., $P = 512$), PLDA+ begins to achieve better speedup than AD-LDA; (2) In fact, if we take the waiting time for synchronization in AD-LDA into consideration, the speedup of AD-LDA could have been even worse. For example, in a busy distributed computing environment, when $P = 128$, AD-LDA may take about 70 seconds for communication in which only about 10 seconds are used for transmitting word-topic matrix and most of time is used to wait for each other.

Appendix: Open Source Software

By the end of 2010, my team have released three pieces of software to the public through the Apache Open Source foundation to assist research communities of signal processing, computer vision, data mining, machine learning, and database to conduct large-scale studies and experiments. The locations of the software are as follows:

- *PSVM* at code.google.com/p/psvm/.
- *PLDA+* at code.google.com/p/plda/.
- *Parallel Spectral Clustering* at code.google.com/p/pspectralclustering/.

Index

The precision and recall of suggested labels are two important metrics for measuring QuestionLabeling performance. Precision measures the correctness of suggested labels, while recall measures the completeness.

Figure 12.18 shows the two parts of QuestionLabeling: offline training and online suggestion. In the offline training part, we employ LDA to model the relationship between words and topic labels. The training data is the existing set of questions with user-submitted labels. First, we merge all questions with the same label l into a meta-document d_l, and form a set of meta-documents $\{d_l\}$ (Figure 12.18, Steps 1-2). Second, we remove all stop words and rare words to reduce the size of each meta-document (Step 3). Third, we use $\{d_l\}$ as the corpus to train LDA models (Steps 5-6). The label corresponds to the document in LDA definition, while the words in the meta-documents correspond to the words. The resulted LDA model decomposes the probability $p(w|l)$ to $\sum_z p(w|z)p(z|l)$—this is similar to the factor model in recommendation algorithms, expressed in terms of probabilities. Instead of a single model, QuestionLabeling trains several LDA models with different number of latent topics. Using multiple LDA models with different k-s is known as *bagging*, which typically outperforms a single model and avoids the difficult task of setting an optimal k, as discussed by Hofmann [28]. In the current QuestionLabeling system, the following numbers of topics are used: $k = 32, 64, 128$ and 256. We collect all LDA models into a set M (Step 7) and save it to disk. The training part works offline. To handle large training data, we use PLDA+ on thousands of machines in order to maintain training time within the range of a few hours.

The online suggestion part assigns labels to a question as the user types it. The bottom half of Figure 12.18 depicts the suggestion algorithm. First, we use each LDA model in M to infer the topic distributions $\{\theta_{q,k}\}$ of the question q (Step 1). Then, we compute the cosine similarity $CosSim(\theta_{q,k}, \theta_{d_l,k})$ between $\theta_{q,k}$ and $\theta_{d_l,k}$ (Step 2). Third, we use the mean similarity over different values of k as the final similarity $S(q,l)$ between a question and a label (Step 3). Finally, we sort all $l \in L$ by $S(q,l)$ in descending order, and take the first N (say ten) labels as recommended ones (Step 4).

Using PLDA+ for QuestionLabeling has two benefits: semantic matching and scalability. PLDA+ decomposes each question and answer into a distribution over a set of latent topics. When encountering ambiguous words, PLDA+ can use the context to decide the correct semantics. For example, QuestionLabeling suggests only labels such as 'mobile' and 'iPhone' to the question *"How to crack an apple?"*, although the word *apple* also means the fruit "apple." In addition, PLDA+ is designed to scale gracefully to more input data by employing more machines.

12.7 Concluding Remarks

In this chapter, we first presented the implementation of AD-LDA based on MPI. We then analyzed the communication bottleneck of AD-LDA. In order to reduce this communication bottleneck, PLDA+ divides processors into two types, namely

P_d processors and P_w processors, and also employs pipeline-based Gibbs Sampling (PGS). Though any distributed scheme may subject to pathological workload, PLDA+ appears to be resilient to substantial deadline misses caused by imbalanced workload. Extensive experiments on large-scale document collections demonstrated that PLDA+ can achieve much higher speedup than AD-LDA, thanks to both its improved load balancing and reduced communication overhead. From the experience with implementing PLDA+ we learned that on top of MapReduce or MPI, advanced strategies such as data placement and pipeline processing should be considered to further smooth out bottlenecks.

References

1. Liu, Z., Zhang, Y., Chang, E.Y., Sun, M. Plda+: Parallel latent dirichlet allocation with data placement and pipeline processing. *ACM Transactions on Intelligent Systems and Technology (ACM TIST)*, 2(3), 2011.
2. Blei, D.M., Ng, A.Y., Jordan, M.I. Latent dirichlet allocation. *Journal of Machine Learning Research*, 3:993–1022, January 2003.
3. Mimno, D.M., McCallum, A. Organizing the oca: learning faceted subjects from a library of digital books. In *Proceedings of ACM/IEEE Joint Conference on Digital Libraries*, pages 376–385, 2007.
4. Newman, D., Asuncion, A., Smyth, P., Welling, M. Distributed inference for latent dirichlet allocation. In *Proceedings of NIPS*, 2007.
5. Asuncion, A., Smyth, P., Welling, M. Asynchronous distributed learning of topic models. In *Proceedings of NIPS*, pages 81–88, 2008.
6. Wang, Y., Bai, H., Stanton, M., Chen, W.Y., Chang, E.Y. Plda: Parallel latent dirichlet allocation for large-scale applications. In *Proceedings of AAIM*, pages 301–314, 2009.
7. Minka, T., Lafferty, J. Expectation-propagation for the generative aspect model. In *Proceedings of UAI*, pages 352–359, 2002.
8. Griffiths, T., Steyvers, M. Finding scientific topics. *Proceedings of the National Academy of Sciences*, 101(90001):5228–5235, 2004.
9. Rosen-Zvi, M., Chemudugunta, C., Griffiths, T., Smyth, P., Steyvers, M. Learning author-topic models from text corpora. *ACM Transactions on Information Systems*, 28(1):1–38, 2010.
10. Li, W., MaCallum, A. Pachinko allocation: DAG-structured mixture models of topic correlations. In *Proceedings of ICML*, pages 577–584, 2006.
11. Chemudugunta, C., Smyth, P., Steyvers, M. Modeling general and specific aspects of documents with a probabilistic topic model. In *Proceedings of NIPS*, 2007.
12. Newman, D., Asuncion, A., Smyth, P., Welling, M. Distributed algorithms for topic models. *Journal of Machine Learning Research*, 10:1801–1828, 2009.
13. Nallapati, R., Cohen, W., Lafferty, J. Parallelized variational em for latent dirichlet allocation: An experimental evaluation of speed and scalability. In *ICDM Workshop*, pages 349–354, 2007.
14. Wolfe, J., Haghighi, A., Klein, D. Fully distributed em for very large datasets. In *Proceedings of ICML*, pages 1184–1191, 2008.
15. Gomes, R., Welling, M., Perona, P. Memory bounded inference in topic models. In *Proceedings of ICML*, pages 344–351, 2008.
16. Porteous, I., Newman, D., Ihler, A., Asuncion, A., Smyth, P., Welling, M. Fast collapsed gibbs sampling for latent dirichlet allocation. In *Proceedings of ACM KDD*, pages 569–577, 2008.
17. Thakur, R., Rabenseifnery, R., Gropp, W. Optimization of collective communication operations in mpich. *International Journal of High Performance Computing Applications*, 19(1):49–66, 2005.

18. Graham, S., Snir, M., Patterson, C. *Getting up to speed: The future of supercomputing*. National Academies Press, 2005.
19. Shen, J.P., Lipasti, M.H. *Modern Processor Design: Fundamentals of Superscalar Processors*. McGraw-Hill Higher Education, 2005.
20. Blinn, J. A trip down the graphics pipeline: Line clipping. *IEEE Computer Graphics and Applications*, 11(1):98–105, 1991.
21. Berenbrink, P., Friedetzky, T., Hu, Z., Martin, R. On weighted balls-into-bins games. *Theoretical Computer Science*, 409(3):511–520, 2008.
22. Chang, E.Y., Garcia-Molina, H., Li, C. 2d bubbleup: Managing parallel disks for media servers. In *Proceedings of FODO*, pages 221–230, 1998.
23. Cohn, D., Chang, H. Learning to probabilistically identify authoritative documents. In *Proceedings of ICML*, pages 167–174, 2000.
24. Popescul, A., Ungar, L., Pennock, D., Lawrence, S. Probabilistic models for unified collaborative and content-based recommendation in sparse-data environments. In *UAI*, pages 437–444, 2001.
25. Li, L.J., Wang, G., Fei-Fei, L. OPTIMOL: automatic online picture collection via incremental model learning. In *Proceedings of CVPR*, 2007.
26. Si, X., Chang, E.Y., Gyöngyi, Z., Sun, M. Confucius and its intelligent disciples: Integrating social with search. In *Proceedings of the VLDB*, pages 1505–1516, 2010.
27. Chen, W.Y., Chu, J.C., Luan, J., Bai, H., Wang, Y., Chang, E.Y. Collaborative filtering for orkut communities: Discovery of user latent behavior. In *Proc. of the 18th International WWW Conference*, pages 681–690, 2009.
28. Hofmann, T. Probabilistic latent semantic analysis. In *Proceedings of UAI*, pages 289–296, 1999.

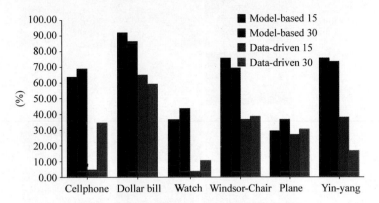

Fig. 2.4: Model-based outperforms data-driven

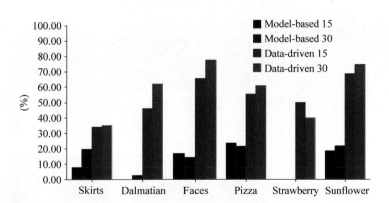

Fig. 2.5: Data-driven outperforms model-based

Table 2.2: Images with a rigid body

Category	Images
Cellphone	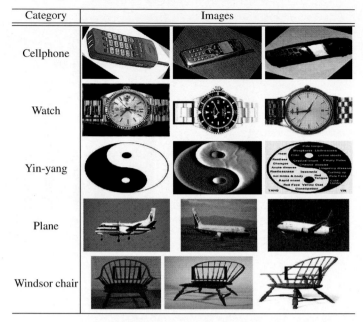
Watch	
Yin-yang	
Plane	
Windsor chair	

Table 2.3: Images of color or texture rich

Class	Category	Images
Visual words (color)	Strawberry	
	Sunflower	
Visual words (texture)	Dalmatian	
	Pizza	

Table 2.4: Texture examples of some images

Class	Category	Images
Rigid body	Watch	
	Windsor chair	
Texture	Dalmatian	
	Pizza	

Table 2.7: Tough categories

Category	Images
Beaver	
Barrel	
Cup	
Mayfly	
Wild cat	
Gramophone	

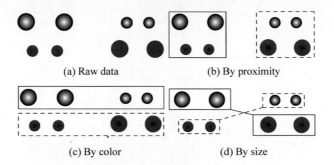

(a) Raw data   (b) By proximity

(c) By color   (d) By size

Fig. 5.1: Clustering by different functions

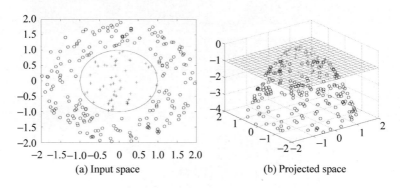

(a) Input space   (b) Projected space

Fig. 5.2: Clustering via the kernel trick

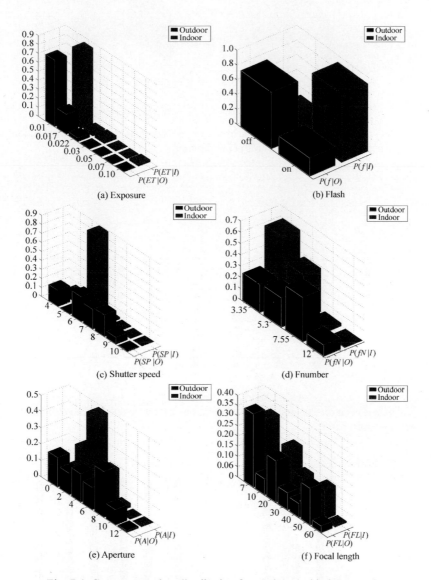

Fig. 7.4: Camera metadata distribution for outdoor and indoor scenes

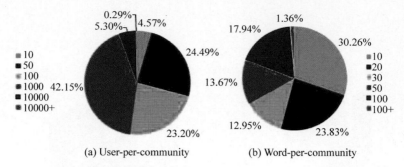

(a) User-per-community   (b) Word-per-community

Fig. 8.6: (a) Distribution of the number of users per community, and (b) distribution of the number of description words per community

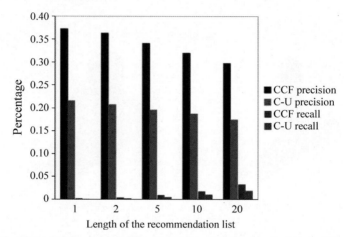

Fig. 8.8: The precision and recall as functions of the length (up to 20) of the recommendation list

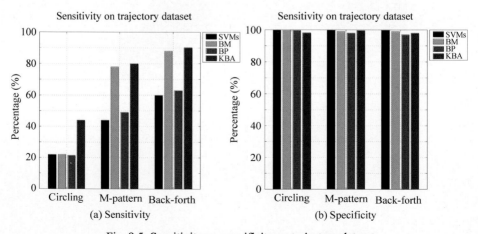

(a) Sensitivity   (b) Specificity

Fig. 9.5: Sensitivity vs. specificity on trajectory dataset

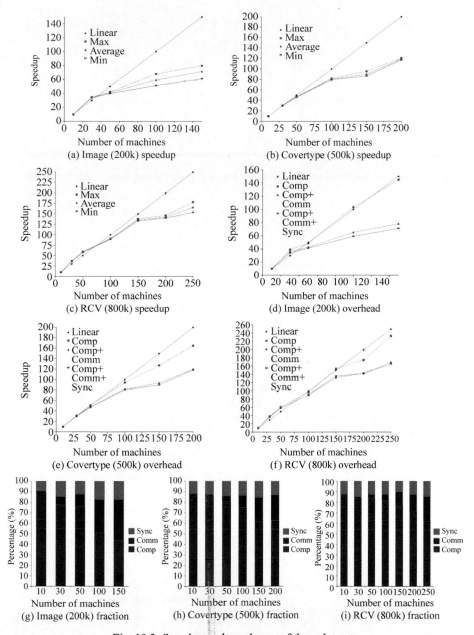

Fig. 10.3: Speedup and overhead three datasets

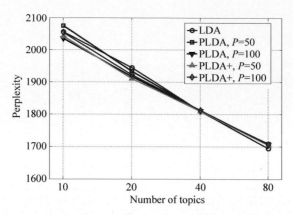

Fig. 12.10: Test perplexity on NIPS versus # topics T when the number of iterations is 400

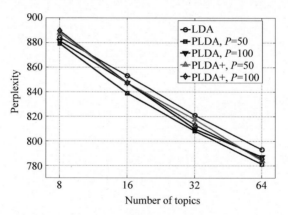

Fig. 12.11: Test perplexity on Dianping versus # topics T when the number of iterations is 400

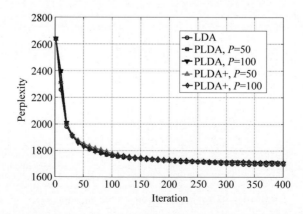

Fig. 12.12: Convergence of test perplexity versus iteration on NIPS with $T = 80$

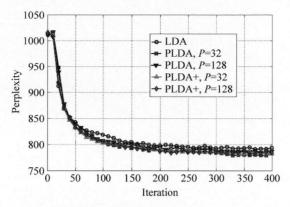

Fig. 12.13: Convergence of test perplexity versus iteration on Dianping with $T = 64$

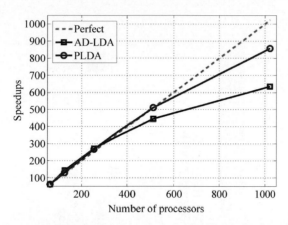

Fig. 12.14: Parallel speedup results for 64 to 1,024 processors on Wiki-20T

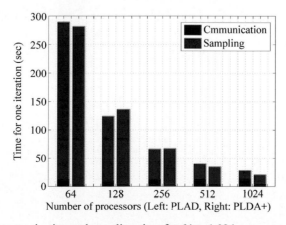

Fig. 12.15: Communication and sampling time for 64 to 1,024 processors on Wiki-20T

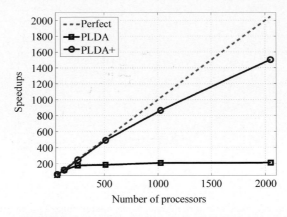

Fig. 12.16: Parallel speedup results for 64 to 2,048 processors on Wiki-200T

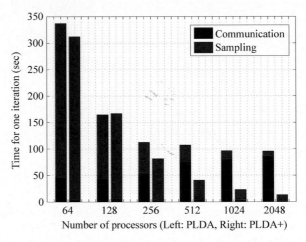

Fig. 12.17: Communication and sampling time for 64 to 2,048 processors on Wiki-200T